Materials, Energy and Environment Engineering

Raj Mohan B. · G. Srinikethan
Bhim Charan Meikap
Editors

Materials, Energy and Environment Engineering

Select Proceedings of ICACE 2015

Springer

Editors
Raj Mohan B.
Department of Chemical Engineering
National Institute of Technology Karnataka
Mangalore, Karnataka
India

Bhim Charan Meikap
Department of Chemical Engineering
Indian Institute of Technology Kharagpur
Kharagpur, West Bengal
India

G. Srinikethan
Department of Chemical Engineering
National Institute of Technology Karnataka
Mangalore, Karnataka
India

ISBN 978-981-10-9678-5 ISBN 978-981-10-2675-1 (eBook)
DOI 10.1007/978-981-10-2675-1

Printed on acid-free paper

This Springer imprint is published by Springer Nature
The registered company is Springer Nature Singapore Pte Ltd.
The registered company address is: 152 Beach Road, #22-06/08 Gateway East, Singapore 189721, Singapore

Preface

The magnificence and destiny of humanity are intricately tied to the various processes that shape its environment. However, virtually all aspects of environmental changes remain under the dominant influence of human activity. The fact that many of the resources available today will be further depleted, and many of the environmental sinks that help assimilate waste products become overburdened from continued accumulation, has accelerated the pace of scientific research in the areas of materials, energy and environment. The field is rapidly growing and the resultant technological change has afforded humanity ever more cleverer ways to expand and tap into its resource base, seek and use limited materials and energy more efficiently, and cut down on the emission of many pollutants. However, the speculations that the "demand" would overpower the existing availability of resources owing to the unprecedented population growth has continued to challenge human race to further explore the technological prowess to develop sustainable solutions in form of cleaner production, smarter materials and cost-effective and eco-friendly fuel generation. These efforts have continually been backed up by the various findings in environmental science and technology that acts as a means to provide the science base to understand human–environment interactions and as a guide/resource for production and consumption decisions, as well as the release of pollutants.

Given the diversity of the challenges at hand, and the historical, social and cultural context within which these challenges are understood and addressed, it is evident that interdisciplinary collaborations and methodological pluralism are prerequisites for anyone who aspires to get their minds and hands around the underlying and emerging complexities. This book is a vivid reminder of the pervasiveness of the complexity and the rich knowledge base that has been generated in the fields of materials, energy and environmental studies in the recent years. The contributors highlight important findings that are generated by an interplay of theory and experimentation and an understanding of human–environment interactions at conceptual level and through case-specific inquiry. The contents of the book focuses on the broad area of *Materials, Energy and Environment* with five dedicated sub-sections namely: *Materials, Biosorption and Degradation, Nanomaterials, Nanomaterials Synthesis and Pollution Control.*

Part I of the book attends to the recent trends in Materials research; including interesting findings in macro-material fabrication, optimization and characterization. The field of materials research has an extended past as well as a long and promising future and interestingly, the scientific study of materials have always generated significant findings that have had a major impact on the evolution of the main stream of modern science, including quantum mechanics and allied fields. The research discussion on materials research in the present book would highlight the continued role of the field as a pacemaker that determine the nature and effectiveness of devices used for the upliftment of mankind.

Part II describes the current research in Biosorption and degradation with thrust on heavy metal bioremediation and degradation on an optimized scale for industrial applications. The findings pave way towards sustainable development by providing important insights for the development and promotion of environmental management and green technologies to treat a wide range of aquatic and terrestrial habitats contaminated by increasing anthropogenic activities with the main sources of contaminants being the effluents from the chemical industries. Since bioremediation and biosorption not only represent an emerging technology but also present a great advantage of being cost-effective when compared to the traditional remediation methods due to the use of indigenous microorganisms with a versatile metabolism, this book presents an up-to-date and comprehensive collection of research findings bioremediation technology research and application.

Parts III and IV represent the upcoming field of nanomaterials—their synthesis and applications. Nanomaterials have recently become one of the most active research fields in the areas of solid state physics, chemistry and engineering. Evidence of this interest was proved by the massive response in the form of research papers devoted to the subject. Nanomaterials display novel and often enhanced properties compared to traditional materials, which opens up possibilities for new technological applications. The dedicated section is intended to satisfy the need for a broad coverage that will provide an introduction, background and recent advances in all areas of nanomaterials and their synthesis.

Part V of the book presents detailed analyses to the environmental enthusiasts regarding the present day challenges in pollution control. The field has been avidly embraced by a large percentage of research community who has been actively analyzing and finding solutions to address the concerns regarding pollution of the various strata of our environment. The publication represents an embodiment of the modest initiatives undertaken to deal with source apportionment and assessment of the pollutant load in air and water and proposes effective means to manage the challenges.

In short, this book spans a rich array of recent research and developments in the field of materials, energy and environment. It portrays the diversity of challenges faced by the present generation on a universal scale and is also a testament to the wealth of insights that can be generated when bridging interdisciplinary research areas to address these issues. The usefulness of the data represented in this publication can be summed up as better science and better environmental solutions. With knowledge of the methods and applications of this book in the hands of its readers, new frontiers open up for inquiries into the further exploration in the field of material sciences, energy and environment; their interdependencies and novel opportunities.

Mangalore, India	Raj Mohan B.
Mangalore, India	G. Srinikethan
Kharagpur, India	Bhim Charan Meikap

Contents

Part III Pollution Control

About the Editors

Dr. Raj Mohan B. is working as Associate Professor in Department of Chemical Engineering, National Institute of Technology Karnataka. He has more than 50 research publications in the field of Air Pollution: Particulate Matter analysis, Control and Abatement and CO_2 sequestration, Bioremediation and Separation Technology, Wastewater Treatment and Quality monitoring, Biosynthesis of nanoparticles, reduction of antioxidants and antimicrobial compounds.

Dr. G. Srinikethan is working in Department of Chemical Engineering, National Institute of Technology Karnataka. He has more than 40 research publications. His research fields are Transfer Operations, Industrial Pollution Control, Hydrodynamics, Environmental Biotechnology.

Dr. Bhim Charan Meikap is working at Chemical Engineering, Indian Institute of Technology Kharagpur. He has more than 100 publications in reputed journals. He is working on the research fields such as Air Pollution: Particulate Matter analysis, Control and Abatement and CO_2 sequestration, Bioremediation and Separation Technology, Wastewater Treatment and Quality monitoring.

Part I
Materials and Nanomaterials

Characterization of Citrus Peels for Bioethanol Production

John Indulekha, M.S. Gokul Siddarth, Ponnusamy Kalaichelvi and Appusamy Arunagiri

1 Introduction

The increasing greenhouse gas emissions, acid rain, climate change and fossil fuel depletion have given rise to interest in clean energy. Biomass has greater potential for renewable and sustainable source of energy. Biomass is the fourth largest energy source around the world. One way to reduce the crude oil consumption as well as environmental pollution is to produce biofuels from biomass (Demirabs 2005). The advantages of biofuels over fossil fuels are that it is obtained from common biomass sources, biodegradable, helps in environmental sustainability and combustion of biofuels is included in CO_2 cycle (Demirabs 2007). Among the liquid biofuels, bioethanol is the most widely used. Ethanol requisites have augmented considerably because of the use of ethanol along with gasoline in high-octane fuels since it act as a fuel oxygenate (Lyons 2004). Lignocellulose is one of the most unutilized and economical biomaterial. The huge availability of these feedstock and capability of producing biofuel makes them very attractive. These are renewable energy resources which are classified into forest products (softwood and hardwood), agricultural residues (e.g., corn stover, sugarcane bagasse, wheat straw), and dedicated crops (salix, switchgrass). The compositions of these lignocelluloses are very important as they affect the conversion of polysaccharides into other useful monosaccharides for bioethanol production. The chemical and structural composition of lignocellulose varies based on the environmental and genetic interactions (Lee et al. 2007). Any lignocellulosic material is composed of cellulose, hemicellulose and lignin. It is always suggested to opt for a biomass which contains low

---

J. Indulekha · M.S. Gokul Siddarth · P. Kalaichelvi · A. Arunagiri (✉)
Department of Chemical Engineering, National Institute
of Technology Tiruchirappalli, Tiruchirappalli, Tamil Nadu, India
e-mail: aagiri@nitt.edu

J. Indulekha
e-mail: indulekhajohn@gmail.com

R. Mohan B. et al. (eds.), *Materials, Energy and Environment Engineering*,
DOI 10.1007/978-981-10-2675-1_1

volatile matter and fixed carbon by difference. The presence of high moisture in biomass requires prior drying to remove the moisture and reduces the energy efficiency of the system. The volatile matter content of these peels are high which revealed that it can also be utilized for the pyrolysis process because biomass with more volatile matter is highly reactive and also get devolatilized easily than that of the lesser volatile matter and producing less fixed carbon. The ash in the biomass is low in comparison to other biomass. Ultimate analysis of orange peel and sweet lime peel shows carbon, hydrogen, nitrogen, sulphur and oxygen by difference. This result proves that the nitrogen and sulfur content in the peel is very little, hence this peel produces the lowest measure of nitrogen and sulfur oxides in thermo chemical transformation process. Good amount of carbon, hydrogen and oxygen can lead to higher ethanol yields. Higher heating value of sweet lime peel is 14,853 kJ/kg and for orange peel is 15,690 kJ/kg.

3.2 *FTIR Spectroscopy*

The FTIR technique is utilized to analyze the chemical components and distinctive functional group. Figures 2 and 3 show the FTIR spectra of orange and sweet lime peels. These spectra give the information about the presence of cellulose, hemicellulose, lignin and pectin (Corrales et al. 2012). The spectral bands at 3175–3490 cm^{-1} indicate the existence of O–H stretching intramolecular hydrogen bonds (Kumar et al. 2014). The spectra of both orange and sweet lime peel are almost same. These two citrus peels have same functional groups but composition may vary (Table 2).

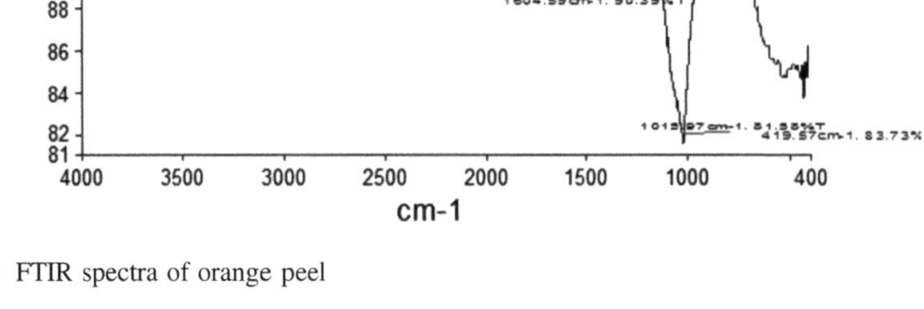

Fig. 2 FTIR spectra of orange peel

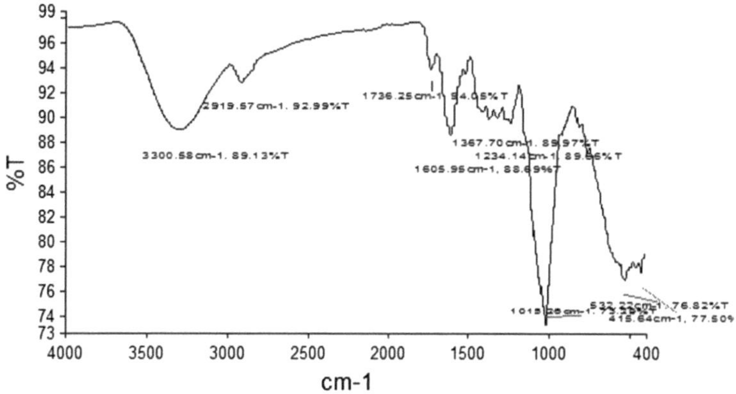

Fig. 3 FTIR spectra of sweet lime peel

Table 2 Functional group present in citrus peels from FTIR spectroscopy

Wavenumber (cm^{-1})	Bond	Composition
3333	O–H stretching	Alcohol and pectic acid
2921	C–H Asymmetrical stretching	Lignin
1633	C=O stretching	Lignin
1371	C–H bending	Cellulose
1016	C–C–O, C=O, Vibrational stretching	Cellulose, Hemicellulose

3.3 *Thermal Analysis*

Thermogravimetric analysis (TGA), which is a function of temperature, was investigated for citrus peels and Fig. 4 shows TGA of sweet lime peel. Both the peels showed similar pattern and hence results are shown for sweet lime peel. The total mass rcduction throughout the temperature range can be disintegrated into the

Fig. 4 Thermogravimetric analysis of sweet lime peel

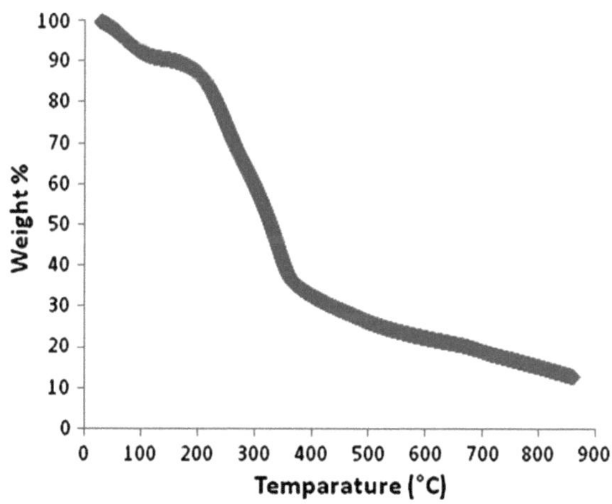

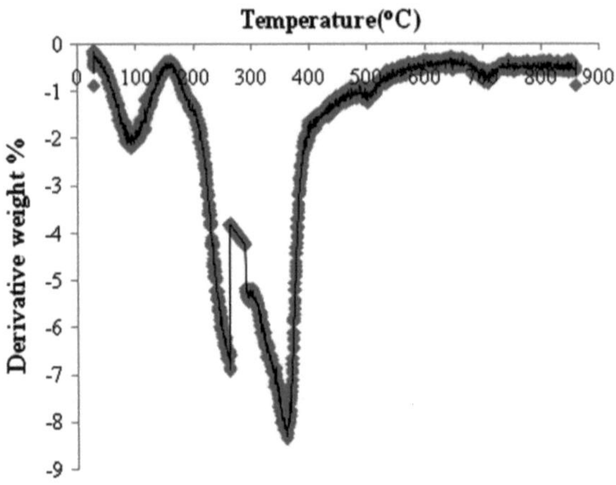

Fig. 5 DTG of sweet lime peel

three sections such as (i) removal of moisture, (ii) degradation of hemicellulose, cellulose and pectin, (iii) lignin degradation, which is clearly shown in derivative thermogravimetric (DTG) of the peels. Pectin as a component of lignocellulosic material is not much explored for thermal analysis. In general, at normal heating rate the degradation of hemicellulose occur at less than 350°C, cellulose degradation between 250 and 500°C, and lignin degradation is gradually distributing throughout the process even after 500°C. There is no definite weight reduction happens since lignin is more thermostable than cellulose and hemicellulose (Li et al. 2004). As indicated in Fig. 5, a weight loss below 200°C is because of the evacuation of moisture and some amount of volatile matter. When the temperature

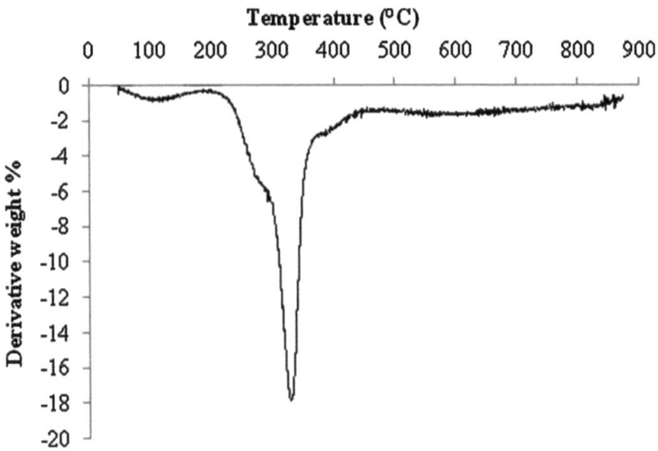

Fig. 6 DTG of acid detergent fiber (ADF)

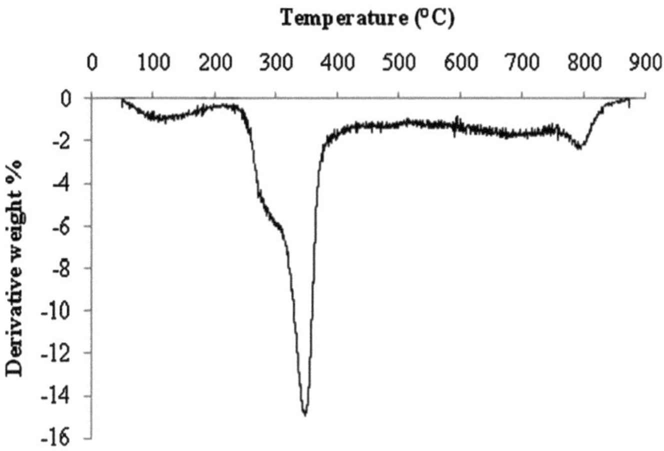

Fig. 7 DTG of neutral detergent fiber (NDF)

varied from 200 to 400°C, a higher mass loss is observed. The degradation of cellulose and hemicellulose occurs at a closer temperature range approximately below 350°C (Fig. 5), and it is impractical to distinguish isolated peaks for cellulose and hemicellulose (Mothe et al. 2009). This is clear from the derivative thermograph of acid detergent fiber (ADF) (Fig. 6) and neutral detergent fiber (NDF) (Fig. 7) because, other than lignin, NDF contains cellulose as well as hemicellulose whereas ADF contains only cellulose. But there is not much difference in the peaks. Moreover, above the temperature of 370°C, decomposition of

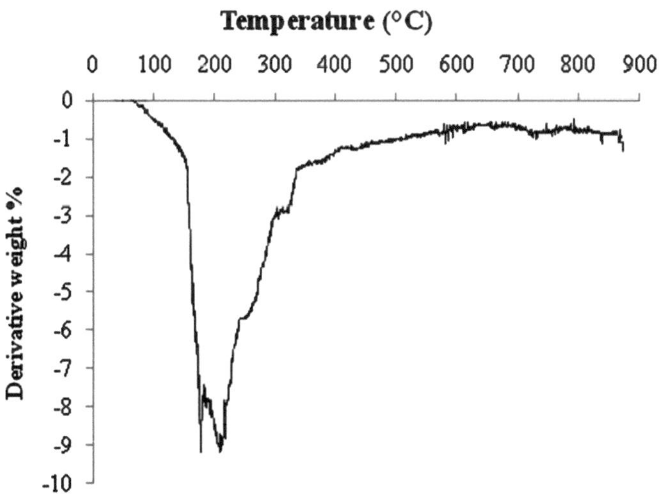

Fig. 8 DTG of pectin

lignin takes place with lower and almost constant rate of mass loss upto 950°C (Gupta et al. 2011). From Fig. 8, pectin degradation occurs at faster rate when compared to other components. The degradation of pectin occurs in the temperature range of 150 to 300 °C. Since ADF and NDF also contain lignin, it is assumed that lignin degradation also starts within that temperature range of 200 to 400°C. The peak obtained within this range in Fig. 9, can also be due to release of carbon dioxide and methane from the biomass (Strezov et al. 2004). Lignin degrades at 320 to 450°C (Gracia et al. 2012) which is proved in this study also. DSC profile of the

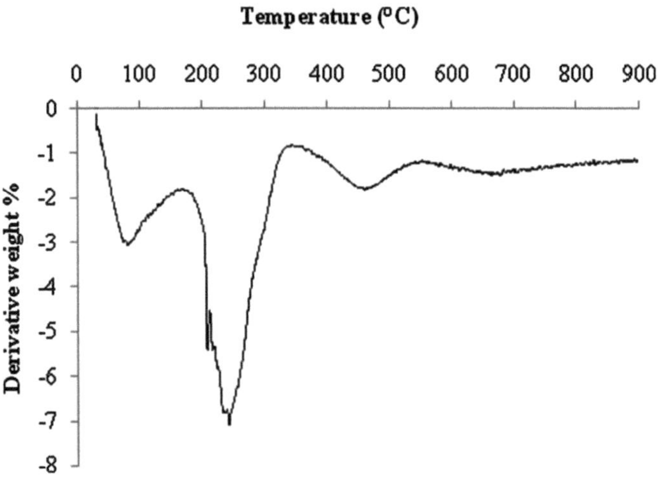

Fig. 9 DTG of acid detergent lignin (ADL)

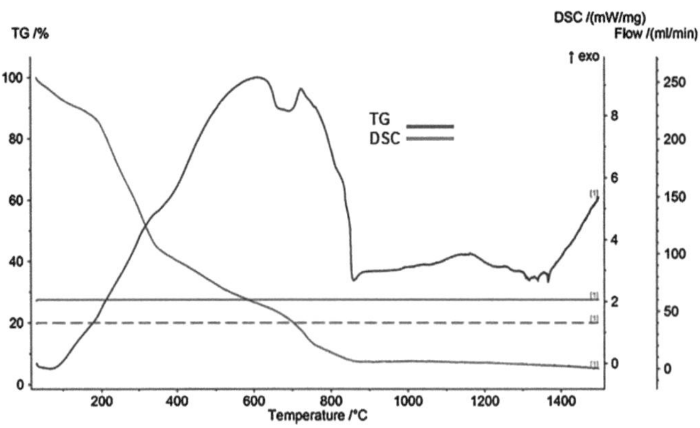

Fig. 10 Differential scanning calorimetry

sweet lime peel showed that the reaction is endothermic initially which is attributed mainly to the removal of moisture. Since the degradation of cellulose, lignin, hemicellulose and pectin proceeds from 200 to 800°C, the reaction is exothermic (Fig. 10).

4 Conclusion

The analysis of citrus peels shows that low moisture, ash, nitrogen, sulfur and high carbon and volatile matter suggest that citrus peels has a good potential for biofuel production. The HHV is comparable to other biomass feedstock for biofuel production. The lignocellulosic fraction in the citrus peels shows that the high amount of cellulose content, gives high sugar yield which can in turn produce more bioethanol. From FTIR analysis, it was confirmed that the presence of elements mainly carbon and oxygen as well as the presence of lignocellulosic fraction such as cellulose, hemicellulose, pectin and lignin in citrus peels. The thermal decomposition behaviour of the citrus peels was also described and used as a method to determine the lignocellulosic composition of these peels. Derivative thermogravimetric analysis of neutral detergent fiber, acid detergent fiber, pectin and acid detergent lignin was compared with that of raw peel and it was confirmed that lignin, hemicellulose, pectin and cellulose are present in the citrus peels like other lignocellulosic biomass. Hence, the characterization studies of citrus peels revealed that it has good potential for the production of bioethanol as well as value added product like pectin.

References

Corrales, R.C.N.R., Mendes, F.M.T., Perrone, C.C., Anna, C.S., de Souza, W., Abud, Y., da Silva Bon, E.P., Ferreira-Leitão, V.: Structural evaluation of sugar cane bagasse steam pretreated in the presence of CO_2 and SO_2. Biotechnol. Biofuels **5**(36), 1–8 (2012)

Demirbas, A.: Bioethanol from cellulosic materials: a renewable motor fuel from biomass. Energy Sources **27**, 327–337 (2005)

Demirbas, A.: Progress and recent trends in biofuels. Prog. Energy Combust. Sci. **33**, 1–18 (2007)

Garcia, R., Pizarro, C., Lavin, A.V., Bueno, J.L.: Characterization of Spanish biomass wastes for energy use. Bioresour. Technol. **103**, 249–258 (2012)

Gupta, N., Tripathi, S., Balomajumder, C.: Characterization of pressmud: a sugar industry waste. Fuel **90**, 389 (2011)

Kumar, A., Negi, Y.S., Choudhary, V., Bhardwaj, N.K.: Characterization of cellulose nanocrystals produced by acid-hydrolysis from sugarcane bagasse as agro-waste. J. Mater. Phys. Chem. **2** (1), 1–8 (2014)

Lee, D., Owens, V.N., Boe, A., Jeranyama, P.: Composition of herbaceous biomass feedstocks, pp 1–07. South Dakota State University Publication, Brookings, SD. (2007)

Li, S., Xu, S., Liu, S., Yang, C., Lu, Q.: Fast pyrolysis of biomass in free-fall reactor for hydrogen-rich gas. Fuel Process. Technol. **85**, 1201–1211 (2004)

Lyons, T.R.: Ethanol around the world: rapid growth in policies, technology and production. In: Jacques, K.A., Lyons, T.P., Kelsall D.R. (eds.) The Alcohol Textbook. 4th ed. pp. 1–8 Nottingham University Press, Nottingham, UK (2004)

Mothe, C.G., de Miranda, I.C.: Characterization of sugarcane and coconut fibers by thermal analysis and FTIR. J. Therm. Anal. Calorim. **97**, 661–665 (2009)

Strezov, V., Moghtaderi, B., Lucas, J.A.: Computational calorimetric investigation of the reactions during thermal conversion of wood biomass. Biomass Bioenergy **27**(5), 459 (2004)

Sudhakar, D.V., Maini, S.B.: Isolation and characterization of mango peel pectins, J. Food Process. Preserv. **24**(3), 209–227 (2000)

Van Soest, P.J.: Nutritional Ecology of the Ruminant, p. 476. Cornell University, USA. (1994)

Study of Mechanical Properties and Microstructure of Aluminium Alloy Reinforced with TiB_2, by in Situ Technique

Akshay Mohan Pujar and Chetan Kulkarni

1 Introduction

A composite material is a material made up of a discrete constituent phase, the reinforcement and continuous phase, the matrix. Based on the physical or chemical nature of the continuous phase (the matrix) these materials are classified as: polymer matrix, metal-matrix and ceramic composites (Surappa 2003).

Composites with metal as the matrix material are very popular for automotive, aircraft and aerospace industry applications (Deuis et al. 1997; Suresha and Sridhara 2010). Metal matrix composites have emerged as the important class of advanced material giving engineers the opportunity to develop the material properties according to their needs (Surappa 2003). This class of material exhibits the ability to withstand high tensile and compressive stresses by the transfer and distribution of the applied load from the matrix phase to the reinforcement phase.

Among metal matrix composites, the discontinuously reinforced aluminium metal matrix composites exhibit properties such as low density, high specific stiffness, high specific strength, controlled co-efficient of thermal expansion, high fatigue resistance and improved stability at high temperatures (Surappa 2003). Aluminium matrix composites are anticipated to replace monolithic materials as well as aluminium alloys, ferrous alloys, titanium alloys and polymer based composites in several applications as these exhibits outstanding combination of properties. These can offer economically viable solutions for a wide variety of commercial applications (Deuis et al. 1997).

A.M. Pujar (✉) · C. Kulkarni (✉)
B. V. Bhoomaraddi College of Engineering and Technology, Vidyanagar, Hubballi, India
e-mail: ampujar0602@gmail.com

C. Kulkarni
e-mail: chetan27993@gmail.com

R. Mohan B. et al. (eds.), *Materials, Energy and Environment Engineering*,
DOI 10.1007/978-981-10-2675-1_2

In aluminium matrix composites (AMC), aluminium/aluminium alloy forms the continuous phase (the matrix). The reinforcement phase typically is non-metallic and ceramic namely silicon carbide, fly ash, titanium boride, aluminium oxide, titanium oxide etc. By changing the volume fraction and nature of the ingredients the properties of these materials can be altered.

Wang et al. 2007 produced Al/TiB_2 AMC by the in situ method and explained the role of TiB_2 particles in grain refinement. Zhao et al. (2007) synthesized Al/(TiB_2 + Al_2O_3) hybrid composite by the in situ reaction of K_2TiF_6, KBF_4 and CuO to molten aluminium and reported the distribution and blending of TiB_2 particles with $CuAl_2$ phase, alongside the grain boundaries. Kumar et al. (2008) synthesized Al–7Si/TiB_2 composite by the in situ reaction of K_2TiF_6 and KBF_4 to molten aluminium and observed significant improvement in wear and mechanical behaviour of the reinforced composites as compared to the base alloy. Ramesh et al. (2010, 2011) prepared AA6063/TiB_2 composite by the in situ reaction of Al–10Ti and Al–3B mater alloys and investigated the consequence of different compositions of reinforcement on the development and mechanical characteristics of the developed composites. Xue et al. (2011) studied that when CeO_2 is added to the in situ reaction of K_2TiF_6 and KBF_4, improvement in the distribution of TiB_2 particles and enhancement in property of Al/TiB_2 composites can be obtained.

Keeping in view, the necessity of light weight material for developing different automobile components and to maximize the efficiency, in this article an effort has been made to develop new and light AA7175 based TiB_2 reinforced composite material. Besides, an attempt has been constituted to examine the effect of TiB_2 reinforcement on the microstructure and mechanical properties of the developed composites.

2 Materials and Methods

In the present work, AA7175 (Al–Zn–Mg–Cu) alloy as well as salts such as potassium hexa flouro titanate (K_2TiF_6), and potassium tetra fluoro borate (KBF_4) have been employed as the basic raw materials. The salts used to be of commercial grade in powder form. Aluminium alloy ingots and inorganic salts were bought from FENFE Metallurgicals, Uttarahalli, Bengaluru and Sigma Aldrich, Bengaluru respectively.

2.1 Composition of Alloy

The composition of AA7175 alloy which comes under 7XXX alloy family is shown in the Table 1.

Table 1 Chemical composition of AA7175 alloy

Element	Zn	Mg	Mn	Cu	Fe	Si	Cr	Ti	Ca	Mo
wt%	5.47	2.45	0.13	1.52	2.23	0.14	0.2	0.043	0.002	0.0003

2.2 *Preparation of Composites by Mixed Salt Route Technique*

The AA7175 (Al–Zn–Mg–Cu) TiB_2 composite was processed through stir casting method, in which the TiB_2 particles were precipitated in situ through an exothermic reaction between K_2TiF_6 and KBF_4 (Mandal 2004; Herbert 2007). This technique of generating particles in situ by a chemical reaction involving salts is called as mixed salt route method (Wood 1993; Bartels 1997; Lakshmi 1998). The requisite amounts of K_2TiF_6 and KBF_4 salts necessary for yielding the desired weight fraction of TiB_2 were weighed, and then dried discretely in an oven at 150 °C for one hour to expel the absorbed moisture present. The salts were mixed carefully to get a uniform mixture, and then wrapped in aluminium foils to form packets of suitable size to be added to the molten alloy. The alloy melt was heated to the preset reaction temperature of 800 °C, in a vertical muffle furnace (VBF-1200X; MTI Corporation) and then degassed by adding hexachloroethane (C_2Cl_6) tablets. The degassing was accompanied by a slight drop in temperature. Once the melt was reheated to 800 °C, the packets containing salts mixture were added and stirred for uniform mixing using a zirconia coated graphite rod. The entire mixture was held at the temperature of 800 °C for an hour after the addition of salts to ensure complete reaction. The molten mixture was stirred at every 10 min interval to ensure uniform distribution of salts. The TiB_2 particles were formed in situ inside the molten aluminium alloy through the reaction of K_2TiF_6 and KBF_4 salts following Eq. (1) (Mandal 2004; Herbert 2007).

$$K_2TiF_6 + KBF_4 + \text{Al alloy} \rightarrow \underset{\text{(Composite)}}{\text{Al alloy} + TiB_2} + \underset{\text{(Dross)}}{KAlF_4} \quad (1)$$

After removing the dross, the superheated melt was stir-cast by top pouring into plate-shaped cast wooden mold. In preparing the samples for microstructure, wear, hardness and tensile testing a square plate of size (0.15 m × 0.15 m × 0.065 m) was cast using the sand casting method.

2.3 *Sample Preparation for Optical Microscopy and SEM*

Samples having dimensions of 0.01 × 0.005 × 0.0005 m were cut from the cast plate of AA7175-TiB_2 composite for carrying out micro structural study by optical microscope and scanning electron microscope. The samples were prepared for micro structural study as per ASTM E3-11 standard.

2.4 Micro Structural Characterisation

Optical microscopy was carried out using the Leica image analyzer (Model: Vertimet CP), while for scanning electron microscopy, analyzer from JEOL limited (Model: JSM-6380) was employed. Optical and SEM micrographs of metallographic ally polished samples of the composite were recorded using a digital camera attached to the microscope, which was interfaced with a computer for further analysis.

2.5 Wear Testing

Sliding wear tests were carried out on AA7175-TiB_2 composite samples, using a pin-on-disc wear machine in accordance with ASTM standard G99-05. For the purpose of wear testing, specimens having dimensions of $0.006 \times 0.006 \times 0.03$ m were cut from the cast plate. These specimens attached to the pin were subjected to wear tests under normal loads of 10, 20 and 30 N. Every test was carried out to a sliding distance of 1200, 1500 and 1800 m in contact with the hardened disc, which was rotated at 240 rpm. The track diameters of 80, 100 and 120 mm were used so that linear circumferential speed in tangential direction was 1, 1.25 and 1.5 m/s respectively.

2.6 Hardness Testing

Vickers micro-hardness tests were carried out on the developed AA7175-TiB_2 samples using micro hardness tester (Model: MVH-I, Omnitech, Pune, India) as per ASTM standard E384-04a. For testing, the samples with metallographic ally polished surfaces were employed. Micro hardness measurements were carried out by applying an indentation load of 25 gf allowing a dwell time of 15 s.

2.7 Tensile Testing

The tensile tests were conducted on cast composite samples using mini universal testing machine (Model:PC-2000, Kudale Instruments (P) Ltd, Pune, India) as per ASTM standard E8 M-08 with crosshead speed of 0.3 mm/min. Samples having dimensions of $0.016 \times 0.004 \times 0.004$ m were used for this test.

3 Results and Discussion

3.1 Microstructure of AA7175-TiB_2 Composite

The optical micrographs of AA7175-TiB_2 composites for varying percentages of TiB_2 are shown in Fig. 1. The micrographs are composed of dendritic structure. From the below figures it is clearly seen that with the increase in percentage of TiB_2 particles the grain size also reduces. The reinforced TiB_2 particulates induce modification in the dendritic structure and the grain refinement. During the solidification process the TiB_2 particles provide resistance to the growth of α-Al.TiB_2 particles act as nucleation centres on which the aluminium grains solidify.

The SEM micrographs of the fabricated composites for varying composition of TiB_2 particles with a magnification of 200 μm are shown in Fig. 2. Uniform distribution of TiB_2 particles throughout the matrix can be noticed from the micrographs, which plays a major role in the betterment of mechanical properties. Distribution of reinforced particulates is influenced by process of solidification. The presence of clear interface between two constituent phases can be clearly seen. This plays a major role in improvising the mechanical and tribological properties of the developed composites. Micrographs also depict single TiB_2 particles and TiB_2

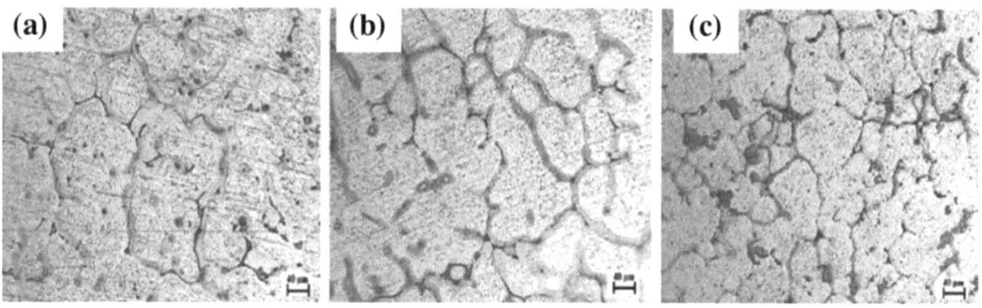

Fig. 1 Optical micrographs of the composites with different compositions of TiB_2 **a** AA7175 + 2.5 % **b** AA7175 + 5 % **c** AA7175 + 7.5 %

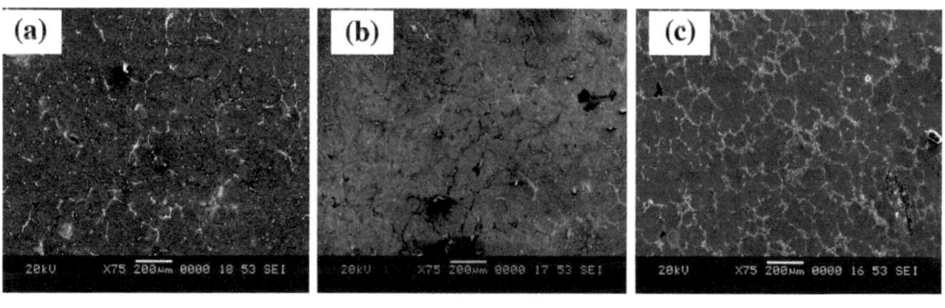

Fig. 2 SEM micrographs of the developed composites with different compositions of TiB_2: **a** AA7175 + 2.5 % **b** AA7175 + 5 % **c** AA7175 + 7.5 %

clusters which are uniformly distributed. With the increase in TiB_2 particles the nucleation sites also get increased and this also offers more resistance to the growth of the grains leading to grain structure modification.

3.2 Wear Behaviour

For analyzing wear behaviour, we have adopted the response surface method (RSM) and applied design of experiments (DOE) approach, as a result of which the following graphs are obtained. From the Fig. 3 and Fig. 4 it is seen that reinforcement of TiB_2 particulates enhances the resistance to wear. It can be clearly concluded from the below graphs that, increase in the reinforcement reduces co-efficient of friction of AA7175-TiB_2 composites. Wear coefficient is of the order ($XX \times 10^{-4}$) which indicates that the wear is mainly influenced by the wear between mating surfaces which is adhesive in nature (Figs. 5 and 6). The most effective results were reported for the composites reinforced with 5 % TiB_2 particles. H.B. Michael Rajan et al. reported similar behaviour when wear performance of aluminium alloy reinforced with TiB_2 was analyzed.

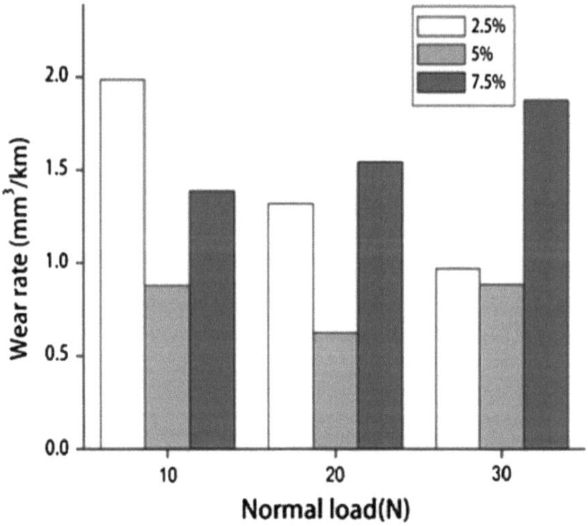

Fig. 3 Wear rate versus normal loads for different compositions of TiB_2

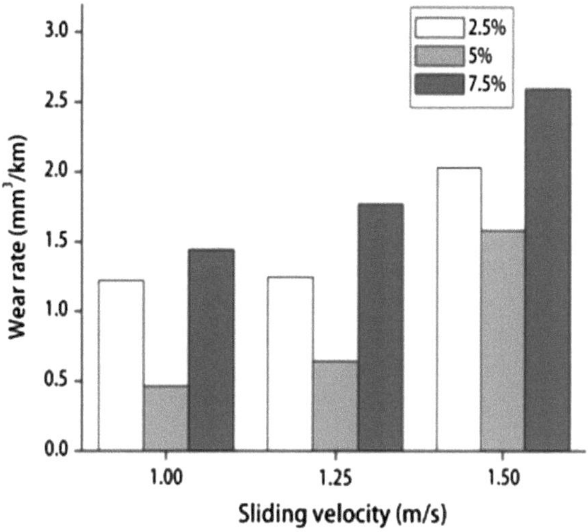

Fig. 4 Wear rate versus sliding velocity for different compositions of TiB_2

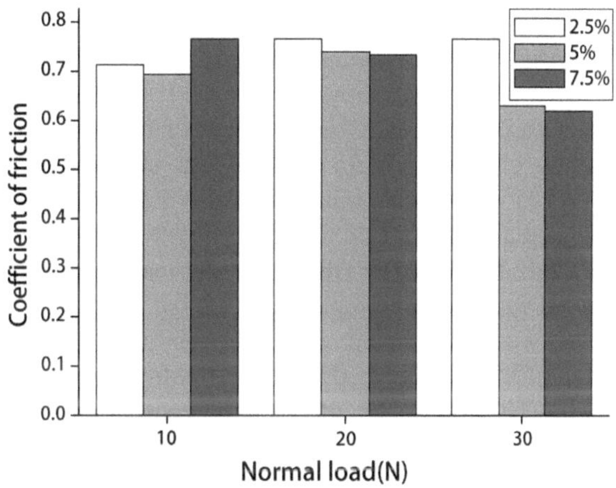

Fig. 5 Coefficient of friction versus normal loads for different compositions of TiB_2

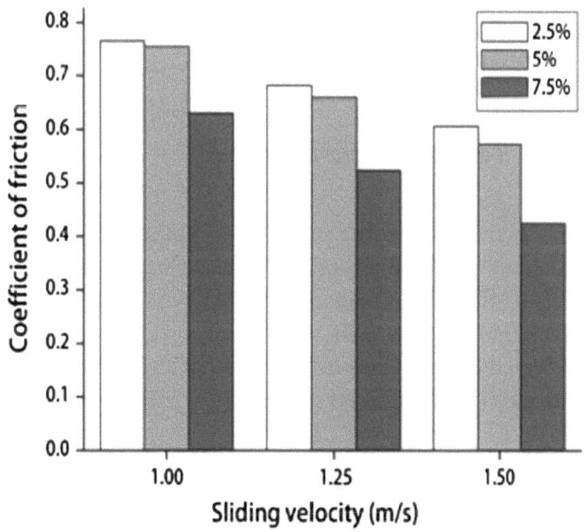

Fig. 6 Coefficient of friction versus sliding velocity for different compositions of TiB_2

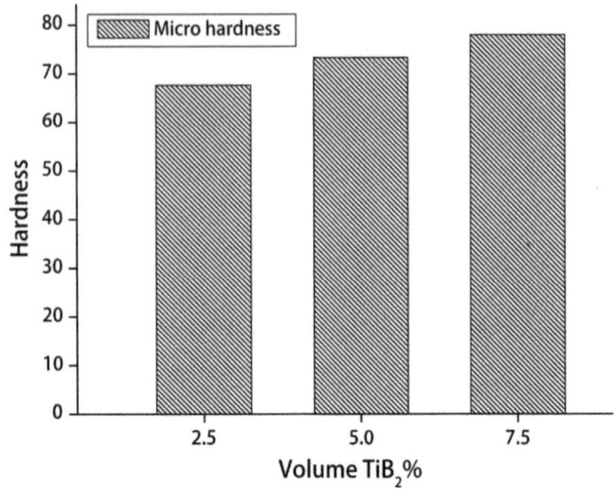

Fig. 7 Micro hardness of composites for various compositions of TiB_2

3.3 *Hardness of Composites*

The Vickers-micro hardness of AA7175-TiB_2 composites for varying volume % of TiB_2 particles is as shown in Fig. 7. It can be observed that as the volume % of TiB_2 particle increase, the hardness of the composites is increased. The hardness values of 67.63, 73.254 and 77.92 are obtained for AA7175 + 2.5 % TiB_2, AA7175 + 5 % TiB_2 and AA7175 + 7.5 % TiB_2 composites respectively, with a net increase of 13.21 %. The minimum value is reported for the composite material, reinforced with 2.5 %TiB_2 while the maximum value of composite material reinforced with 7.5 %TiB_2 particles. Similar behaviour was reported by Michael Rajan et al. when aluminium alloy was reinforced with titanium boride particles. This strengthening effect is due to the reinforcement of TiB_2 particles which are very hard. The propagation of cracks is resisted by the interaction between the dislocations and the reinforced TiB_2 particles. The grain refining action resulted due to the reinforcement of titanium boride particles also contributes towards improved hardness of the in situ fabricated composites. Similar results were reported by Han et al.

3.4 *Tensile Behaviour*

The Fig. 8 shows the variation of yield strength of the developed composites for different percentages of TiB_2 particles. Yield strength of 137.361, 171.84 and 186.1 MPa are obtained for AA7175 + 2.5 % TiB_2, AA7175 + 5 % TiB_2 and AA7175 + 7.5 % TiB_2 composites respectively.Net increase of 26.19 % was obtained in the values of yield strength of the developed composites as compared to the material reinforced with 2.5 % TiB_2 particles. Variation of % elongation of

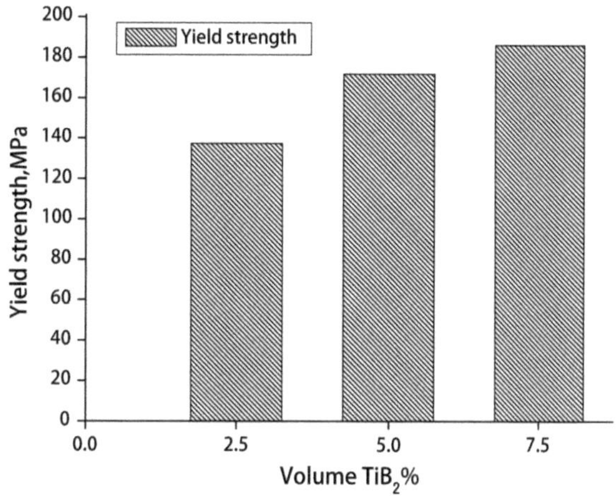

Fig. 8 Variation of yield strength of composites for various percentages of TiB_2

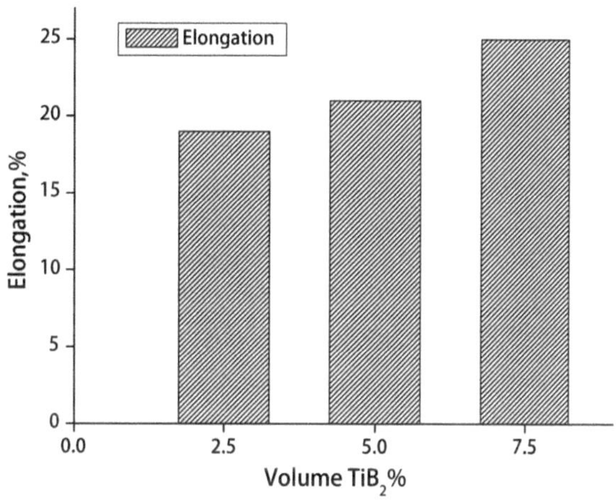

Fig. 9 Variation of % elongation of composites for various percentages of TiB_2

composites for various percentages of TiB_2 particulates is as shown in Fig. 9. Similarly variation of ultimate tensile strength of composites for various percentages of TiB_2 is as depicted in Fig. 10. As the percentage of reinforcement increases, increment in percentage elongation as well as in ultimate tensile strength was observed. % Elongation of 19 for 2.5 %, 21 for 5 % and 25 for 7.5 %TiB_2 reinforced composite material was noted; thus resulting net increment of 24 %. While Ultimate tensile strength of 151.38 MPa for 2.5 %, 198.62 MPa for 5 % and 216.47 MPa for 7.5 %TiB_2 reinforced composite material was observed, yielding net increment of 30.01 % as compared to material with 2.5 % reinforcement. Wang et al. reported similar behaviour when TiB_2 particles were reinforced with 3XXX series based aluminium alloy. Similarly when TiO_2 was reinforced with AA7075 alloy, improvement in tensile behaviour was reported by Murali et al. as compared to base alloy.

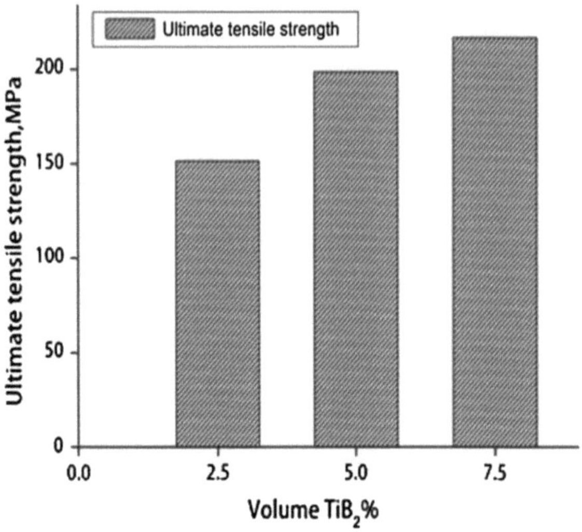

Fig. 10 Variation of ultimate tensile strength of composites for various percentages of TiB_2

Improvement in tensile behaviour of the developed composites is attributed to the reinforcement of titanium boride particulates. The addition of TiB_2 particles not only refines grain structure but also reduces the grain size as the percentage of reinforcement increase; which is one of the prime factors in enhancing the tensile behaviour of the prepared composites. In addition to this, homogeneous distribution of the reinforced particles in the matrix also contributes to the enhanced tensile behaviour of in situ synthesised composites.

4 Conclusions

AA7175/TiB_2 AMCs were successfully synthesized by the in situ reaction of inorganic salts such as K_2TiF_6 and KBF_4 to molten aluminium. The in situ reaction resulted in the formation of TiB_2 particles. In the current study, refinement of grain structure and modification of the mechanical properties due to the reinforcement of TiB_2 particulates in the developed AA7175-TiB_2 composites, as compared to the base alloy were studied.

1. From the optical and SEM micro-graphs it can be inferred that reinforcement of TiB_2 particulates to the aluminium alloy leads to grain refinement and grain structure modification. Also the homogeneous distribution of reinforced particulates can be observed from the micro-graphs.
2. The wear properties of the AA7175 alloy were considerably improved by the addition of TiB_2 particulates and the wear resistance of the composites was much higher than that of the unreinforced aluminium alloy. The wear resistance of composites increased with decreasing particle size of TIB_2 particulates.

Reinforcement of TiB_2 has led to apparent increase in the wear resistance and the coefficient of friction (CoF) was found to be decreasing with increasing percentage of TiB_2.

3. Addition of hard TiB_2 particulates lead to improved micro hardness and this also increased with increase in percentage of TiB_2 reinforcement. Net increment of 13.21 % was observed as compared to composites with 2.5 % TiB_2.
4. Enhancement of the tensile behaviour of in situ synthesised composites was observed with the reinforcement of titanium boride particulates and this increased with the increasing percentage of reinforcement phase. Thus an average increment of 26.73 % was reported. Thus, on closure it can be concluded that the newly developed material can be effectively used for automotive and aerospace applications.

References

Bartels, C., Raabe, D., Gottstein, G., Huber, U.: Investigation of the precipitation kinetics in an Al6061/TiB2 metal matrix composite. Mater Sci Eng. A. **237**, 12–23 (1997)

Deuis, R.L., Subramanian, C., Yellup, J.M.: Dry sliding wear of aluminium composites—a review. Compos. Sci. Technol. **57**, 415–435 (1997)

Herbert, M.A., Sarkar, C., Mitra, R., Chakraborty, M.: Microstructural Evolution, Hardness, and Alligatoring in the mushy state rolled cast Al-4.5Cu Alloy and in-situ Al-4.5Cu-5TiB2 composite. Metall. Trans. **38A**, 2110–2126 (2007)

Kumar, S., Chakraborty, M., Sarma, V.S., Murty, B.S.: Tensile and wear behaviour of in situ Al–7Si/TiB_2 particulate composites. Wear **265**, 134–142 (2008)

Lakshmi, S., Lu, L., Gupta, M.: In-situ preparation of TiB2 reinforced Al based composites. J Mater. Proc. Tech. **73**, 160–166 (1998)

Mandal, A., Maiti, R., Chakraborty, M., Murty, B.S.: Effect of TiB2 particles on aging response of Al-Cu alloy. Mater. Sci. Eng. A. **386**, 296–300 (2004)

Ramesh, C.S., Ahamed, A., Channabasappa, B.H., Keshavamurthy, R.: Development of Al 6063–TiB_2 in situ composites. Mater. Des. **31**, 2230–2236 (2010)

Ramesh, C.S., Pramod, S., Keshavamurthy, R.: A study on microstructure and mechanical properties of Al 6061–TiB_2 in situ composites. Mater. Sci. Eng. A. **528**, 4125–4132 (2011)

Surappa, M.K.: Aluminium matrix composites-Challenges and opportunities. Sadhana **28**, 319–334 (2003)

Suresha, S., Sridhara, B.K.: Effect of addition of graphite particulates on the wear behaviour in aluminium–silicon carbide–graphite composites. Mater. Des. **31**, 1804–1812 (2010)

Wang, C., Wang, M., Yu, B., Chen, D., Qin, P., Fenga, M.: The grain refinement behavior of TiB_2 particles prepared with in situ technology. Mater. Sci. Eng., A **459**, 238–243 (2007)

Wood, J.V., Davies P., Kellie, J.L.F.: Properties of reactively cast Aluminium-TiB2 alloys. Mater. Sci. Tech. **9**, 833–840 (1993)

Xue, J., Wang, J., Han, Y., Li, P., Sun, B.: Effects of CeO_2 additive on the microstructure and mechanical properties of in situ TiB_2/Al composite. J. Alloy. Compd. **509**, 1573–1578 (2011)

Zhao, D.G., Liu, X.F., Pan, Y.C., Bian, X.F., Liu, X.J.: Microstructure and mechanical properties of in situ synthesized (TiB_2 + $Al2O_3$)/Al–Cu composites. J. Mater. Process. Technol. **189**, 237–241 (2007)

Development of Bio-Based Epoxide from Plant Oil

Srikanta Dinda, Nikhil S.V. Reddy, U. Appala Naidu and S. Girish

Nomenclature

A_O, A_{OH}, A_i	Atomic weight of oxygen, hydroxyl group and iodine respectively g mol^{-1}
G_{exp}, G_{the}	Experimentally and theoretically obtainable α-glycol respectively, mole/100 g sample
IV_0	Initial iodine value, g of I_2/100 g sample
OO_{exp}, OO_{the}	Experimentally and theoretically obtained epoxide oxygen, g/100 g oil sample

1 Introduction

Transformation of petroleum to renewable resources is becoming difficult due to gradual depletion oil reserves and external environmental alarms. In search of renewable resources for day-to-day products, people are looking at valuable alternatives from natural sources (Namhoon et al. 2015). Oil derived from plants has been a major asset as a non-renewable source. Products from plant oil are known to be environment friendly, clean, and bio-degradable in nature. Conversion of unsaturated –C=C– double-bond into epoxide ring via epoxidation process received special care to synthesize bio-lubricant feed stock, which can be further cured with a suitable curing agent to make adhesives (Borugadda and Goud 2014). Due to increasing demand for petroleum-based epoxy resin, the epoxidized vegetable oil came into existence with its tremendous applications in the fields of paints and coating applications, besides acting as a good stabilizing agent (Ewumbua et al. 2015).

S. Dinda (✉) · N.S.V. Reddy · U. Appala Naidu · S. Girish
Department of Chemical Engineering, BITS Pilani, Hyderabad Campus,
Hyderabad 500078, India
e-mail: srikantadinda@gmail.com

R. Mohan B. et al. (eds.), *Materials, Energy and Environment Engineering*,
DOI 10.1007/978-981-10-2675-1_3

Epoxides are well known as stable intermediates which can be converted to a lot of products like glycols, carbonyl compounds, polyesters, polyurethanes, etc. (Padmasir et al. 2009). Epoxidation process depends on many parameters, such as, type of feedstock, catalyst used, temperature, etc. Okieimen et al. (2002) have studied various aspects of epoxidation of rubber seed oil using acetic acid and hydrogen peroxide. The literature shows that around 50 % epoxide conversion was obtained after 8.5 h at 50 °C. Epoxidized soybean oil is produced at a large scale for synthesis of plasticizers and polymer stabilizers (Biermann et al. 2000). Dinda et al. (2008, 2011) have studied the kinetics of epoxidation of cottonseed oil by peroxyacetic acid in the presence of H_2SO_4 and acidic ion exchange resin Amberlite IR 120H as acid catalyst. The literature shows that the selectivity of epoxide product is better in presence of Amberlite compared to H_2SO_4 as a catalyst under similar conditions. Goud et al. (2010) have studied the epoxidation reaction of karanja and jatropha oil by hydrogen peroxide and acetic acid catalyzed by Amberlite IR 120H. The study shows that around 72 % epoxide conversion was possible after 6 h at 60 °C.

Economic viability of a bio-based epoxide or epoxide derivative depends highly on availability and cost of raw oil. Based on the literature survey, no article was found on epoxidation of Nahor oil. Nahor trees are very common in tropical countries like India, Sri Lanka, Nepal, Burma, Malaysia etc. Nahor oil, with a composition of around 60 % oleic acid and 15 % linoleic acid (http://www.chempro.in/fattyacid.htm), is an appropriate raw material for epoxidation reaction. Epoxidised nahor oil can be a promising raw material for making bio-lubricant or an intermediate feedstock for chemical industries. The objective of the present study was to prepare epoxide product from nahor oil (Mesua ferrea Linn). In the present work, epoxidation of nahor oil was carried out by using different carboxylic acids, in the presence of different catalysts to obtain an optimum epoxide yield.

2 Experimental Details

2.1 Materials

Nahor oil (*Mesua ferrea* Linn) was procured from a local vendor (Sanjeevani Herbal Health Society, Hyderabad, India). Chemicals, such as, glacial acetic acid (CH_3COOH), formic acid ($HCOOH$), propanoic acid (CH_3CH_2COOH), 50 % aqueous hydrogen peroxide (H_2O_2), Amberlite IR 120H, and Sulphuric acid (H_2SO_4) for epoxidation reaction were procured from S D Fine-chem Ltd. (Mumbai, India). Dowex 50 WX8 was procured from Sigma Aldrich.

2.2 Experimental Procedure

Epoxidation reactions were performed in a mechanically agitated glass reactor. The reactor was fixed in a water bath to ensure a constant reaction temperature. The epoxidation method reported by Dinda et al. in 2008 was used for all the experimental runs dealing with different parameters. Fixed amount of nahor oil, organic carboxylic acid and acid catalyst were taken into the reactor and the mixture was stirred at 1600 rev/min for 20 min at 30 °C. Thereafter, the calculated amount of aqueous H_2O_2 was slowly added to the reaction mixture. The extent of reaction was monitored by analyzing the sample at regular intervals of time.

2.3 Chemical and Instrumental Analysis

The degree of unsaturation was determined according to Wijs method. Following expression was used to calculate the iodine value.

$$\text{Iodine value}\ (\mathrm{gI_2/100\,g\,oil}) = \frac{12.69 \times (V_B - V_S) \times N}{W} \tag{1}$$

Epoxide yield was determined by Pequot method (Pequot 1979). The relative percentage yield of epoxide was estimated as follows:

$$\text{Relative percentage yield of epoxide} = \left(OO_{exp}/OO_{the}\right) \times 100 \tag{2}$$

where

$$OO_{the} = \left[\frac{(IV_0/2A_i)}{100 + (IV_0/2A_i)A_0}\right] A_0 \times 100 \tag{3}$$

α-Glycol yield was determined according to May method (May 1987) in the presence of a non-aqueous medium.

$$\text{Relative percentage yield of } \alpha\text{-glycol} = \left(G_{exp}/G_{the}\right) \times 100 \tag{4}$$

where,

$$G_{the} = \left[\frac{(IV_0/2A_i)}{100 + (IV_0/2A_i)2A_{0H}}\right] \times 100 \tag{5}$$

Fourier transforms Infra-Red (FTIR) spectroscopy analysis of nahor oil and ENO was performed using PerkinElmer Spectrum GX (Model No. 4200) instrument to know the presence of functional groups qualitatively.

3 Results and Discussion

To investigate the epoxide formation, epoxidation reactions were performed at a constant temperature of 50 ± 1 °C in the presence of acid catalyst and oxygen carrier. Catalysts used for the reaction were H_2SO_4, Amberlite IR 120H, and Dowex 50WX8. The organic acids, namely, acetic acid (AA), formic acid (FA), and propanoic acid (PA) were used as oxygen carriers in the present study. H_2O_2 concentration expressed as H_2O_2-to-ethylenic unsaturation mole ratio of 2.0 was used for the present study. Carboxylic acid-to-ethylenic unsaturation mole ratio of 0.4 was maintained for each run. Repeatability analysis shows that the deviation between two experimental results is less than 3 %. Properties of the nahor oil as experimentally determined are: specific gravity 0.95 at 30 °C; iodine value 90 g I_2/100 g oil.

3.1 Epoxidation Reactions

A two-step process may be considered for the in situ epoxidation reactions: (i) aqueous phase formation of peroxycarboxylic acid by reacting between H_2O_2 and organic carboxylic acid in presence of acid catalyst (Eq. 6); (ii) reaction of peroxycarboxylic acid with double bond to form epoxide in organic phase (Eq. 7).

$$CH_3COOH + H_2O_2 + H^+ \rightleftharpoons CH_3COOOH + H_2O \tag{6}$$

$$-CH=CH- + CH_3COOOH \longrightarrow -HC\overset{O}{\frown}CH- \; + CH_3COOH \tag{7}$$

3.2 Comparison of Different Acid Catalysts on Epoxide Yield

To investigate the effect of acid catalysts on epoxide yield, epoxidation reactions were performed with acetic acid in the presence of different acid catalysts, namely, H_2SO_4, Amberlite IR 120H, and Dowex-50 under identical conditions. In each case, 48.7 g H_2O_2, and 8.77 g CH_3COOH were mixed with 100 g nahor oil at 50 °C temperature. Around 15 wt% catalyst loading was used for ion exchange resin catalyst, and in case of H_2SO_4, 2 wt% concentration was used in the reaction. The effectiveness of various catalysts on epoxide yield is shown in Fig. 1.

The figure shows that epoxide yield increased with increase in reaction time (within the experimental time limit) in the presence of Amberlite and Dowex ion exchange resin catalysts. However, in case of H_2SO_4, epoxide yield initially increased with reaction time, attained a maximum value (≈65 %), and then started to decrease with time. Among the catalysts studied, Amberlite IR 120H shows highest yield of epoxide as compared to the other two catalysts. A maximum of

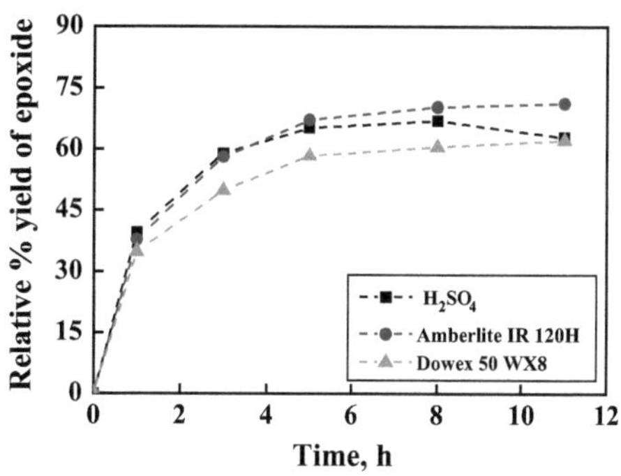

Fig. 1 Effect of catalyst on epoxide yield

70 % epoxide yield was obtained with Amberlite IR 120H at 50 °C after a reaction time of 5 h. For H_2SO_4, the decrease of epoxide yield with time is mainly due to the hydrolysis and other consecutive reactions of epoxide to glycol products. The yield of α-glycol (α-G) and iodine value conversion corresponding to epoxide yield with three different catalysts is tabulated in Table 1. Most probable ring cleavage reactions are shown by Eqs. (8) and (9).

$$-\underset{\diagdown O \diagup}{HC-CH}- + H_2O + H^+ \longrightarrow -CH(OH)-CH(OH)- \quad (8)$$

$$-\underset{\diagdown O \diagup}{HC-CH}- + CH_3COOH \longrightarrow -CH(OH)-CH(OCOCH_3)- \quad (9)$$

3.3 *Comparison of Different Carboxylic Acids on Epoxide Yield*

To compare the effect of organic acids on epoxide yield, epoxidation reactions were carried out with three different acids, namely, formic acid (FA), acetic acid (AA), and propanoic acid (PA) under identical conditions. In each case, 48.7 g H_2O_2, and 15 g Amberlite IR 120H were treated with 100 g nahor oil at 50 °C temperature under a stirring speed of 1600 rev/min. Organic acid-to-EU mole ratio of 0.4 was maintained for all three experiments, and results are shown in Fig. 2. The plot shows that, initially the epoxide conversion was relatively higher with FA as compared to

Table 1 Percentage of iodine value conversion and α-G yield for different catalysts

Catalyst		Time (h)			
		3	5	8	11
H_2SO_4	% conv. of IV	66.9	75.1	81.5	85.1
	Relative % yield of α-G	2.9	4.8	9.9	16.1
Amberlite IR 120H	% conv. of IV	64.1	73.5	79.2	83.2
	Relative % yield of α-G	1.7	1.8	3.9	6.9
Dowex 50 WX8	% conv. of IV	56.4	65.1	68.6	72.1
	Relative % yield of α-G	1.6	1.7	3.2	5.5

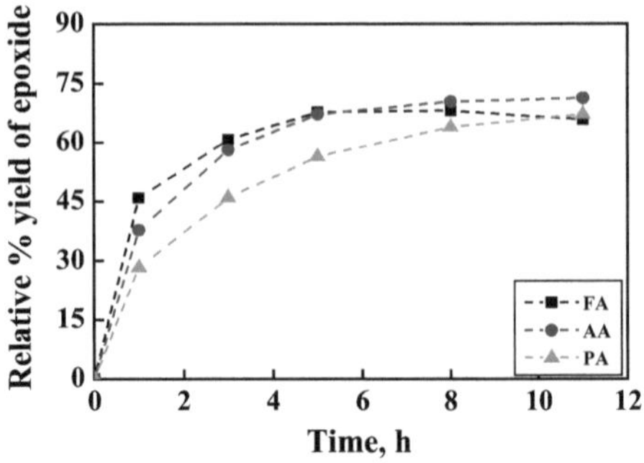

Fig. 2 Effect of carboxylic acids on epoxide yield

AA or PA. Epoxide yields with FA, AA, and PA after 3 h of reaction are 61, 57 and 46 % respectively. However, as time progressed, epoxide yield started to decrease after reaching a maximum value of around 68 %, whereas, the iodine value conversion increased gradually with time in presence of FA. Conversely, in case of AA and PA, both iodine value conversion and epoxide yield increased gradually within the experimental time limit. The result shows that, after 11 h of epoxidation, the epoxide yields are 71 and 67 % with AA and PA respectively. In case of AA, no significant (≈1 %) change in epoxide yields was observed between 8 and 11 h. Iodine value conversion and α-glycol yield corresponding to epoxide yield are given in Table 2. The order of effectiveness of carboxylic acids was found to be FA > AA > PA. The decreasing order of reactivity from FA to PA may be due to the gradual increase of positive inductive effect (+I) of $-CH_3$ and $-C_2H_5$ group attached to the carboxylic group in AA and PA respectively.

3.4 *FTIR Analysis of Nahor Oil and Products*

To identify the functional groups present in the nahor oil and epoxidised nahor oil (ENO), FTIR analysis was performed for both the oil and ENO samples. A typical FTIR spectrum of the oil and ENO is shown in Fig. 3. A characteristic band at 1583 cm^{-1} (in Fig. 3a) indicates the presence of C=C bond in the raw oil. A broad band at 3429 cm^{-1} and a band at 829 cm^{-1} in Fig. 3b indicate the formation of

Table 2 Percentage of iodine value conversion and α-G yield for different carboxylic acids

Carboxylic acid		Time (h)			
		3	5	8	11
Formic acid (FA)	% conv. of IV	68.9	78.3	84.7	86.9
	Relative % yield of α-G	4.1	6.8	11.8	18.1
Acetic acid (AA)	% conv. of IV	64.1	73.5	79.2	83.2
	Relative % yield of α-G	1.7	1.8	3.9	6.9
Propanoic acid (PA)	% conv. of IV	49.6	61.2	71.2	75.7
	Relative % yield of α-G	1.1	1.5	2.8	4.8

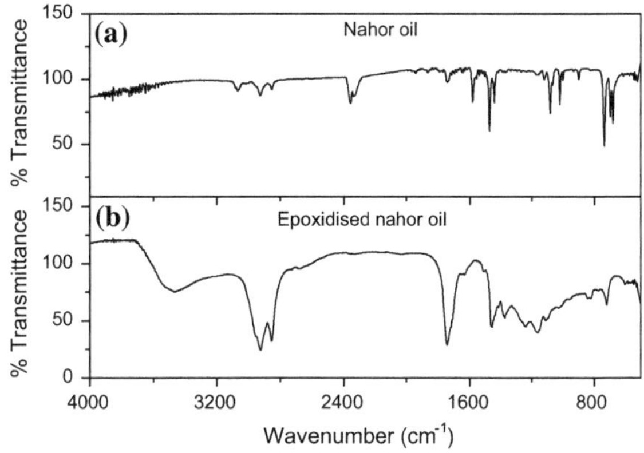

Fig. 3 FTIR spectra of **a** nahor oil; **b** epoxidised nahor oil (ENO)

O–H group and C–O group respectively in the product. The band at around 1740 cm^{-1} corresponds to C=O stretching. The presence of O–H group in the ENO (Fig. 3b) confirmed the cleavage of epoxide ring during the course of reaction.

4 Conclusion

The present study reveals that nahor oil could be epoxidised effectively by peroxyacid generated in-situ from carboxylic acid, and H_2O_2 at moderate temperature range of 50 °C. Different carboxylic acids, namely, formic acid, acetic acid, and propanoic acid on epoxidation reaction were examined as an oxygen carrier in the present study. The effectiveness of carboxylic acids was found to be in the order of formic acid > acetic acid > propanoic acid. The effect of different acid catalysts namely H_2SO_4, Amberlite IR 120H, and Dowex 50 WX8 on epoxide yield was examined. Amberlite IR 120H shows better epoxide yield compared to H_2SO_4 and Dowex 50 WX8. It was possible to obtain around 70 % epoxide yield within 8 h at moderate temperature of 50 °C. FTIR analysis confirmed the formation of glycol product during the course of reaction. The findings of the present study will provide useful information for making bio-plasticizers and lubricants.

Acknowledgments We are grateful to BITS Pilani Hyderabad Campus for extending the necessary supports and facilities for the above studies.

References

Biermann, U., Friedt, W., Lang, S., Luhs, W., Machmuller, G., Metzger, J.O., Klaas, M.R., Schafer, H.J., Schneider, M.P.: New syntheses with oils and fats as renewable raw materials for the chemical industry. Angew. Chem. Int. Ed. **39**, 2206–2224 (2000)

Borugadda, V.B., Goud, V.V.: Epoxidation of castor oil fatty acid methyl esters (COFAME) as a lubricant base stock using heterogeneous ion-exchange resin(IR–120) as a catalyst. Energy Procedia **54**, 75–84 (2014)

Dinda, S., Patwardhan, A.V., Goud, V.V., Pradhan, N.C.: Epoxidation of cottonseed oil by aqueous hydrogen peroxide catalysed by liquid inorganic acids. Bioresour. Technol. **99**, 3737–3744 (2008)

Dinda, S., Goud, V.V., Patwardhan, A.V., Pradhan, N.C.: Selective epoxidation of natural triglycerides using acidic ion exchange resin as catalyst. Asia-Pac. J. Chem. Eng. **6**, 870–878 (2011)

Ewumbua, M.M., Darrin, M.H., Dennis, P.W.: Characterizing the epoxidation process conditions of canola oil for reactor scale-up. Ind. Crops Prod. **67**, 364–372 (2015)

Goud, V.V., Dinda, S., Patwardhan, A.V., Pradhan, N.C.: Epoxidation of Jatropha (Jatropha curcas) oil by peroxyacids. Asia-Pac. J. Chem. Eng. **5**, 346–354 (2010)

http://www.chempro.in/fattyacid.htm. Top-notch technology in production of oils and fats

May, C.A. (1987). Epoxy Resins: Chemistry and Technology, 2nd edn. Marcel Dekker Inc, New York

Namhoon, K., Yonghui, L., Xiuzhi, S.S.: Epoxidation of Camelina sativa oil and peel adhesion properties. Ind. Crops Prod. **64**, 1–8 (2015)

Okieimen, F.E., Bakare, O.I., Okieimen, C.O.: Studies on the epoxidation of rubber seed oil. Ind. Crops Prod. **15**, 139–144 (2002)

Padmasiri, K.G., Michael, O.B., Laleen, K.: Epoxidation of some vegetable oils and their hydrolysed products with peroxy formic acid—optimized to industrial scale. J. Nat. Sci. Found Sri Lanka **37**, 229–240 (2009)

Pequot, C.: Standard Methods for the Analysis of Oils, Fats and Derivatives, Part-1, 5th edn. Pergamon Press, Germany (1979)

Experimental and FEM Analysis on the Mechanical Properties of Al-8011 Alloy Reinforced with Fly-Ash and E-Glass Fibers

Chetan Kulkarni, Akshay Mohan Pujar and Balappa Hadagali

1 Introduction

Metal matrix composites (MMCs) are progressively turning into another class of material in automobile, aviation and mineral processing applications in light of the fact that their properties can be customized through the addition of selected reinforcements (Hamouda et al. 2007). Due to their lower weight and excellent thermal conductivity properties, aluminium alloys are utilized in many applications as an engineering material. Due to their blend of high strength to weight and high stiffness to weight ratios research in recent years has moved towards particulate reinforced metal matrix composites (Sudarshan and Surappa 2008). Particulate reinforced metal matrix composites are acquiring significance due to their specific strength and specific stiffness at room or elevated temperatures. Aluminium matrix composites (AMCs) are potential materials for different applications because of their comparatively better physical and mechanical properties. When compared to traditional engineering materials the properties such as wear, stiffness, creep, specific strength and fatigue can be improved by the addition of suitable reinforcements. Fly ash is a discrete discontinuous reinforcement among the low density and low-priced reinforcements and is available in considerable quantity. It is a by-product in solid form gained from combustion of coal in thermal power plants (Ganesh Kumar et al. 2014). Embedding and binding of fibers by a continuous matrix forms fibrous composites. Due to their higher strength-weight ratio,

C. Kulkarni (✉) · A.M. Pujar (✉)
B.V. Bhoomaraddi College of Engineering and Technology, Vidyanagar, Hubballi, India
e-mail: chetan27993@gmail.com

A.M. Pujar
e-mail: ampujar0602@gmail.com

B. Hadagali
Samarth Group of Institutions, College of Engineering, Belhe, India
e-mail: balappa.hadagali2015@gmail.com

R. Mohan B. et al. (eds.), *Materials, Energy and Environment Engineering*,
DOI 10.1007/978-981-10-2675-1_4

fatigue strength-weight ratio, fatigue damage tolerance, modulus-weight ratio, and being lighter in weight, synthetic fibers are extraordinarily better than those of metallic materials (Pavanu Sai et al. 2014). As a result of its simplicity, adaptability, and usability to a large capacity production, Stir casting is most generally utilized and acknowledged for commercially practicable method. Additionally, it is most practical method to fabricate metal matrix composites. To achieve homogeneous distribution of metal matrix composites in an economical way, a mechanized stirring setup was developed. The volume fraction and the size of the reinforcement affect the mechanical properties of the composites. Aluminium metal composites have excellent mechanical properties in industries by its competent materials (Sakthivel et al. 2008). Aluminium alloy reinforced with glass fibers possess a unique combination of high specific tensile and compressive strength, low coefficient of thermal expansion, good wear and fatigue resistance than the corresponding non-reinforced matrix alloy system (Pavanu Sai et al. 2014). Fly ash reinforced aluminium matrix composites have many implementations because of their hardness, wear resistance, higher tensile and compression than non-reinforced AMCs (Shanmughasundaram et al. 2011). Various compositions of Al-8011 hybrid composites are fabricated by stir-casting method by varying the weight percentage of E-glass fiber with constant fly ash. Tensile and compression strength of the base Al 7075 alloy are enhanced by reinforcing with E-glass and fly ash particulates (Kammer et al. 2012). Al 6061 reinforced with fly ash and E-glass rises the mechanical properties of the composite (Kumar and Swamy et al. 2011). Similarly a study on Al-8011 reinforced with fly ash and E-glass fiber for different composition is carried by Yogananda et al. (2013). In this direction an attempt has been made to study the tensile and compression behaviour of Al-8011/E-glass fiber/Fly ash composites developed. The objective of the present work is to study the influence of reinforcements of fly ash and E-glass fiber on mechanical properties of Al-8011 using experimental and analytical studies.

2 Experimental

2.1 Raw Materials and Their Properties

Al-8011 is used as a matrix material and it is an Al–Fe–Si alloy commonly used as foil and closures as well as heat exchanger fin stock due to its unique combination of attractive properties such as its low weight, corrosion resistance, and easy maintenance of final product. The reinforcing material Fly ash and E-glass fiber were used for the present work. Fly ash is by-product acquired by coal combustion, and is made out of the fine particles that are driven out of the boiler with the vent gases. The particles of fly ash are spherical in form and their size ranges from less than 1 to 100 μm and the specific surface area ranges between 250 and 600 m^2/kg. The specific gravity of fly ash alters in the scale of 0.6–2.8 g/cc. E-glass fiber is a low alkali glass having density of 2.58 g/cc^3. E glass is chemically resistant, has relatively low density, high strength and stiffness properties.

2.2 *Fabrication of Composites*

The Al-8081, fly ash and E-glass fiber reinforced hybrid composite was processed through stir casting method. The requisite amounts of Al-8011 alloy was super-heated using Vertical Muffle furnace (VBF-1200X; MTI Corporation; Richmond, USA) up to a temperature of 750 °C and then degassed by purging 5 g of hexachloroethane (C_2Cl_6) tablets. The degassing was accompanied by a slight drop in temperature. Once the melt was reheated to the set temperature, fly-ash and E-glass fiber were added in required proportions followed by stirring for uniform mixing using a zirconia coated graphite rod to prepare the hybrid composite. Later the red hot liquid mixture is poured into the plate-shaped cast iron mould. Upon cooling the hybrid composites of different compositions were machined to ASTM standard test specimen size for carrying out tensile and compression tests. Hybrid composites of different compositions were prepared by keeping fly-ash constant and varying E-glass fiber by weight (%). The composition and code prepared for Al-8011 alloy and composites are shown in Table 1.

2.3 *Brinell Hardness Test*

Hardness of the composites was tested as per the ASTM standard E10-08 using the Brinell hardness tester. The specimens with dimensions 20 mm × 20 mm were cut from the cast obtained and then polished. The test was carried out by applying the constant load of 500 kgf for the time interval of 15 s using the indenter.

2.4 *Tensile and Compression Tests*

In the present research work universal testing machine is used for determining the tensile strength, percentage elongation and compression strength of the specimen as

Table 1 Different compositions of prepared hybrid composite

Sl. no.	Composition	Code
1	Al-8011 alloy (base)	Al
2	Al-8011+6 % fly ash+1 % E-glass fiber	C1
3	Al-8011+6 % fly ash+3 % E-glass fiber	C2
4	Al-8011+6 % fly ash+5 % E-glass fiber	C3
5	Al-8011+6 % fly ash+7 % E-glass fiber	C4

per ASTM standard E8-08 and ASTM standard E9-89a respectively. The tensile specimen with gauge length of 62.5 mm and diameter of 12.5 mm is prepared for the test. Similarly, specimen with gauge length of 60 mm and diameter 20 mm is prepared for the compression test. Five specimens for each composition are employed for the tests. Test speed of 4 mm/min is used in carrying out above tests.

3 Results and Discussions

The results and discussion obtained from the experimental investigations of physical and mechanical properties are presented in this section and are analysed along with analytical results for comparison.

3.1 Hardness

Figure 1 shows Brinell hardness number (BHN) for varying E-glass fiber percentage with constant 6 % fly-ash. In common it can be seen from the graph that hardness number increases with the increase in E-glass fiber percentage. From the graph, hardness value is found to be minimum for base alloy, and is maximum for Al-8011 +6 % fly ash+7 % E-glass. The increase in hardness could be due to the fact that the reinforcement material is much harder than the matrix material. And also inclusion of fly ash improves the brittle nature of alloy thereby improving the hardness of the developed composites. Similar results were reported by Ramesh et al. (2010). The hardness of composite has increased from 57.2 BHN (for base alloy) to 73.41 BHN (Al-8011+6 %fly ash+7 %E-glass) thus showing 22.08 % increment.

Fig. 1 Brinell hardness number versus varying % E-glass fiber for constant 6 % fly-ash

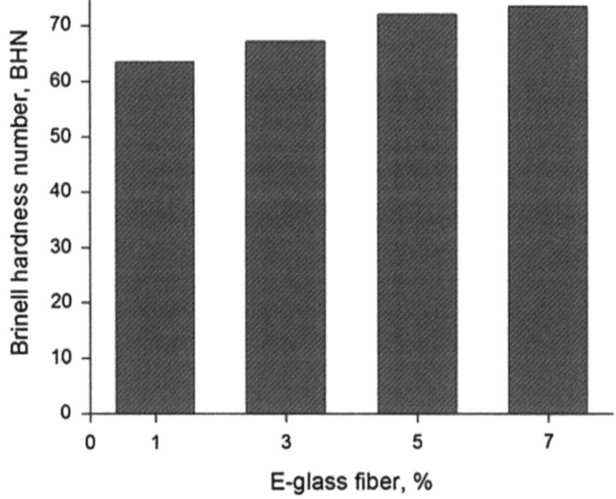

3.2 *Tensile Strength*

Figure 2a illustrates the variation of ultimate tensile strength as a function of varying % E-glass fiber and constant 6 % Fly-ash. From Fig. 2a, it follows that the tensile strength is minimum for Al-8011 alloy and maximum for Al-8011+6 % Fly ash+5 % E-glass fiber. The tensile properties vary according to the percentage of reinforcements added. The tensile strength of composite was increased from 139 to 178.79 MPa as shown in Fig. 2a. The yield strength of composite was increased from 125.3 to 168.74 Mpa as shown in Fig. 2b. Figure 2c shows the variation of percentage elongation with different percentages of E-glass fiber, the ductility of Al-8011+6 % fly ash+5 % E-glass fiber composite is high as compared to Al alloy. The percentage elongation of the composites increases with the increase in weight percentage of reinforcement up to 5 % E-glass fiber and decreases for 7 % E-glass fiber. This is because of the existence of fly ash particles and E-glass fibers, which limit the plastic deformation of the composite. More loads were carried for plastic deformation due to limitation by E glass fibre and fly ash particles, which also

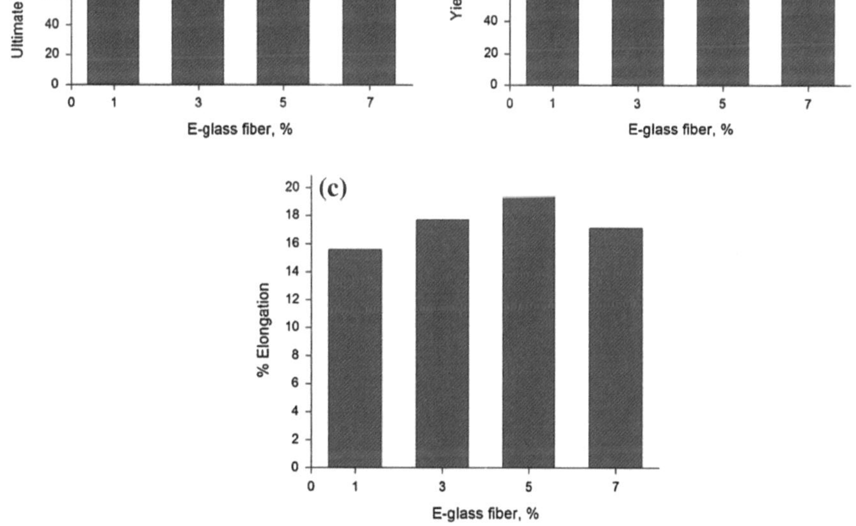

Fig. 2 **a** Ultimate tensile strength versus varying % E-glass fiber for constant 6 % fly-ash **b** yield strength versus varying % E-glass fiber for constant 6 % fly-ash **c** % elongation versus varying % E-glass fibers for constant 6 % fly ash

resulted in a higher tensile strength of the composite. Fly ash particles help in strengthening the matrix, behave as obstacles to the dislocations when taking up the load applied from this value, it is concluded that the tensile strengths of the composites are greater than that of the aluminium base alloy.

The tensile strength was decreased from 178.79 to 168.54 MPa (5.73 %) for 7 % E-glass fiber composition. This is due to increase in interactions between the fibers that may also cause difficulties in separation resulting in lowering of mechanical properties. Similar behaviour was demonstrated by Prajit et al. (2014).

3.3 Compression Strength

Figure 3 shows the results acquired from uniaxial compression as a function of E-glass fiber at constant 6 % fly-ash. From the above graph it is observed that compression strength are minimum for base alloy, and is maximum for Al-8011 +6 % fly ash+5 % E-glass. Increase in percentage of E-glass fiber increases the compression strength of composites. The enhancement in compression strength is because of the rise in the density of the composite material. This proves that the inclusion of fly ash and E-glass to a soft matrix increases its density and thus increasing the compressive strength. It may also be because of hardening of base alloy by the addition of E-glass fiber and fly ash particulates. Because compressive forces are acting normal to the section hence the loads are uniformly distributed in the section thus the reinforcement material carries the load. Identical results were revealed in numerous investigations carried on the compressive strength of composite materials by Prajit et. al (2014).

The compressive strength of composite was increased from 601.2 to 669.74 MPa as shown in figure which is 10.23 % increase in strength.

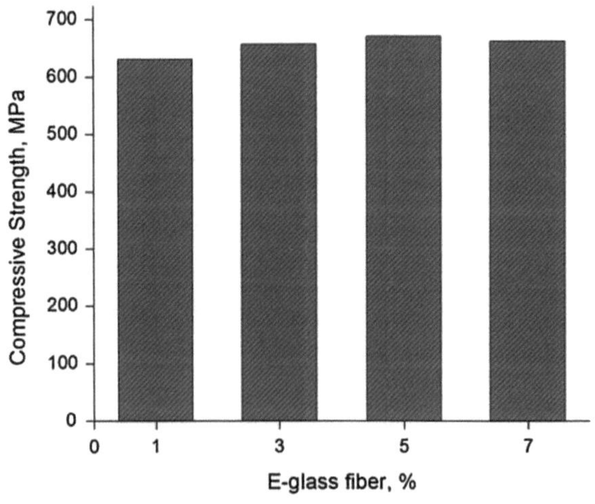

Fig. 3 Compressive strength versus varying % E-glass fiber for constant 6 % fly-ash

4 FEM Approach

With the advancement of computers, finite element analysis has become one of the most important tools available to an engineer for design analysis. In order to validate the experimental results Finite Element Analysis (FEA) approach was adopted. The meshing of specimen is done using tetrahedral element and the properties are given to the specimen using Halpin Tsai equation (Halpin and Kardos 1976). The number of elements and nodes are 10804 and 47343 respectively for tensile specimen analysis and number of elements and nodes are 4560 and 20229 respectively for compression specimen analysis (Figs. 4 and 5; Table 2).

The finite element method is one of the most general procedures for solving complex analysis problems. For performing finite element analysis the material was considered to be isotropic in nature and the boundary condition and load conditions applied were similar to the experimental condition. Results revealed that tensile results for numerical analysis are better than experimental results. This deviation of results occurred due to manufacturing defects of composites like blow holes, porosity etc. Further in analysis part isotropic behaviour was considered, but it is not possible to achieve isotropic behaviour practically using random oriented fibres due to stress concentrations at fibre ends.

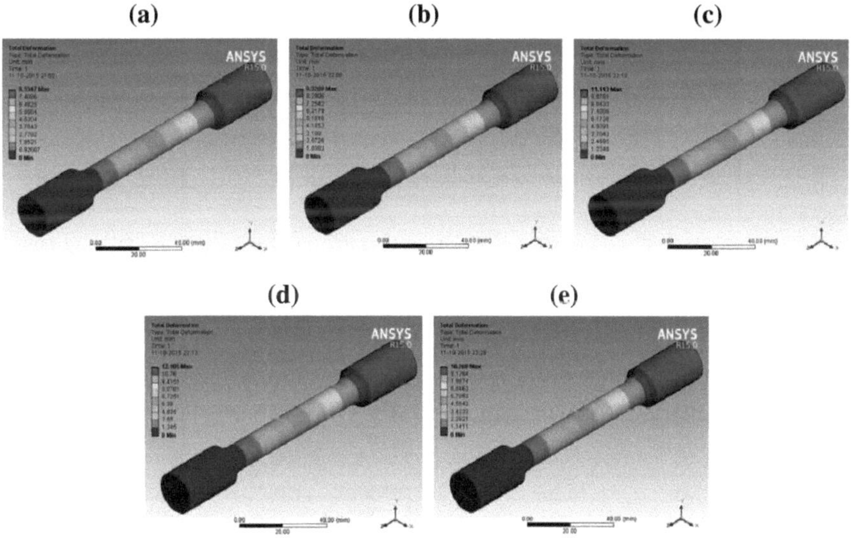

Fig. 4 Deformation of tensile specimens **a** A1 alloy **b** C1 **c** C2 **d** C3 **e** C4 composites respectively, using FEA analysis

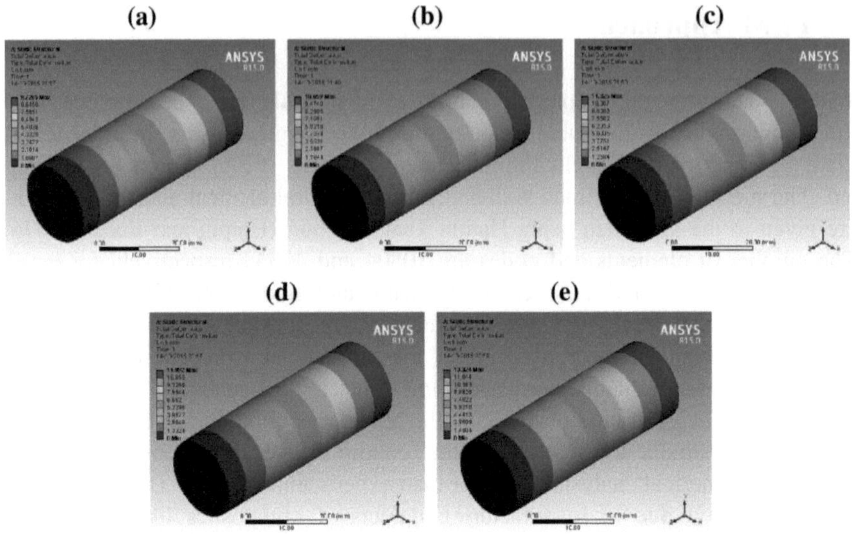

Fig. 5 Deformation of compressive specimens **a** Al alloy **b** C1 **c** C2 **d** C3 **e** C4 composites respectively, using FEA analysis

Table 2 Comparison of experimental and numerical (FEA) results

	Tensile			Compression			
E-glass fiber (%)	Experimental (mm)	Numerical (mm)	Error (%)	Load (N)	Experimental (mm)	Numerical (mm)	Error (%)
0 % E-glass fiber (Al-8011 alloy)	8.125	8.3347	2.52	184532	10.21	9.7265	4.73
1 % E-glass fiber	8.935	9.3269	4.22	195170	11.58	10.659	7.953
3 % E-glass fiber	10.546	11.113	5.10	210490	12.06	11.325	6.095
5 % E-glass fiber	11.254	12.105	7.03	225740	12.98	11.992	7.611
7 % E-glass fiber	9.7941	10.269	4.75	230619	14.19	13.324	6.102

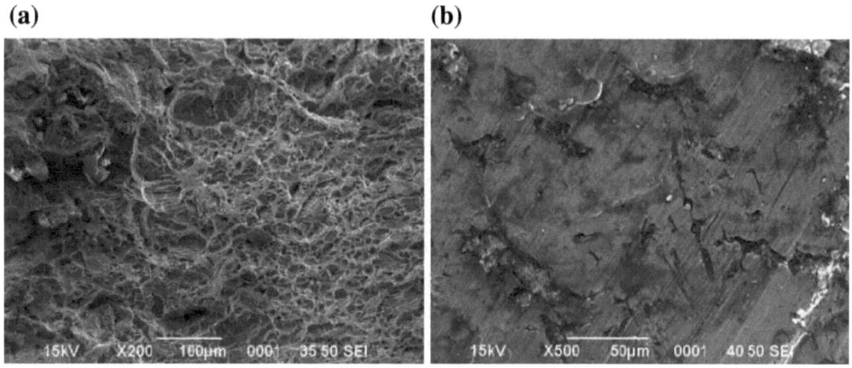

Fig. 6 Fracture surfaces for Al-8011 with 6 % fly ash and 5 % E-glass fiber composite of **a** tensile specimen **b** compression specimen

5 Scanning Electron Microscope Analysis

Each type of the composite materials reacts differently to the externally applied load, and the failure pattern is unique to the same class of composites. The fracture surfaces of tensile and compression specimens are examined for fracto graphs under scanning electron microscope. Figure 6a, b shows micrographs of the fractured surfaces of tensile and compression specimens for Al-8011+6 % fly ash+5 % E-glass fibre composite.

The SEM micrographs of test specimens are taken from the fractured surfaces of tensile and compression specimen of the developed composites. Figure 6a shows SEM micrographs of fractured surfaces of tensile specimen consisting of nearly dimple pattern, therefore confirming ductile fracture. The presence of silica of fly ash induces a white colour in the images and black phase in the micrographs is because of the presence of graphite flakes. From the above micrographs, a close interfacial adhesion between the Al-8011 matrix with fly ash particulates can be observed. Figure 6b shows SEM images of fractured surfaces of compression specimen; it can be revealed that incorporation of E-glass acts as a tailoring agent in holding the matrix, also they act as closing agents for several over lapping layers in the matrix. The agglomeration of structures is due to the amorphous crystalline nature of aluminium matrix.

6 Conclusions

From the exhaustive experimental and analytical results obtained the following conclusions were made:

1. Stir casting method can be successfully employed in for preparing the hybrid composites reinforced with fly ash and E-glass fibre.

2. Ultimate tensile strength of the developed hybrid composites has been increased as compared to base alloy showing net increase of 22.25 %. The deformation of tensile specimens were analysed experimentally and analytically, the results obtained were found to be in good agreement with an average error of 4.724 %.
3. On the addition of E-glass fibre and fly ash, 10.23 % increase in compression strength of the composites has been observed as compared to as cast Al-8011 alloy. Similarly, an average error of 6.49 % was obtained when deformation of compression specimens were evaluated experimentally and analytically.
4. Increment of 22.08 % in hardness of the hybrid composites reinforced with fly ash and E-glass fibre has been obtained in comparison with base alloy.
5. From the SEM analysis the nearly dimple pattern confirms ductile fracture and close interfacial adhesion between the Al-8011 matrix with E-glass fibres and fly ash particulates confirms compressive fracture.

On the closure it can be concluded that the newly developed composite could be used for automotive and aerospace applications.

References

Arunkumar, M.B., Swamy, R.P.: Evaluation of mechanical properties of Al6061, fly ash and E-glass fibre fiber reinforced hybrid metal matrix composites. ARPN J. Eng. Appl. Sci. **6**, 1–7 (2011)

Ganesh Kumar, A.G., Ranganath, G., Pounraj, B., Shylin Jose, H., Sakthivel, M.: Experimental analysis of E-glass fibre fiber and fly ash reinforced to E-waste aluminium. J. Chem. Pharm. Sci. **4**, 138–140 (2014)

Halpin, J.C., Kardos, J.L.: The Halpin-Tsai equations: a review. Polym. Eng. Sci. **16**, 344–352 (1976)

Hamouda, A.M.S., Sulaiman, S., Vijayaram, T.R., Sayuti, M., Ahmad, M.H.M.: Processing and characterization of particulate reinforced aluminium silicon matrix composite. J. Achievements Mat. Manufact. Eng. **25**, 11–16 (2007)

Pavanu Sai, J., Srinivasa Rao, A., Dr. Hari Babu, N.: Experimental investigation of effect of aluminum filler material on thermal properties of palmyra fiber reinforced composite. Int. J. Eng. Res. Ind. Appl. **4**, 23–29 (2014)

Prabhakar, K., Shivanand, H.K., Santhosh Kumar, S.: Experimental studies on mechanical properties of E-glass short fibers and fly ash reinforced Al 7075 hybrid metal matrix composites. Int. J. Appl. Res. Mech. Eng. **1**(4), 111–115 (2012)

Prajit, R., Santhosh Kumar, V., Sathwik, V., Anantha Narayanan, R.: Evaluation of mechanical properties of Al 6061, fly ash reinforced with E-glass fiber composites. IOSR J. Mech. Civ. Eng. **11**, 33–39 (2014)

Ramesh, D., Swamy, R.P., Chandrashekar, T.K.: Effect of weight percentage on mechanical properties of frit particulate reinforced Al 6061 composite. ARPN J. Eng. Appl. Sci. **5**, 32–36 (2010)

Sakthivel, A., Palaninathan, R., Velmuruganand, R., Rao, P.R.: Production and mechanical properties of SiC particle reinforced 2618 aluminium alloy composites. J Mater Sci. **43**, 7047–7056 (2008)

Shanmughasundaram, P., Subramanian, R., Prabhu, G.: Some studies on aluminium—fly ash composites fabricated by two step stir casting method. Eur. J. Appl. Eng. Sci. Res. **63**, 204–218 (2011)

Sudarshan, Surappa, M.K.: Synthesis of fly ash particle reinforced A356 Al composites and their characterization. J. Mat. Sci. Eng. A. **480**, 117–124 (2008)

Yogananda, A., Shivanand, H.K., Santhosh Kumar, S.: Investigation on mechanical properties of E-glass fibre and fly ash reinforced Al 8011 based hybrid composites. Int. J. Mech. Eng. Technol. **4**, 78–83 (2013)

Effects of Single, Double, Triple and Quadruple Window Glazing of Various Glass Materials on Heat Gain in Green Energy Buildings

Kirankumar Gorantla, Saboor Shaik and Ashok Babu Talanki Puttaranga Setty

1 Introduction

Modern residential and commercial buildings use glass materials for building enclosures. Extensive use of glass in building envelopes increases the heat gain, which in turn increases the energy requirement to maintain the building at comfortable conditions. In general solar radiation enters in buildings through walls, floor, roof and windows. Most of the direct solar radiation enters through window glass materials only. Selection of appropriate glazing type can reduce the cooling loads in buildings and result in significant power savings for Air-conditioning. Thermal and optical properties of clear and bronze glass windows with multiple glazing filled with air and argon were evaluated by Singh et al. (2011). The influence of the insulation location inside the flat roof on thermal performance of buildings was reported in the literature (Saboor et al. 2015). This work presents the thermal performance of buildings with various window glasses such as single, double, triple and quadruple glasses of clear, bronze, green and bronze-reflective glasses.

2 Experimental Methodology

The Perkin- Elmer lambda 950 Spectrophotometer was used to measure the spectral optical properties of clear, bronze, green and bronze-reflective glass materials as per the standards (ASTM E424 1971). The experiments were conducted with four

K. Gorantla · S. Shaik (✉) · A.B.T.P. Setty
Department of Mechanical Engineering, National Institute of Technology Karnataka, Surathkal, Mangalore 575025, Karnataka, India
e-mail: saboor.nitk@gmail.com

R. Mohan B. et al. (eds.), *Materials, Energy and Environment Engineering*,
DOI 10.1007/978-981-10-2675-1_5

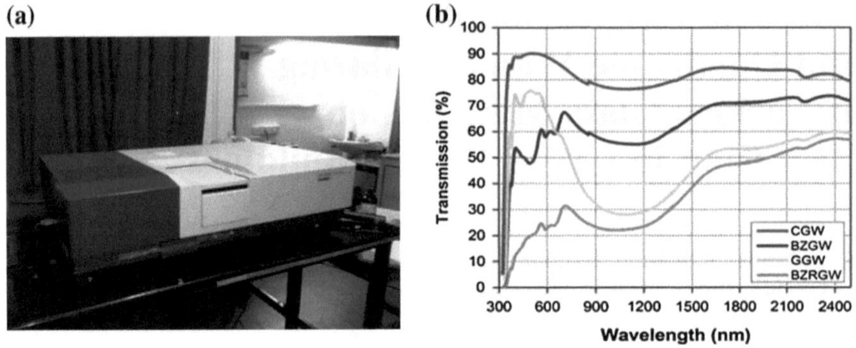

Fig. 1 **a** Perkin-Elmer lambda spectrophotometer **b** spectral transmission of different glass materials

Table 1 Solar thermal properties of glass materials

Glass material	Transmittance (τ) (%)	Reflectance (ρ) (%)	Absorptance (α) (%)
Clear glass	73	11	16
Bronze glass	53	9	38
Green glass	47	9	44
Bronze reflective glass	38	34	28

different glass materials of thickness 5 mm at 0° angle of incidence and at various wavelengths ranging from 300 to 2500 nm. Figure 1a shows the Perkin-Elmer lambda 950 Spectrophotometer.

Figure 1b shows the spectral transmission of clear, bronze, green and bronze-reflective glass materials. From the measured spectral optical property data, the solar thermal properties such as transmittance and reflectance of glass materials can be obtained as per BS: EN 410 standard method. Equations (1) and (2) are used to compute solar thermal properties (BS: EN 410 1998) from experimentally measured spectral optical property data. Table 1 shows the solar thermal properties of glass materials.

$$\tau_s = \frac{\sum_{300}^{2500} S_\lambda \tau(\lambda) \Delta\lambda}{\sum_{300}^{2500} S_\lambda \Delta\lambda} \tag{1}$$

$$\rho_s = \frac{\sum_{300}^{2500} S_\lambda \rho(\lambda) \Delta\lambda}{\sum_{300}^{2500} S_\lambda \Delta\lambda} \tag{2}$$

Table 2 Thermo-physical properties of building materials

Building material	Thermal conductivity (k) (W/mk)	Specific heat (C_p) (J/kgk)	Density (D) (kg/m^3)
Burnt brick	0.811	880	1820
RCC	1.58	880	2288
Dense concrete	1.74	880	2410
Cement plaster	0.721	840	1762

3 Thermal Analysis

In this study, the building models of size 5 m × 5 m × 5 m were designed in Design builder 4.3.0.039 licensed version. The walls of the building models are built with burnt brick of thickness 0.22 m and plastered on both sides of the brick wall. The thickness of the cement plaster used on each side is 0.0125 m. The roof is built with reinforced cement concrete of thickness 0.15 m with cement plaster on external and internal side of the roof. The dense concrete of thickness 0.15 m is used for the floor of the building model. The floor is plastered on inside of the building. The window is placed on the one of the four walls of the building model. The window to wall ratio maintained is 40 % WWR as per the GRIHA standard. Single glazed window, double glazed window, triple glazed window and quadruple glazed windows of clear, bronze, green and bronze- reflective glass materials in four orientations such as east, west, north and south were used one by one. Total sixty four building models were designed and thermal analysis was carried out using Energy plus 8.1 simulation tool. The heat gain in building through walls, roof, floor and windows were computed for four climatic regions of India such as Hot and dry (Ahmedabad), temperate (Bangalore), warm and humid (Bombay) and composite (New Delhi). The peak summer day of each region is considered as the reference design day. May 15th, April 15th, May 15th and June 21st are the peak summer days in Ahmedabad, Bangalore, Bombay and New Delhi regions, respectively as per IS 11907 standard (1986). The thermo-physical properties of building material are taken as per the Indian standard (SP: 41 1987). Table 2 shows thermo-physical properties of building materials.

4 Results and Discussions

4.1 *Heat Gain in Buildings of Hot and Dry (Ahmedabad) and Temperate (Bangalore) Climatic Regions*

Figure 2a shows heat gain in buildings of the Ahmedabad climatic region. From the results, it is observed that heat gain in south orientation is the least as compared to the other orientations of the window glass. Hence, south orientation is the best for the location of windows from the least heat gain point of view for reduced cooling loads. The total heat gain in buildings when the single glazing window is located in south orientation of the Ahmedabad climatic region is 38.80, 37.93, 37.49 and 36.62 kWh for clear, bronze, green and bronze-reflective glasses, respectively. The total heat gain in buildings when double glazing window is located in south orientation of the Ahmedabad climatic region is 37.38, 36.34, 35.92 and 35.27 kWh for clear, bronze, green and bronze-reflective glasses, respectively. The total heat gain in buildings when triple glazing window is located in south orientation of the Ahmedabad climatic region is 36.56, 35.51, 35.14 and 34.64 kWh for clear, bronze, green and bronze-reflective glasses, respectively. The total heat gain in buildings when quadruple glazing window is located in south orientation of the Ahmedabad climatic region is 35.99, 34.99, 34.67 and 34.28 kWh for clear, bronze, green and bronze-reflective glasses, respectively.

Figure 2b shows heat gain in buildings of the Bangalore climatic region. The total heat gain in buildings when the single glazing window is located in south orientation of the Bangalore climatic region is 25.55, 24.59, 24.08 and 23.10 kWh for clear, bronze, green and bronze-reflective glasses, respectively. The total heat gain in buildings when double glazing window is located in south orientation of the Bangalore climatic region is 23.94, 22.79, 22.32 and 21.58 kWh for clear, bronze, green and bronze-reflective glasses, respectively. The total heat gain in buildings when triple glazing window is located in south orientation of the Bangalore climatic region is 23, 21.84, 21.44 and 20.87 kWh for clear, bronze, green and bronze-reflective glasses, respectively. The total heat gain in buildings when quadruple glazing window is located in south orientation of the Bangalore climatic region is 22.34, 21.25, 20.89 and 20.45 kWh for clear, bronze, green and bronze-reflective glasses, respectively.

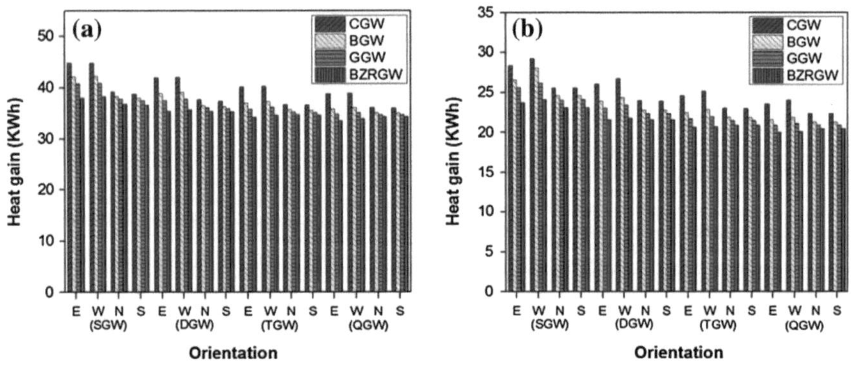

Fig. 2 **a** Heat gain in buildings of Ahmadabad climatic region (23.07°N, 72.63°E) **b** heat gain in buildings of Bangalore climatic region (12.97°N, 77.58°E)

4.2 Heat Gain in Buildings of Warm and Humid (Bombay) and Composite (New-Delhi) Climatic Regions

Figure 3a shows heat gain in buildings of the Bombay climatic region. The total heat gain in buildings when the single glazing window is located in south orientation of the Bombay climatic region is 28.88, 27.78, 27.21 and 26.08 kWh for clear, bronze, green and bronze-reflective glasses, respectively. The total heat gain in buildings when double glazing window is located in south orientation of the Bombay climatic region is 27.54, 26.26, 25.72 and 24.86 kWh for clear, bronze, green and bronze-reflective glasses, respectively. The total heat gain in buildings when triple glazing window is located in south orientation of the Bombay climatic region is 26.71, 25.43, 24.96 and 24.30 kWh for clear, bronze, green and bronze-reflective glasses, respectively. The total heat gain in buildings when quadruple glazing window is located in south orientation of the Bombay climatic region is 26.10, 24.89, 24.48 and 23.96 kWh for clear, bronze, green and bronze-reflective glasses, respectively.

Figure 3b shows heat gain in buildings of New Delhi climatic region. The total heat gain in buildings when the single glazing window is located in south orientation of the New Delhi climatic region is 44.54, 43.34, 42.74 and 41.54 kWh for clear, bronze, green and bronze-reflective glasses, respectively. The total heat gain in buildings when double glazing window is located in south orientation of the New Delhi climatic region is 43.08, 41.67, 41.11 and 40.18 kWh for clear, bronze, green and bronze-reflective glasses, respectively. The total heat gain in buildings when triple glazing window is located in south orientation of the New Delhi climatic region is 42.18, 40.76, 40.27 and 39.55 kWh for clear, bronze, green and bronze-reflective glasses, respectively. The total heat gain in buildings when quadruple glazing window is located in south orientation of the New Delhi climatic region is 41.51, 40.18, 39.73 and 39.19 kWh for clear, bronze, green and

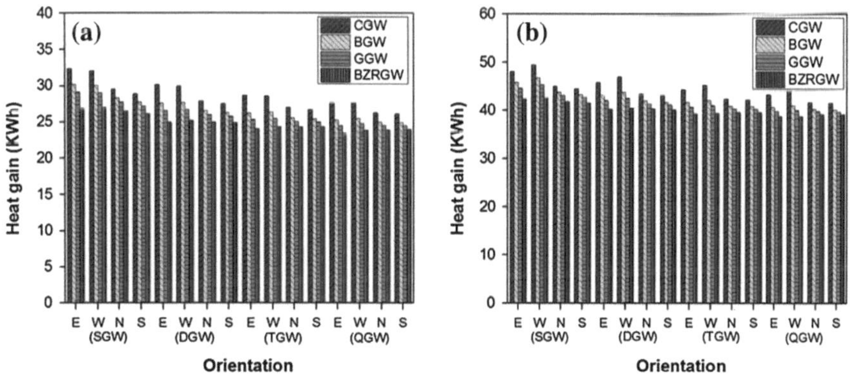

Fig. 3 **a** Heat gain in buildings of Bombay climatic region (19.12°N, 72.85°E) **b** heat gain in buildings of New Delhi climatic region (28.57°N, 77.12°E)

bronze-reflective glasses, respectively. From the results it is noticed that the heat gain decreases with the increase in the no. of air spaces between the glass layers and bronze-reflective glazing is observed to be energy efficient from the lower heat gain point of view in all four different climatic conditions of India.

5 Conclusion

The heat gain in buildings decreases with the increase in the pane glazing layers. The south orientation is found to be energy efficient for window placing from the lower heat gain point of view for cooling loads. The bronze-reflective window glass materials with single, double, triple and quadruple glazing are found to be energy efficient among four glass materials studied with four pane glazing layers in four different climatic conditions of India.

References

ASTM E424.: Test for solar energy transmittance and reflectance (terrestrial) of sheet materials. pp. 1320–1326. Washington DC, USA (1971)

BS EN 410.: Glass in Building-Determination of Luminous and Solar Characteristics of the Glazing, pp. 1–24. British Standards, UK (1998)

IS:11907 Recommendations for calculation of solar radiation on buildings CED 12.: Functional Requirements in Buildings, pp. 1–20. Bureau of Indian Standards, New Delhi, India (1986)

Saboor, S., Ashok Babu, T.P.: Optimizing the position of insulating materials in flat roofs exposed to sunshine to gain minimum heat into buildings under periodic heat transfer conditions. Environ. Sci. Pollut. Res., 3–13 (2015). http://dx.doi.org/10.1007/s11356-015-5316-7

Singh, I., Bansal, N.K.: Thermal and optical properties of different window systems in India. Int. J. Ambient Energy. **23**(4), 201–211 (2011) http://dx.doi.org/10.1080/01430750.2002.967481

SP: 41. (S&T) Handbook on functional Requirement of Buildings Other than Industrial Buildings, pp. 33–40. Bureau of Indian Standards, India (1987)

Synthesis of Ruthenium Nanoparticles by Microwave Assisted Solvothermal Technique

Isha Misra, Riya Parikh, Alisa Chakraborty, Yogeshwar R. Suryawanshi and Mousumi Chakraborty

1 Introduction

Metal nanoparticles have drawn lots of attention in the recent decades due to their size and shape-dependent physical and chemical properties. Nanoscale metal clusters and colloids have unique physical and chemical properties. In recent years, different techniques have been employed in the preparation of several nanoparticles (Harpeness et al. 2005; Zawadzaki et al. 2008). Chemical, physical or electro-chemical methods have been used to synthesize uniformly distributed nanoparticles. Different synthesis techniques such as microemulsion technique, sol-gel process and solvothermal process (Nandanwar et al. 2011a, b), chemical reduction technique (Patharkar et al. 2013) etc. have been employed in the preparation of several nanosized materials. Conventional heating methods for the preparation of metal nanoparticles generally take an hour's due to a multistage temperature program for absolute reduction of metal salts in the solutions. The application of microwave heating to heat and drive chemical reactions is a fast growing research area (Moseley and Kappe 2011). MWI (a) provides a homogeneous and rapid heating to the entire sample (b) initiates reaction rates fast to facilitating the formation of uniform nucleation growth and (c) because of energy efficient and environmental friendly behaviour it obeys the green chemistry principles (Galletti et al. 2008). MWI triggers heating by two main mechanisms, namely dipolar polarization and ionic conduction. When microwave is applied to the reaction mixture, the molecular dipoles accordingly attempt to re-align themselves along the alternating electric-field streamlines and as a result, heat is produced through molecular friction and dielectric loss (dielectric heating) (Moseley and Kappe 2011; Suryawanshi et al. 2013).

I. Misra · R. Parikh · A. Chakraborty · Y.R. Suryawanshi · M. Chakraborty (✉)
Department of Chemical Engineering, S. V. National Institute of Technology,
Surat 395 007, Gujarat, India
e-mail: mch@ched.svnit.ac.in

R. Mohan B. et al. (eds.), *Materials, Energy and Environment Engineering*,
DOI 10.1007/978-981-10-2675-1_6

Many efforts have been made to synthesize Ru nanoparticles using MWI method (Ni et al. 2012; Antonetti et al. 2012). However, the synthesis of ruthenium nanoparticles with optimized reaction protocols was rarely reported. Present study enlightens the use of MWI technique mainly because it allows attaining greener synthesis of Ru nanoparticles by shorter reaction times (15 min including initial induction and reaction period) and reduced energy consumption (low irradiation power, 200 W). Control of uniform nucleation and growth rates are critical to nanomaterials quality. So exploration of different parameters like molar ratio of stabilizer to ruthenium and MW irradiation power are investigated for controlling nucleation and growth rate which affect size and size distribution of the synthesized nanoparticles.

2 Experimental Procedures

2.1 *Materials*

Ruthenium trichloride ($RuCl_3{\cdot}nH_2O$, Ru Content ≥ 37 %) from Finar chemicals, India, Poly(N-vinyl-2-Pyrrolidone) (PVP as stabilizer, average molecular weight 40,000) from Heny fine chemicals, India, Ethylene glycol (solvent) from Merck chemicals, India were purchased. All organic solvents like acetone, ethylene glycol of analytical grade were used as purchased.

2.2 *Synthesis of Ru Nanoparticles in Pressurized Vial*

An advanced microwave synthesis labstation (MILESTONE, India) operates at 2.45 GHz frequency was used in this study. PVP (1.43×10^{-3} mol) and $RuCl_3{\cdot}xH_2O$ (1.72×10^{-4} mol) were dissolved in ethylene glycol to maintain PVP/$RuCl_3$ ratio in the solvent under vigorous stirring. Red wine coloured reaction mixture was produced after complete dissolution of $RuCl_3$ in the solvent. Reaction mixture was charged into the reaction vial [PTFE (polytetrafluroethylene) lined vessels of 100 ml capacity, withstand 300 °C temperature and 50 bar maximum pressure] and heated up to 200 °C for 15 min (optimized reaction temperature and time) varying MW irradiation power.

2.3 *Characterisation of Ru Nanoparticles*

UV-Vis spectrophotometer (HACH, Germany) was used to analyse spectra at reaction time span. Particle size analyzer (Malvern Zetasizer, Nano ZS 90, U.K.)

was used for measuring the size of nanoparticles. Shape and surface morphology was analysed by Tecnai G^2-20 U-Twin Transmission Electron Microscope (TEM, FEI, Netherlands). Nanoparticle stability was analyzed using Turbiscan classic MA2000 at light rays 880 nm wavelength (Formulaction, France).

3 Result and Discussion

3.1 *Formation of Ru Nanoparticles*

Calculated amount for metal, stabilising agent were added into the solvent and treated with MWI. Reaction mixture follows stepwise reduction of Ru^{3+} to Ru^0.

$$\begin{array}{ll} Ru^{3+} \rightarrow Ru^{2+} + 1e^- & \text{(from wine red to yellow)}, \\ Ru^{2+} \rightarrow Ru^{1+} + 1e^- & \text{(from wine yellow to green)}, \\ Ru^{1+} \rightarrow Ru^{0} + 1e^- & \text{(from wine green to dark brown)}. \end{array}$$

Formation of Ru nanoparticle protected by stabilising agent in polyol medium was depicted in (Fig. 1).

MWI treated reaction mixture (200 W, 200 °C) was collected at different time interval (5–15 min) and cooled down by ice chilled water. Progress of the reaction was monitored by UV-Vis absorption spectroscopy. Initially at 0 min, absorption peaks around 370 nm was observed which could be attributed to the charge transfer from the ligand of Cl^- to Ru^{3+} ions (Fig. 2). After 15 min of the MWI, no significant absorption peak was found in the spectra, indicating that all of the metal ions were reduced completely.

3.2 *Effect of PVP/RuCl$_3$ Molar Ratio (MR) on Particle Size*

In this study, PVP was used to restrict agglomeration of the nanosized particles. The average diameters of Ru nanoparticle are strongly influenced by molar ratio of PVP

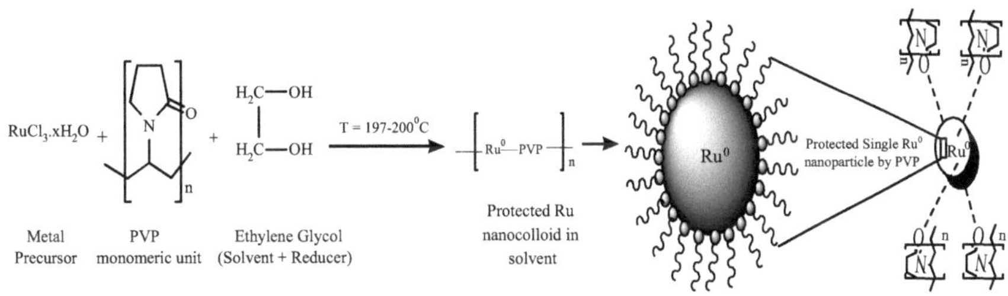

Fig. 1 Protected Ru nanoparticle in polyol medium

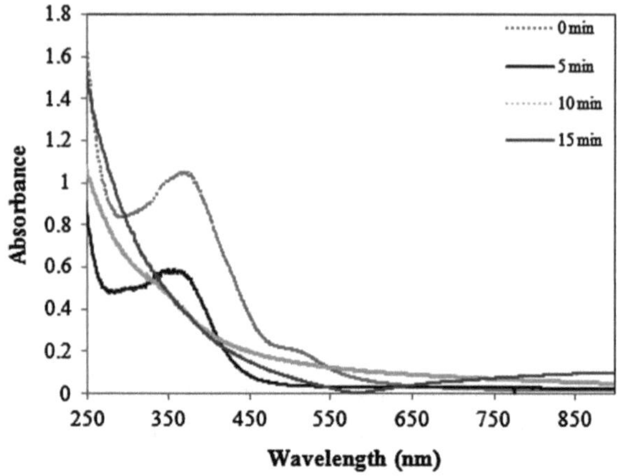

Fig. 2 UV-Vis spectra of colloidal Ru nanoparticles

to $RuCl_3$. It was observed (Fig. 3) that, with increasing molar ratio (0.6–1), particle size decreased and smallest nanoparticle (avg. particle size 14 nm) was obtained at higher molar ratio (MR = 1). At higher PVP/$RuCl_3$ molar ratio, no agglomeration would occur which helped to form smaller size particles.

3.3 *Effect of MWI Power on Average Size of Ru Nanoparticles*

Keeping temperature at 200 °C, reaction time 15 min and PVP/$RuCl_3$ at 1, experiments were conducted varying MWI power. It was observed from Fig. 4 that increasing irradiation power supply from 100 W to 200 W, decreasing particle size (52–14 nm) but generating larger particles above 200 W. Larger size nanoparticles were obtained at lower power supply (100 and 150 W). For the synthesis of nanoparticles, time requirement to attain reaction temperature (induction period) would be high at low power supply, so actual reaction time would be reduced which led to incomplete conversion of metal precursor. At higher power supply (200 W), rapid temperature rise would increase nucleation, growth rate and resulted smallest size particles (14 nm). The main reason is that reaction mixture can get a fast heating rate at high microwave power and precursor can be reduced in a short time. With further increase in MWI power (250–300 W) nucleation and growth rate would be very high and stabilizer protection was weakened due to lower stabilizer concentration compared to large number of ruthenium nuclei generated in the reaction mixture. Thus the tiny particles agglomerated to form larger particles (Fig. 4).

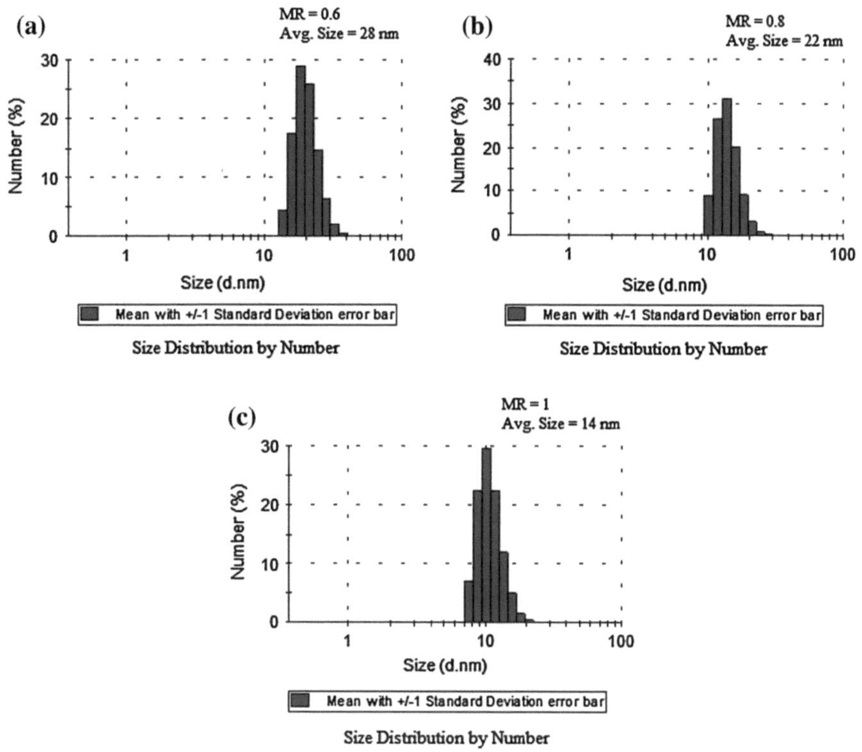

Fig. 3 Size distribution of synthesized Ru nanoparticles at different PVP/$RuCl_3$ molar ratio **a** 0.6, **b** 0.8, **c** 1

Fig. 4 Effect of MWI power on average size of synthesized Ru nanoparticles

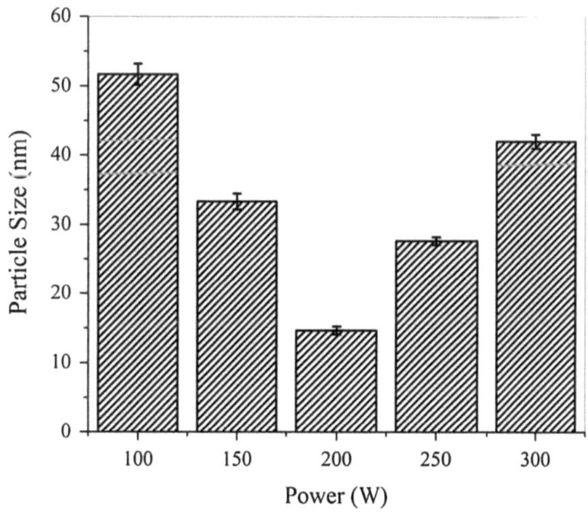

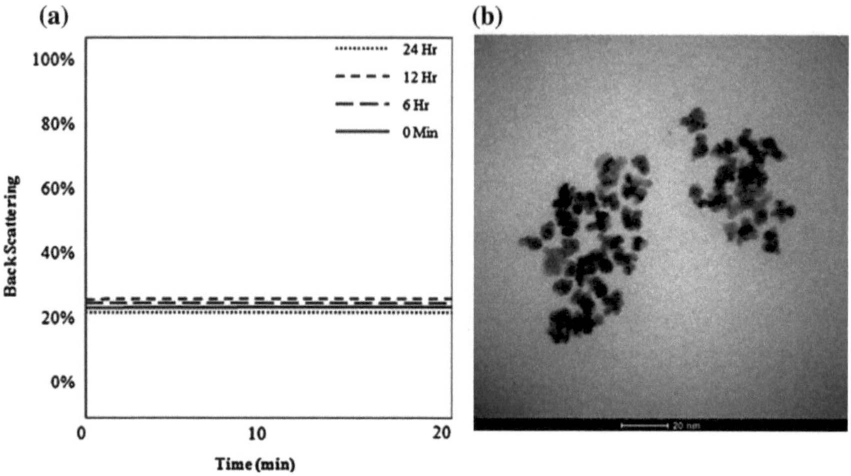

Fig. 5 **a** BS profile, **b** TEM image of synthesized Ru nanoparticles

3.4 Stability of Ru Nanoparticle and TEM Analysis

The nanoparticles were stabilized due to attractive and repulsive electrostatic forces created by capping agent present in the reaction system. Synthesized colloidal nanoparticle solution was scanned by light ray of 880 nm wavelength using Turbiscan. It was observed that Back Scattering (BS) profiles (Fig. 5a) at different time intervals were superimposing which indicated that the structure and average particle size were not changing up to 24 h and from TEM image nanoparticle size (without capping agent) was found to be in the range of 4–8 nm.

4 Conclusion

PVP protected colloidal Ru nanoparticles were successfully synthesized by Microwave Assisted Solvothermal Technique. Uniform nanoparticles were obtained by optimizing different parameters such as reaction temperature and time (200 °C and 15 min), PVP/$RuCl_3$ molar ratio (MR = 1) and MWI power (200 W) Synthesized colloidal nanoparticle size was found to be in the range of 4–20 nm and TEM image (nanoparticles without capping agent) indicated nanoparticle size in the range of 4–8 nm. Superimposed BS profiles suggested that synthesized colloidal Ru nanoparticles remained stable up to 24 h.

References

Antonetti, C., Oubenali, M., Galletti, A.M.R., Serp, P., Vannucci, G.: Novel microwave synthesis of ruthenium nanoparticles supported on carbon nanotubes active in the elective hydrogenation of p-chloronitrobenzene top-chloroaniline. Applied Catalysis A: General, pp. 99–107. Elsevier, The Netherlands (2012)

Galletti, A.M.R., Antonetti, C., Longo, I., Capannelli, G., Venezia, A.M.: A novel microwave assisted process for the synthesis of nanostructured ruthenium catalysts active in the hydrogenation of phenol to cyclohexanone. Applied Catalysis A: General, pp. 46–52. Elsevier, The Netherlands (2008)

Harpeness, R., Penga, Z., Liu, X., Pol, V.G., Koltypin, Y., Gedanken, A.: Controlling the agglomeration of anisotropic Ru nanoparticlesby the microwave–polyol process. Journal of Colloid and Interface Science, pp. 678–684. Elsevier, The Netherlands (2005)

Moseley, J.D., Kappe, C.O.: A critical assessment of the greenness and energy efficiency of microwave-assisted organic synthesis. Green Chemistry, pp. 794–806. Royal Society of Chemistry, London (2011)

Nandanwar, S.U., Chakraborty, M., Mukhopadhyay, S., Shenoy, K.T.: Stability of ruthenium nanoparticles synthesized by solvothermal method. Crystal Research and Technology, pp. 393–399. Wiley, United States (2011)

Nandanwar, S.U., Chakraborty, M., Murthy, Z.V.P.: Formation of ruthenium nanoparticles by the mixing of two reactive microemulsions. Industrial and Engineering Chemistry Research, pp. 11445–11451. American Chemical Society, Washington, DC (2011)

Ni, X., Zhang, B., Li, C., Pang, M., Su, D., Williams, C.T., Liang, C.: Microwave-assisted green synthesis of uniform Ru nanoparticles supported on non-functional carbon nanotubes for cinnamaldehyde hydrogenation. Catalysis Communications, pp. 65–69. Elsevier, The Netherlands (2012)

Patharkar, R.G., Nandanwar, S.U., Chakraborty, M.: Synthesis of Colloidal Ruthenium Nanocatalyst by Chemical Reduction Method. Journal of Chemistry, pp. 1–5. Hindawi Publishing Corporation, Egypt (2013)

Suryawanshi, Y.R., Chakraborty, M., Jauhari, S., Mukhopadhyay, S., Shenoy, K.T., Shridharkrishna, R.: Microwave irradiation solvothermal technique: an optimized protocol for size-control synthesis of Ru nanoparticles. Crystal Research and Technology, pp. 69–74. Wiley, United States (2013)

Zawadzki, M., Okal, J.: Synthesis and structure characterization of Ru nanoparticles stabilized by PVP or γ-Al_2O_3. Materials Research Bulletin, pp. 3111–3121. Elsevier, The Netherlands (2008)

Sonochemical Synthesis of Poly (Styrene-co-Methylmethacrylate)-HNT's Nanocomposites by Mini-emulsion Polymerisation

Buruga Kezia and T.K. Jagannathan

1 Introduction

The synergetic properties of polymer nanocomposites have attracted extensive research around the world. Clays are widely investigated and used as nanofillers in preparation of polymer nanocomposites and are receiving scientific and technological interest as a result of unique combination of properties of both fillers and polymers. Even at low clay loading there is significant enhancement in flame retardancy, mechanical strength and barrier properties (Alexander et al. 2000). Recently, Halloysite Nanotubes (HNTs) have gained much attention as these minerals exhibit the geometry of nanotubes, chemistry of kaolinite and are naturally available. They hold inherent interesting properties like hollow tubular structure, high aspect ratio, mechanical strength, similar to carbon nanotubes in morphology, and available at relatively lower cost. Different methods for synthesis of polymer nanocomposites have been reported in literature which include solution intercalation (Morgan et al. 2004), polymer intercalation by the melt method (Wang et al. 1998), and in situ polymerization (Sedlakova et al. 2009). Few researchers have employed emulsion (Qutubuddin et al. 2002), suspension (Wei et al. 2003), and living free radical (Bottcher et al. 2002) polymerization route for development of polymer-clay nanocomposites. Synthesis of poly(styrene-co-butyl acrylate)/Cloisite-30B core shell nanocomposites (Mirzataheri et al. 2009) and poly (styrene-methylmethacrylate)/Al_2O_3 (Mahdavian et al. 2009) by mini-emulsion polymerization technique were also reported.

B. Kezia (✉) · T.K. Jagannathan
Department of Chemical Engineering, National Institute of Technology Karnataka, Surathkal Mangalore 575025, Karnataka, India
e-mail: kezia.nitk@gmail.com

R. Mohan B. et al. (eds.), *Materials, Energy and Environment Engineering*,
DOI 10.1007/978-981-10-2675-1_7

Mini-emulsion polymerisation has the potential to create nanoparticles which can be continuously dispersed in aqueous phase for encapsulation and hence it is considered advantageous over other methods of synthesis (Moraes et al. 2006). Mini-emulsions are relatively stable aqueous dispersions that are dispersed as mini-droplets with a size ranging from 50 to 500 nm. They are fabricated at high shears in a system containing oil, water, and surfactants. A compound which is highly insoluble in water such as hexadecane is employed as co-surfactant to impede Ostwald ripening of the droplets (Landfester et al. 1999) and aids in attaining particles in aspired size range. Composition of oil phase and incorporation of hydrophobic compounds are the prominent features of mini-emulsion polymerisation (Landfester et al. 1999). This method is environmentally favourable as it exempts use of volatile organic solvents. Utilising ultrasound for preparation of mini-emulsions is promising as the acoustic cavitations generated during sonication results in enormous high temperatures, pressures and rigorous environment and hence initiates chemical reactions (Cheng et al. 1999). Under these extreme conditions, free-radicals are generated due to decomposition of monomer, water, surfactant, or rupture of polymer chains which initiates polymerization. In the present work we have successfully synthesised poly(styrene-co-methylmethacrylate)-HNT nanocomposites by ultrasound assisted mini-emulsion polymerisation, as per the reported literature no work has been carried out using these combination of materials. Sonication time and clay loading are the parameters which affect the stability of mini-emulsion, hence the effect of these parameters on structure, morphology and thermal properties of the synthesised nanocomposites were studied. These nanocomposites can be used for applications such as commercial packings of edibles, barrier coatings, and pharmaceutical purposes.

2 Research Methodology

2.1 Materials

Copolymer was synthesised by using Styrene and Methylmethacrylate as monomers and received from Sigma-Aldrich. HNTs were used as filler material and procured from Sigma-Aldrich. Sodium dodecyl benzene sulfonate (SDBS) and Span-60 were used as surfactants, hexadecane was used as co-surfactant and was received from Merck. The initiator 2,2-azo-isobutyronitrile (AIBN) was procured from kemphasol. Deionised water was used for all purposes; all the reactions were carried out under N_2 atmosphere.

2.2 *Mini-emulsion Copolymerization of Poly (Styrene-co-Methylmethacrylate)-HNT's Nanocomposites*

Phase-I & II was prepared as listed in Table 1 and magnetically stirred for 1 h, cooled to 0 °C and sonicated for 1 h. Phase-I & II were stirred vigorously using magnetic stirring for 30 min, and homogenized by sonication for 1 h. This mini-emulsion was used for subsequent polymerization. A similar experiment was conducted in the absence of ultrasound to examine the effect of ultrasound on final properties of nanocomposites.

2.3 *Polymerization of (Styrene-co-Methylmethacrylate)-HNT's*

The experimental setup consisted of a four-necked round bottom glass reactor as depicted in Fig. 1 for carrying out mini emulsion polymerization. It was armed with a magnetic stirrer for stirring, condenser for water circulation, thermometer for temperature sensing and nitrogen inlet to avoid contact with oxygen. Mini-emulsion was transferred to reactor and was consequently flushed with N_2 at room temperature for 30 min. The reaction was then heated to 70 °C and 0.14 g of AIBN was added to initiate polymerization. Polymerization was done at this temperature for 7 h with continuous magnetic stirring; samples were withdrawn at an interval of 1 h to study about polymerisation and dispersion of filler into polymer-matrix. 1 % hydroquinone solution was added to terminate to reaction. After reactor reaches room temperature the thickened milky mini-emulsion latex was filtered to remove any aggregations present. A series of nanocomposites were prepared by incorporating 10, 20, 30 % of HNTs in the mini-emulsion.

Table 1 Composition of water rich and monomer rich phases

Phase-I					Phase II	
Contents	0 wt%	10 wt%	20 wt%	30 wt%	Contents	Weight (g)
HNT	0	0.1	0.2	0.3	Water	75
Styrene	10	10	10	10	SDBS	0.15
Methymethacrylate	5	5	5	5	Span-60	0.15
					Hexadecane	0.15

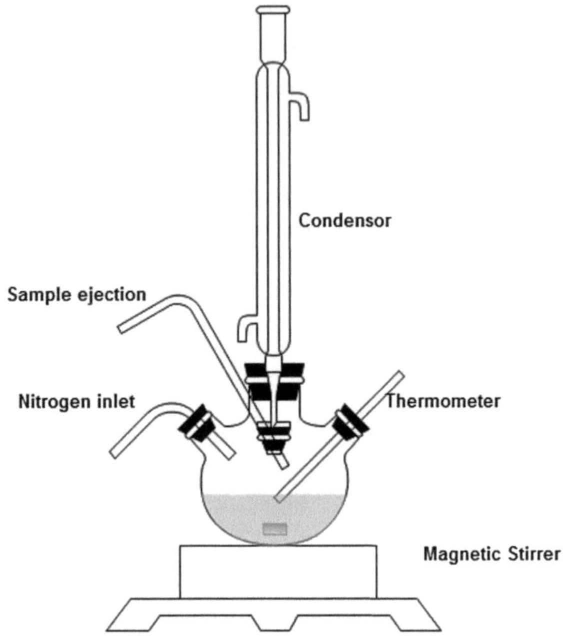

Fig. 1 Experimental setup for mini-emulsion polymerisation

3 Characterisation of Nanocomposites

XRD analysis was performed using RIGAKU Ultima-IV diffractometer and intensity data were collected in 2θ range of 10°–60° and count-time of 2^{-s} using Cu-Kα (λ = 1.542 A°) radiation. SEM analysis was carried out using JEOL-JSM-6360LV.FT-IR analysis was performed using BRUKER spectrometer in region between 4000–450 cm^{-1}. DSC was done using Perkin–Elmer Pyrus at heating rate of 10 °C min^{-1} under nitrogen flow from 0 to 200 °C.

4 Results and Discussions

4.1 Effect of Sonication and Clay Loading on Structure of Nanocomposites

Figure 2a describes interlayer structure of nanocomposites with and without sonication. A significant broad peak appeared at 32° on synthesising nanocomposites without sonication, whereas on using sonication this peak appeared at 17°. This shows an increase in interlayer spacing upon sonication that can be made available for polymer for better interaction with clay thereby enhancing the properties, further

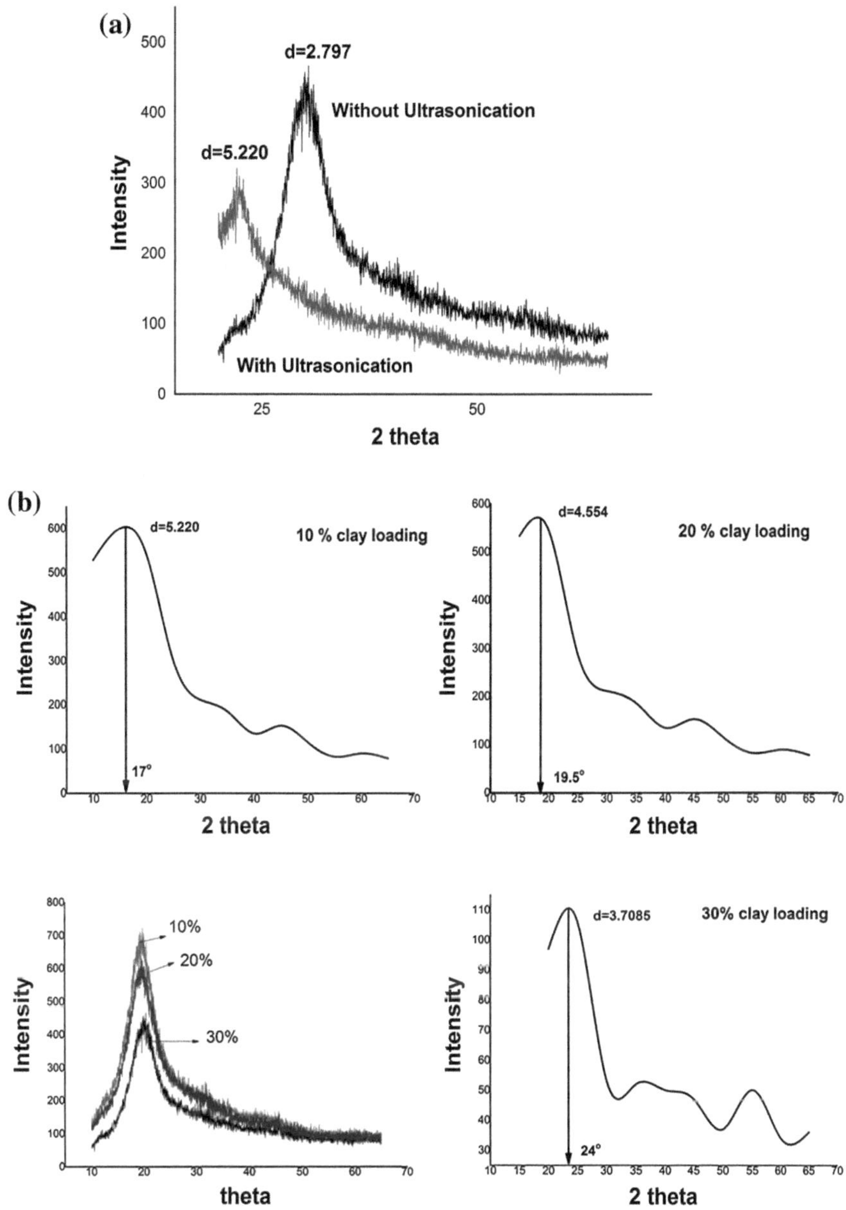

Fig. 2 XRD **a** plot indicating structure with and without ultrasonication, **b** effect of clay loading on interlayer spacing

it can be noted that no significant peaks are observed meaning formation of completely exfoliated structure. Hence, sonication plays a vital role in dispersion quality and chargers in layers.

From Fig. 2b it can be analysed that 10 % clay loading resulted in highest basal spacing and sharp peak meaning it could be better dispersed in comparison with other loadings, implying lower percentage of clay in the polymer matrix results in completely exfoliated nanocomposites. HNTs consists of aluminols located on their interior surface, and siloxanes on their exterior surface hence, they are positively and negatively charged respectively. The high anionic charge density of SDBS facilitates its movement into lumens and interlayers of HNT's due to development of ionic networks and modifier component. This increases the hydrophobicity of HNTs due to inculcation of alkyl long chains which are present in SDBS. This is the key aspect for intercalation of clay (Qutubuddin et al. 2002), which promotes exfoliation process by embodiment of monomer molecules followed by polymerization. Hydroxly group present on the surface of the HNT's are absorbed by SDBS and hence there is decrease in the clay particle hydrophobicity, resulting in enhanced colloidal stability thereby increasing exfoliation of clay platelets. The high shear through acoustic irradiation aids in exfoliation, thus SDBS helps in reaching the co-monomers within the layer and they widen up the gap between the interlayers. Polar carboxyl groups present in methylmethacrylate interacts with polar groups existing in the intergalleries which further facilitates in continous addition of styrene monomers into intergalleries and compels the HNTs to exfoliate.

4.2 *Studies on Morphology of Nanocomposites*

Figure 3. describes SEM analysis of nanocomposites, incorporation of 10 % HNTs into polymer- matrix resulted in nearly mono-dispersed morphology and had smooth surface without any agglomerations in polymer-matrix. The particles were rod shaped and size was in nanorange. It is can also be observed that the particles

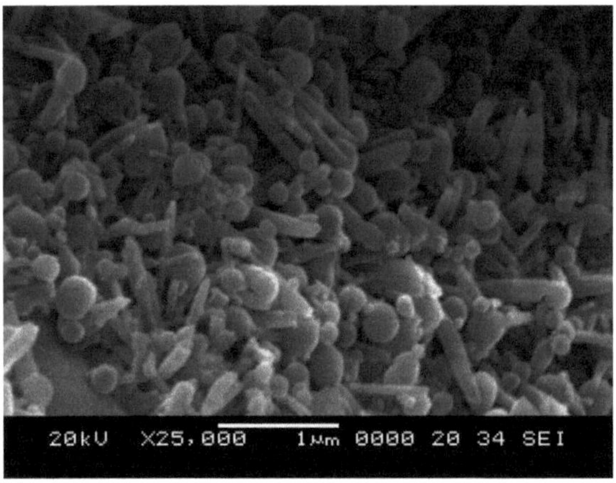

Fig. 3 SEM analysis of poly(styrene-co-methylmethacrylate)-HNT nanocomposites

are irregularly shaped and contains many edges with different sizes, these factors play an important role in the interaction between filler, matrix, and interfacial adhesions thereby enhancing the properties.

4.3 Effect of HNTs Inclusion on Polymer Structure

The stretching vibrations of inner surface of Al-OH are represented by peaks at 3692.79 and 3623.21 cm^{-1} in Fig. 4c. Bending vibrations at 1638.39 cm^{-1} indicate presence of water in interlayer's of HNT's. The stretching of Si–O apical is demonstrated by the peak at 1030.10 cm^{-1}. The bending vibration of inner surface hydroxyl groups can be observed by peak at 909.18 cm^{-1}. Peaks at 750.02 and 687.79 cm^{-1} represents the bending vibrations of Al–O–Si and Si–O–Si respectively. The presence of HNT major absorbance peaks in poly(styrene-co-methylmethacrylate) spectrum can be clearly seen from Fig. 3b. which confirms the existence of HNT's inside the latex particles and indicates encapsulation of HNTs into the final nanocomposites. However, the peak at 3692 cm^{-1} shifts to 3694.43 cm^{-1} in the nanocomposite which attributes interaction of Al layer of HNTs with the polymer surface. Thus from FT-IR study we may infer successful incorporation of HNTs in the polymer matrix.

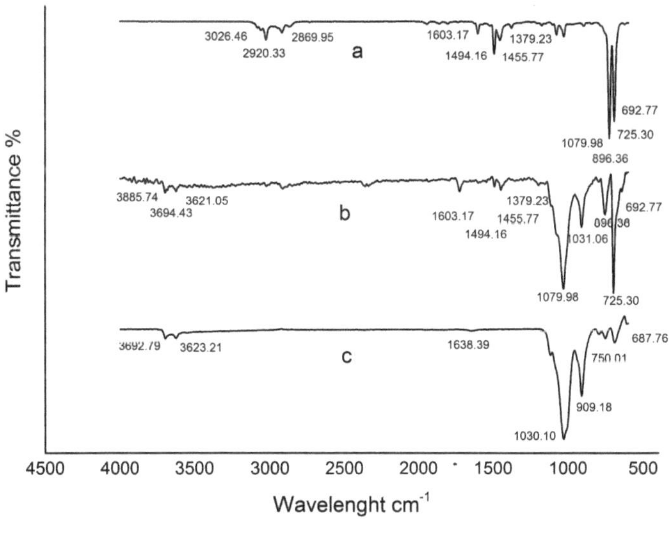

Fig. 4 FTIR of **a** poly(styrene-co-methylmethacrylate), **b** poly(styrene-co-methylmethacrylate)-HNT, **c** HNT

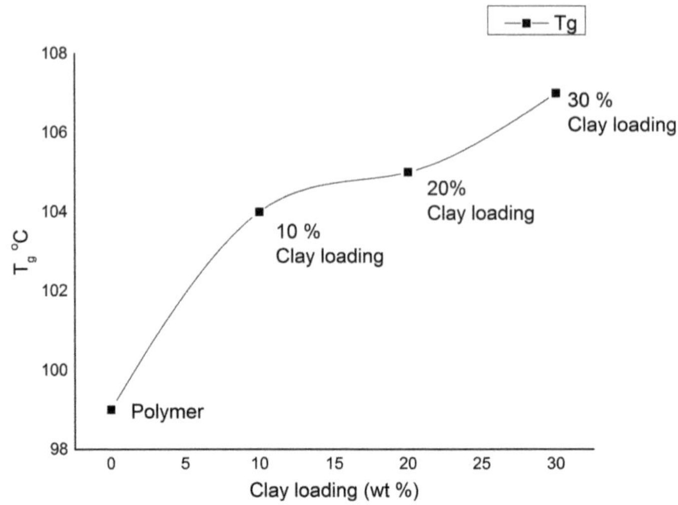

Fig. 5 Effect of clay loading on T_g of nanocomposite

4.4 Effect of Clay Loading on Thermal Stability of Nanocomposites

Figure 5 shows DSC traces of polymer and polymer-nanocomposite with 10, 20 and 30 wt% clay loading. Poly(styrene-co-methylmethacrylate)-HNT nanocomposites showed higher T_g than that which was prepared without HNTs thereby indicating better thermal stability. On incorporation of 10 wt%, the T_g evidently improves and is believed to restrict the movement of intercalated polymer chains between the interlayers of the HNTs which restricts the segmental motions of the polymer chains. On the other hand T_g did not increase appreciably with increasing amount of HNTs, which confirms that the 10 wt% itself can be effective in improving the T_g and hence thermal stability, incorporation of 20 and 30 wt% resulted in non-uniform dispersion and hence T_g did not increase appreciably. The increase in thermal stability may result due to dispersion of layers into the polymer chain. The layered structure of the clay can retard heat-transfer and also strengthen the polymer chain to resist heat degradation.

5 Summary/Conclusion

Nanocomposites of poly(styrene-co-methylmethacrylate)-HNT were successfully synthesized using mini-emulsion polymerisation method. Ultrasonic irradiation has an excellent effect on dispersion of HNTs into polymer matrix in presence of surfactant SDBS. Exceptional high temperatures and pressures are developed inside the cavity due to sonication and under this rigorous condition polymerization gets initiated. SDBS interacts with the clay and makes its way into the intergalleries of clay facilitating entry of the co-monomers within the interlayers causing exfoliation; this results in expansion of the clay galleries to make room for accommodation of polymer chains. XRD results indicated increase in d-spacing on using sonication; also 10 wt% clay could cause sufficient exfoliation. Some of the nanofillers which were originally hollow tube shaped have changed to different shape due to collisions with each other during sonication which aided in better intercalation and the size of particles was less than 150 nm. FTIR results confirms the intercalation of filler into polymer matrix. DSC analysis confirms increase in T_g of nanocomposites which proves increase in thermal stability, hence these can have improved potential application such as barriers properties.

References

Alexander, M., Dubios, P.: Polymer-layered silicate nanocomposites: preparation, properties and uses of new class of materials. Mater. Sci. Eng. **28**, 1–63 (2000)

Bottcher, H., Hallensleben, M.L., Stefan, N., Wurm, H., Bauer, J., Behrens, P.: Organic/inorganic hybrids by 'living'/controlled ATRP grafting from layered silicates. J. Mater. Chem. **12**, 1351–1354 (2002)

Cheng-yi, L., Chun-Hsien, C., An-I, Y., Vivian, L.: Preliminary study on the degradation kinetics of agarose and carrageenans by ultrasound. Food Hydrocolloids **13**, 477–481 (1999)

Landfester, K., Bechthold, N., Tiarks, F., Antonietti, M.: Formulation and stability mechanisms of polymerizable miniemulsions. Macromolecules **32**, 5222–5228 (1999)

Mahdavian, A.R., Sarrafi, Y., Shabankareh, M.: Nanocomposite particles with core-shell morphology III: preparation and characterization of nano Al_2O_3-poly (styrene-methylmethacrylate) particles via miniemulsion polymerization. Polym. Bull. **63**, 329–340 (2009)

Mirzataheri, M., Mahdavian, A.R., Atai, Md: Nanocomposite particles with core-shell morphology IV: an efficient approach to the encapsulation of Cloisite 30B by poly(styrene-co-butyl acrylate) and preparation of its nanocomposite latex via miniemulsion polymerization. Colloid Polym. Sci. **287**, 725–732 (2009)

Moraes, R.P., Santos, A.M., Oliveira, P.C., Souza, F., Amaral, M., Valera, T.S., Demarquette, N. R.: Poly(styrene-co-butyl acrylate)-Brazilian Montmorillonite Nanocomposites, synthesis of Hybrid Latexes via Miniemulsion Polymerization. Macromol. Symp. **245**, 106–115 (2006)

Morgan, A.B., Harris, J.D.: Exfoliated polystyrene-clay nanocomposites synthesized by solvent blending with sonication. J. Polym. **44**, 8695–8703 (2004)

Qutubuddin, S., Fu, X., Tajuddin, Y.: Synthesis of polystyrene-clay nanocomposites via emulsion polymerization using a reactive surfactant. Polym. Bull. **48**, 143–149 (2002)

Sedlakova, Z., Plestil, J., Baldrian, J., Slouf, M., Holub, P.: Polymer-clay nanocomposites prepared via in situ emulsion polymerization. J. Polym. Bull **63**, 365–384 (2009)

Wang, S., Long, C., Wang, X., Li, Q., Qi, Z.: Synthesis and properties of silicone rubber/organomontmorillonite hybrid nanocomposites. J. Appl. Polym. Sci. **69**, 1557–1561 (1998)

Wei, X., Jyh, M.H., George, J., Thandi, M.B., Wei, P.: A study of effect of surfactant on the properties of polystyrene-montmorillonite nanocomposites. Polym. Eng. Sci. **43**, 214–222 (2003)

A Novel Single Step Sonochemical Synthesis of Micro-Nano Size Palladium-Metal Oxides

S. Sivasankaran and M.J. Kishor Kumar

1 Introduction

Transition metals and their oxides have very wide applications in various unit operations and unit processes of chemical engineering. They are used in the engineering applications as adsorbents, catalysts. Specifically noble metal-metal oxides, either as bimetal or multimetal oxides are being explored for their catalytic properties in the emerging research areas such as photo chemical reactions and the conversion of CO_2 to fuels apart from their applications in the health care field for diagnosis. In the conventional wet-chemical method the noble metal-non noble metal had been synthesised in a two step process. The wet chemical methods of synthesis of materials is being replace by sonochemical methods as super critical condition of high temperature and pressures required for reaction could be easily achieved (Suslick 1989) synthesis of nano size noble metal-metal oxide composite particles has been reported in literature (Yamamoto et al. 2004). But by using the novel single-step sonochemical process reported in this paper, both the nano and micro size particles could be produced simultaneously and hence the process could be considered more versatile relatively a cheaper than the existing process. Typically palladium–copper oxide was synthesised (Sivasankaran 2012; Sivasankaran et al. 2013). The histograms of the size analysis and the spectra of XRD analysis of the synthesised product confirmed the bimodal distribution of both nano and micro size particles of Pd (noble metal)-Copper oxide (metal oxide).

S. Sivasankaran (✉) · M.J. Kishor Kumar
Department of Chemical Engineering, Manipal Institute of Technology,
Manipal University, Manipal, India
e-mail: shivashankaran.s@manipal.edu

M.J. Kishor Kumar
e-mail: kish.divine@gmail.com

R. Mohan B. et al. (eds.), *Materials, Energy and Environment Engineering*,
DOI 10.1007/978-981-10-2675-1_8

2 Experimental

The experimental set up and the process is very simple. A 100 mL glass vessel was used for conducting the reaction which may be called as batch sonochemical reactor. 1 mmol of palladium acetate and 1 mmol of copper acetate were taken in the vessel and mixed with 70 mL water and 20 mL ethanol. The solution was directly irradiated with 20 kHz ultrasound wave from a 750 W SONICS ultrasonic processor for 1 h by using a ½ in. probe operating at 45 % amplitude. The black colour slurry of the product obtained at the end of the experiment was centrifuged at 7000 RPM for 15 min. The black colour slurry containing the smaller (nano) particles of the product was separated from the larger (micro) particles deposited on the centrifuge tube. Both the slurry and the deposited solids were analysed for particle size distribution and also chemical composition. The XRD analysis of the dried product confirmed the presence Pd-CuO in both the fractions. The particle size analysis by using Zeta sizer confirmed that the slurry contained nano particles and the solid deposit contained micro particles. Further size analysis of the whole product slurry from the reactor was done by using the dynamic light scattering method. There are two methods for the synthesis of Pd-CuO were mentioned in Fig. 1. Different steps involved in the synthesis process were mentioned in process flow diagram Fig. 2. The step by step chemical reaction involved during the

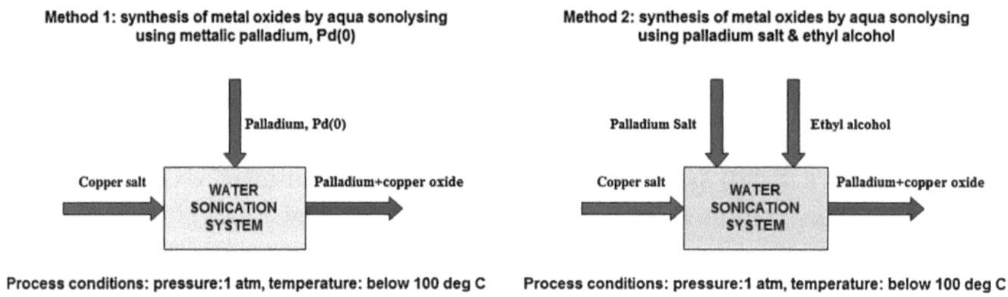

Fig. 1 Two different methods for the synthesis of Pd-CuO

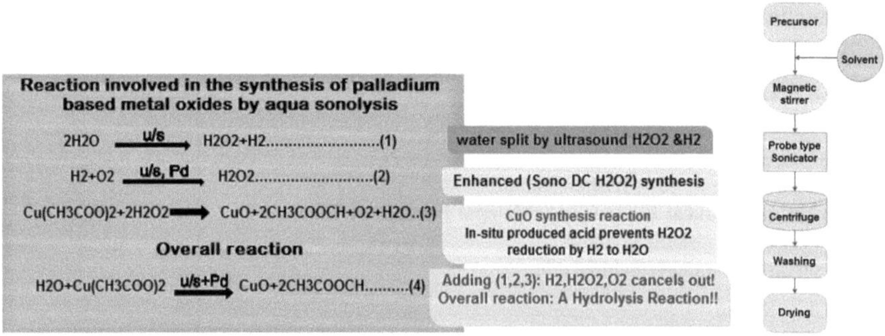

Fig. 2 Different steps involved in the synthesis process of Pd-CuO

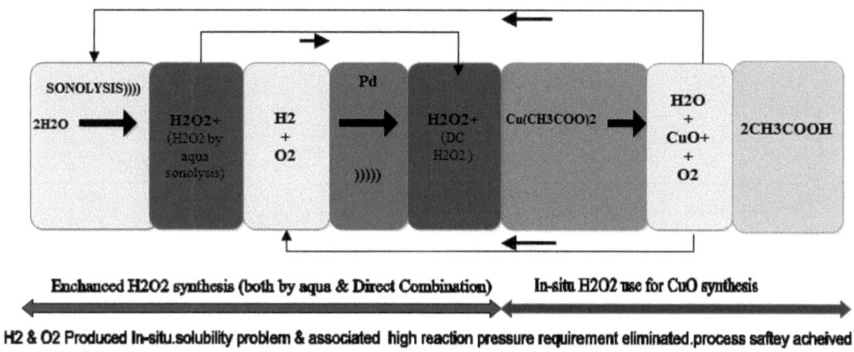

Fig. 3 The chemical reactions proposed during the synthesis of Pd-CuO

synthesis of Pd-CuO was given in Fig. 3. The histogram obtained showed bimodal distribution of the particles in both the nano and micro size ranges (Fig. 6).

3 Results and Discussions

3.1 X-Ray Diffraction and Size Distribution Analysis

XRD analysis of the product (Fig. 4a, b) confirms the chemical composition of the product is palladium cupric oxide. The carbon present in the product comes from acetate precursors used and also depending on the sonication conditions employed. Nitrate precursors will give only palladium and copper oxide. The size analysis (Fig. 5a, b) by using Zeta sizer shows the average size of the nano fraction is 81.58 nm whereas the average size of particles in the micro fraction is 1.548 μm. Particle size analysis (Fig. 6) shows the bimodal size distribution which confirms the presence of nano and micro fractions.

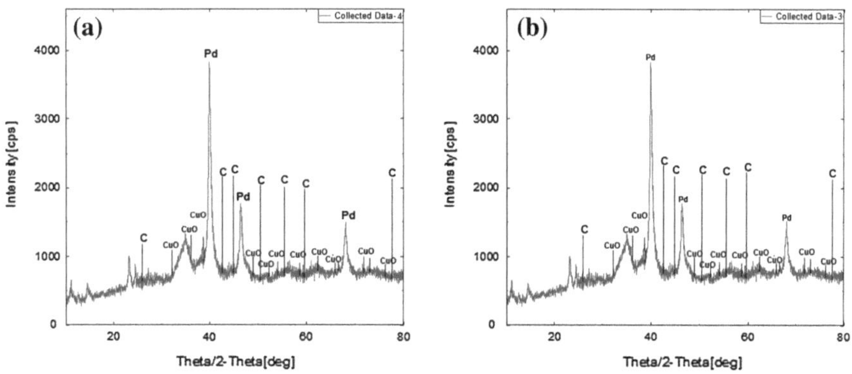

Fig. 4 **a** The nano fraction Pd-CuO. **b** The micro fraction Pd-CuO

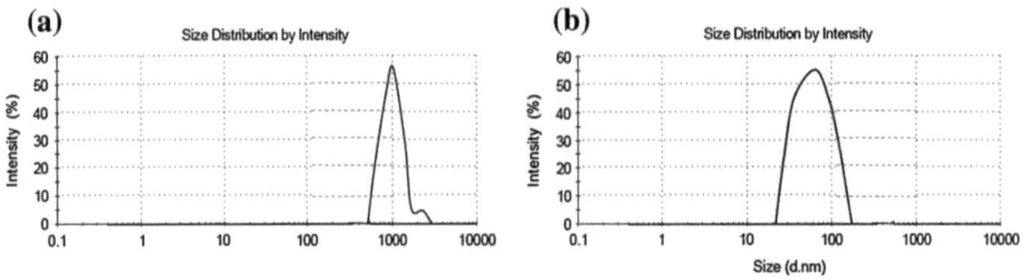

Fig. 5 **a** The micro fraction of Pd-CuO (average size; 1.548 μm). **b** The nano fraction of Pd-CuO (average size; 81.58 nm)

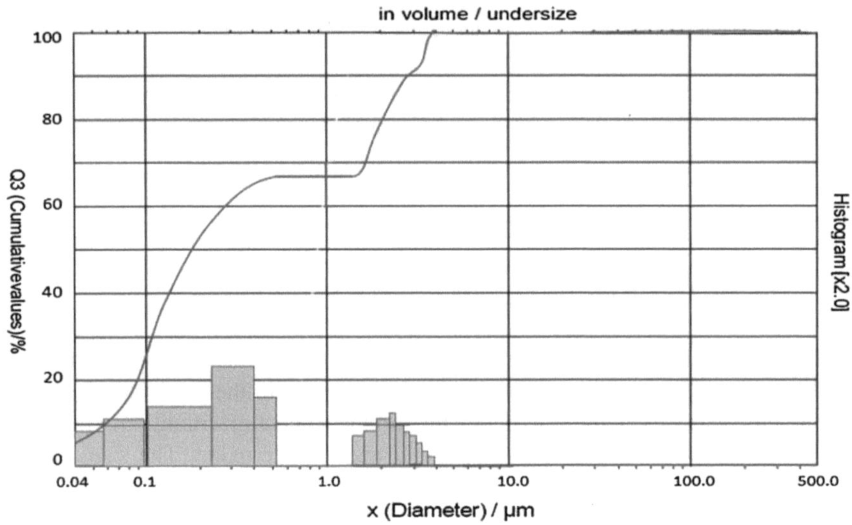

Fig. 6 Bimodal size distribution. The average particle size is 0.94 μm

3.2 *Microscopy and Elemental Analysis*

We have performed the EDX elemental analysis to confirm the presence of Pd–Cu–O–C results showed that as synthesized sample contains the mass% of Pd (34.28 %), Cu (55.33 %) and O (10.039 %) Fig. 7. Micro and nano size samples were subjected to SEM analysis; images were taken at 10,000 magnification Fig. 8a reveals that the synthesized sample was having particle size in micrometer and Fig. 8b shows that the size of particle was less than 1 μm. Transmission electron microscopy (TEM) analysis is most suitable method to know the size and shape of the material. We have taken TEM images at 20 and 50 nm scale by Fig. 9 we can observe the cluster of particles TEM image is one of the evidence that the synthesized sample contains particle size less than 50 nm.

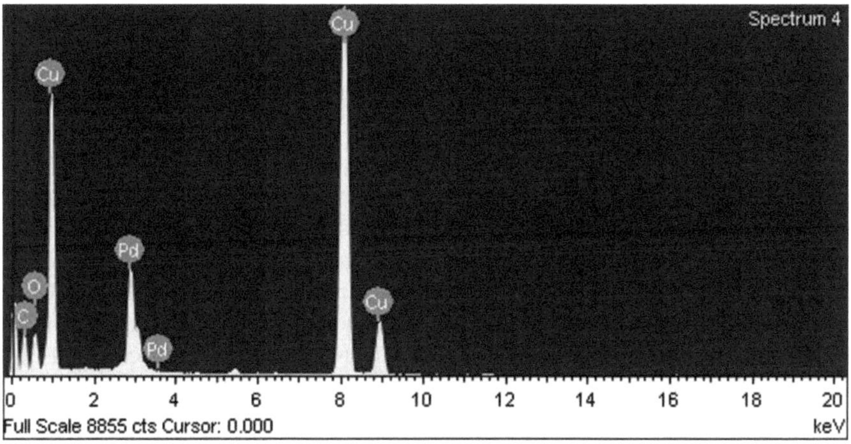

Fig. 7 Elemental composition Pd–Cu–O–C

Fig. 8 SEM images of as synthesized Pd-CuO powder **a** micro fraction, **b** nano fraction

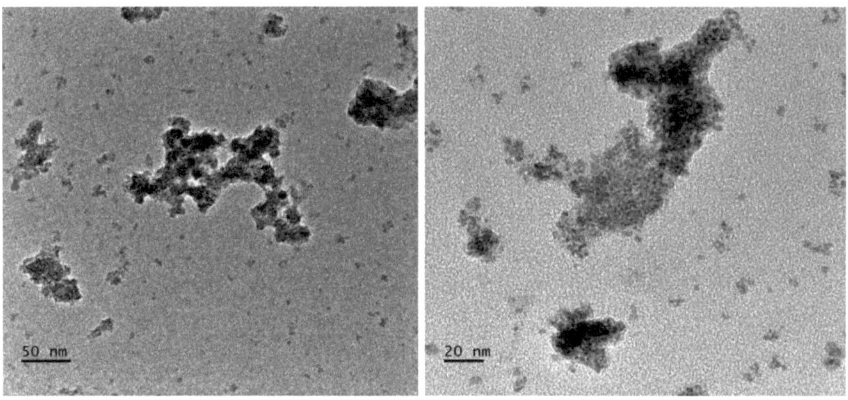

Fig. 9 TEM images of as synthesized Pd-CuO powder

4 Conclusion

Particle size analysis shows the bimodal distribution of Pd-CuO; which confirms that the synthesized sample contains both nano and micro fraction. The process developed for Pd-CuO could be extended to the synthesis of Pd-CoO (Sivasankaran and Mrinal 2014) and Ruthenium-metal oxides easily (Sivasankaran and Sarvesh 2014). Thus general processes for the synthesis Pd-metal oxides and Ru-metal oxides have been successfully developed. It is also possible to synthesis carbon material (activated carbon, CNT, graphene etc.) supported noble metal-metal oxide catalysts easily in a single step method. These materials have very wide applications in the engineering field as adsorbents, catalysts and in the medical fields for diagnosis. The batch sonochemical process (reactor) could be scaled up and also made a continuous sonochemical process (reactor) by using proper reactor design. Thus a continuous sonochemical process (reactor) for the simultaneous synthesis of both the nano and micro size Pd-CuO could be developed.

References

Sivasankaran, S.: Synthesis of Palladium Based Metal Oxides by Sonication. Publication no. WO 2012172396 A1/US patent US20130004412 (2012)

Sivasankaran, S., Sankaranarayanan, S., Ramakrishnan, S.: A novel sonochemical synthesis of metal oxides based bhasmas. Mater. Sci. Forum **754**, 89–97 (2013)

Sivasankaran, S., Ranjan, M.: Synthesis of nanometre sized Pd-CoO a transition bimetal oxide using sonochemical method and its characterization. IJSR, 872–875 (2014)

Sivasankaran, S., Dhingra, S.: A hybrid ultrasound assisted advance oxidation process using ruthenium based catalyst and its application to waste water treatment. IJSR, 920–923 (2014)

Suslick, K.S.: The chemical effects of ultrasound. Sci. Am. **260**, 80–86 (1989)

Yamamoto, N., Okitsu, S.: Nobel metal-magnetic metal oxide composite particle and method for producing same. European Patent. Publication no. EP No. 1604953A1, 2004

A Novel Single Step Ultrasound Assisted Synthesis of Nano Size Metal Oxides Metal Carbides and Metal Nitrides

S. Sivasankaran and M.J. Kishor Kumar

1 Introduction

The major challenges involved in the field of nanomaterials are reproducible, large scale and economical synthesis of stable nanomaterials which could handled easily. The nature of chemicals used in the synthesis and the process conditions such as temperature, reaction time, pH, mode of addition of chemicals, mixing etc. could control the shape, size, crystallinity which have direct impact on the properties of nanomaterials. The energy required for the chemical reactions may be supplied directly as thermal and pressure energies or indirectly through microwave, ultrasound energies. Microwaves produces heating only but ultrasound energy produces heat, pressure and turbulence leading to enhanced mass transfer rates in the liquid being irradiated leading the facile creation of super critical conditions of high temperature and pressure cannot be realized by other methods easily. Therefore the use of ultrasound could be seriously considered for the chemical reactions used in the bottom up method of nano particle synthesis. The phenomenon involved in ultrasonic irradiation is acoustic cavitation (i.e., the formation, growth, and implosive collapse of bubbles in liquids), driven by high intensity ultrasonic waves which is responsible for the chemical effects of ultrasound (Suslick et al. 1990). When liquids are irradiated with ultrasound, bubbles start oscillating because of the alternating expansive and compressive acoustic waves. This oscillating bubble can overgrow and at a subsequent time bubble collapse, releases the concentrated energy which is stored in the bubble within a very short time with typical temperature of 5000 K and pressure of 1000 bar (with a heating and cooling rate of

S. Sivasankaran (✉) · M.J. Kishor Kumar
Department of Chemical Engineering, Manipal Institute of Technology, Manipal University, Manipal, India
e-mail: shivashankaran.s@manipal.edu

M.J. Kishor Kumar
e-mail: kish.divine@gmail.com

R. Mohan B. et al. (eds.), *Materials, Energy and Environment Engineering*,
DOI 10.1007/978-981-10-2675-1_9

>10^{10}Ks^{-1}) (Suslick 1990). Cavitation could be a future tool for creating high Temperature, Turbulence and Pressure (TTP) in a facile, greener and inherently safer manner over conventional methods for the synthesis of nano size metal oxides, metal carbides and metal nitrides for uniform size distribution, a higher and rough surface area, faster reaction time, and improved phase purity. Some of the examples of successful sonochemical synthesis include Palladium-copper oxide (Sivasankaran 2012; Sivasankaran et al. 2013), Titanium dioxide (Yu et al. 2001), Zinc oxide (Qian et al. 2003). Several methods have been reported to synthesize metal carbides and metal nitrides namely sol-gel route (Hasegawa et al. 1997), microwave carbothermal reduction (Hassine et al. 1995), thermal plasma (Tong and Ramana 2005), laser method (Schaaf et al. 2005) and microwave synthesis (Satapathy et al. 2005). Typically the conventional methods for the synthesis of metal oxides, metal carbides and metal nitrides are multiple step processes of longer duration requiring many chemicals and conducted at higher temperature and pressure. Hence there is need for a simple, single step and environmentally benign process for the synthesis of these materials. However there are no reports available on the sonochemical synthesis of silicon carbide, silicon nitride and titanium carbide. In this present work we have successfully synthesized nano size silicon dioxide (SiO_2), silicon carbide (SiC), and titanium carbide (TiC) and silicon nitride (Si_3N_4) in a single step sonochemical method.

2 Experimental

A SONICS make 750 W ultrasonic processor with 13.2 mm diameter probe operating at 20 kHz frequency was used for probe type Sonication. In general procedure precursor was dissolved in suitable solvent; add 50–60 ml of distilled water kept under magnetic stirrer for 15 min then subjected to the probe type sonication; sonication parameters were mentioned below. Centrifugation was done to separate the nano-size materials and washed thoroughly with distilled water. Finally samples were dried in hot air oven at 120 °C for 4 h. Details about probe type sonicator (Fig. 1), process flow diagram (Fig. 2) and materials used (Table 1) were mentioned below.

Sonication parameters

- Frequency 20 kHz
- Pulse: 05 s ON - 02 s OFF
- Amplitude: 45 %
- Time: 60 min
- Set point temperature 70 °C
- Probe temperature 32 °C

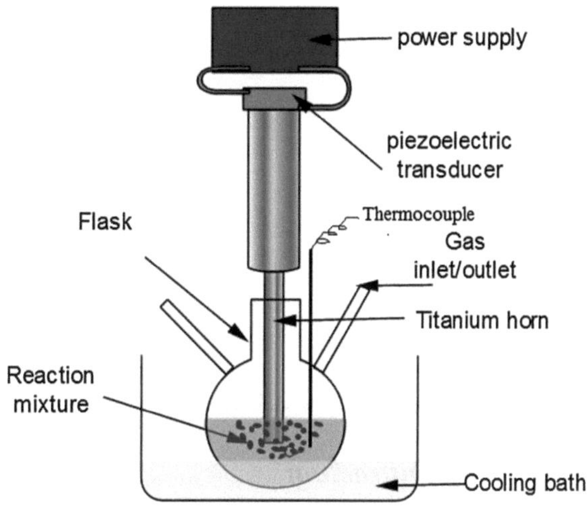

Fig. 1 Probe type sonicator (schematic)

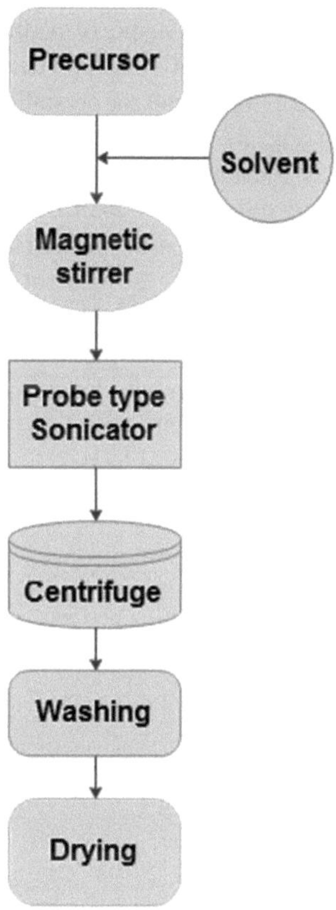

Fig. 2 Flow diagram

Table 1 Materials used for the synthesis of nanopowders

S. no.	Final product	Precursor	Quantity
1	Silicon dioxide	Sodium silicate	5 g
		Hydrochloric acid	10 ml
2	Silicon carbide	Silicon dioxide	1 g
		Activated carbon	3 g
3	Titanium carbide	Titanium dioxide	1 g
		Activated carbon	3 g
4	Silicon nitride	Silicon dioxide	3 g
		Activated carbon	6 g
		Aqueous ammonia	10 ml

3 Results and Discussions

3.1 X-ray Diffraction

XRD pattern of the synthesized powder confirms the presence of silicon dioxide, sharp peak found at diffraction angle (2θ) 21.56°, no other major peaks were found. It confirms the amorphous nature of synthesized SiO_2 powder shown in (Fig. 3). The sharp peaks of synthesized SiC nano powder found at diffraction angles (2θ) of 32.63°, 60.0°, 71.61°, 83.22° it confirms the crystalline nature of β-SiC; carbon peak was found at 36.5° shown in (Fig. 4). The characteristic silicon nitride planes of (1 0 0) (1 1 0) (2 0 0) (1 0 1) and (1 2 0) locating at 23°, 30°, 34.8°, 41°, and 50° for β-Si_3N_4 powder have been obviously observed (JCPDS card no. 33-1160). There are no reflections corresponding to SiO_2 and other impurities in the diffraction pattern (Fig. 6). The typical XRD patterns of the as-prepared TiC the peaks at 35.86°, 41.64°, 60.36°, 72.24° and 76.00° can be indexed to the reflection from (1 1 1), (2 0 0), (2 2 0), (3 1 1) and (2 2 2) planes of the TiC and it could be categorized as cubic titanium carbide (TiC) (JCPDS card no. 065-7994) 2θ are calculated with wavelength = 1.54059 (Fig. 5).

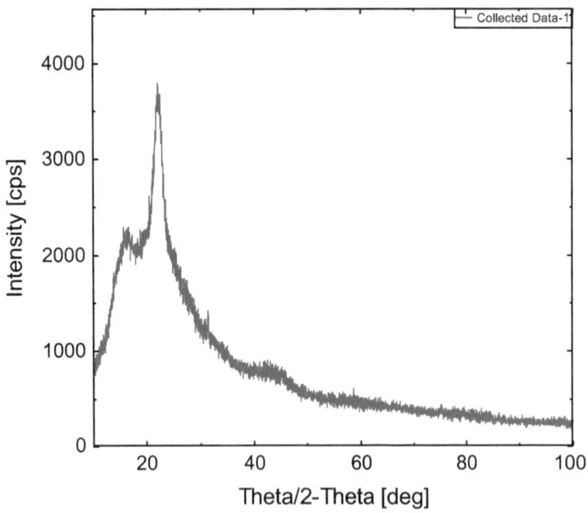

Fig. 3 XRD pattern of SiO_2

Fig. 4 XRD pattern of SiC

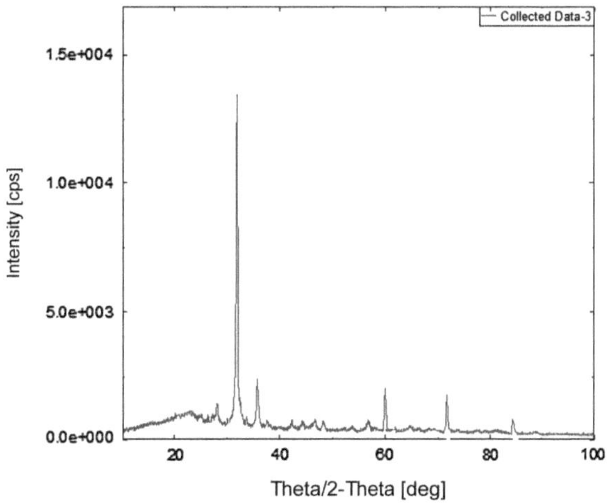

Fig. 5 XRD pattern of TiC

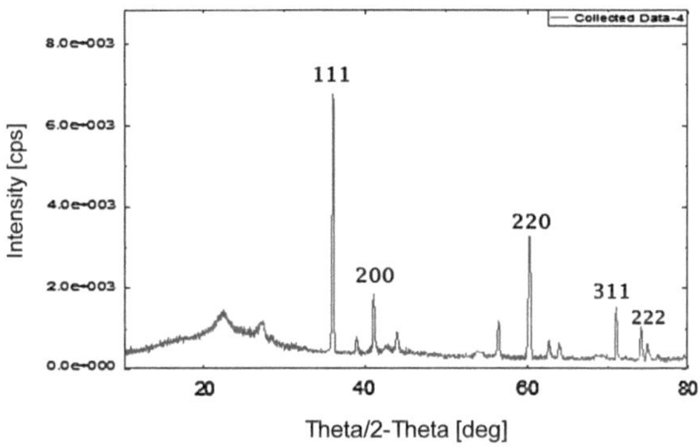

Fig. 6 XRD pattern of Si_3N_4

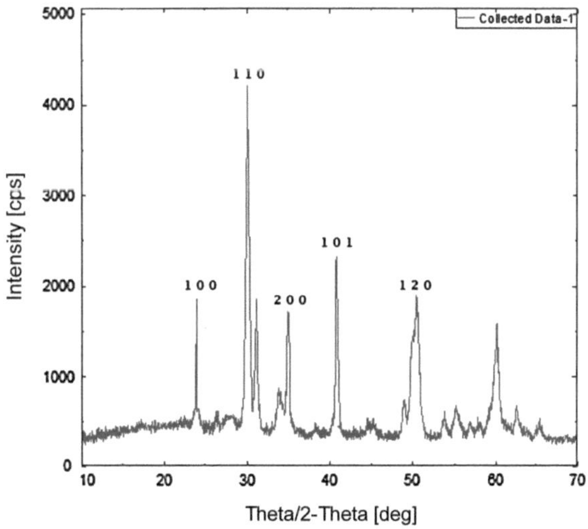

3.2 BET Surface Area Analysis

Specific surface area of the synthesized SiO_2, SiC, β-Si_3N_4 and TiC nano powders were calculated by Brunauer–Emmett–Teller (BET) equation from the adsorption data; specific surface area of the nanopowders were 207.3, 878.92, 803.67, 731.78 m^2/gm respectively.

3.3 Size Distribution Analysis

We have done the size distribution analysis by intensity using Malvern Zeta-sizer which uses the dynamic light scattering (DLS) principle, from the Fig. 7 we concluded that average particle diameter of the synthesized powders were in nanometre. The Z-average particle diameter of β-Si_3N_4 was 76 nm (Fig. 7a); TiC 126.95 nm (Fig. 7b); SiC 86.7 nm (Fig. 7c) and SiO_2 64 nm (Fig. 7d).

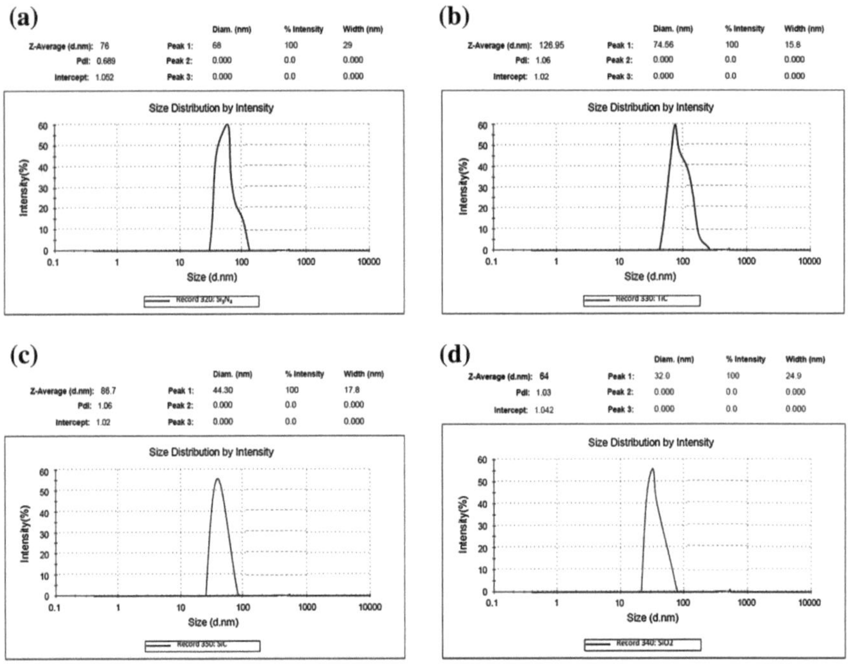

Fig. 7 Average particle diameter of the synthesized powders

Fig. 8 SEM images of as synthesized powder

3.4 Scanning Electron Microscopy (SEM) Analysis

SEM is used very effectively in microanalysis and morphological studies. Scanning electron microscopy is performed at high magnifications, generates high-resolution images. We performed SEM analysis for prepared nanopowders at 50.00 KX (100 nm range) magnification. SEM images clearly showed the size and shape of the synthesized material. We confirmed that the prepared powders containing the particles which are less then 200 nm shown in Fig. 8.

4 Conclusion

We have successfully synthesized nano-size SiO_2, SiC, Si_3N_4 and TiC powders by a novel, general and single-step sonochemical method in a facile, greener, faster, and safer way. All the synthesized materials were having particle size below 200 nm. Phase composition and crystal structure were confirmed by X-ray diffraction. By dynamic light scattering technique the average size distribution was

Table 2 summerized results of the sonochemical method

S. No	Final product	BET surface area (m^2/gm)	Average size distribution (nanometres)	Crystal structure
1	Silicon dioxide	207.3	64	Amorphous
2	Silicon carbide	878.92	86.7	Cubic
3	Silicon nitride	803.67	76	Hexagonal
4	Titanium carbide	731.78	123.95	Cubic

found out. Summarized results of the sonochemical method was mentioned below in Table 2. This novel sonochemical method could also be attempted for the synthesis of metal borides, boron carbide (B_4C), titanium nitride (TiN), boron nitride (BN) etc. with the suitable precursors. In case of SiC, rice husk could be used as the silicon source to reduce the cost of synthesis and make the process sustainable.

References

Hasegawa, I., Nakamura, T., Motojima, S.: Synthesis of silicon carbide fibers by sol-gel processing. J. Sol-Gel Sci. Technol. **8**, 577–579 (1997)

Hassine, N.A., Binner, J.G.P., Cross, T.E.: Synthesis of refractory metal carbide powders via microwave carbothermal reduction. Int. J. Refract Metal Hard Mater. **13**(6), 353–358 (1995)

Qian, D., Jiang, J.Z., Hansen, P.L.: Preparation of ZnO nanocrystals via ultrasonic irradiation. Chem. Commun. 1078 (2003)

Satapathy, L.N., Ramesh, P.D., Agrawal, D., Roy, R.: Microwave synthesis of phase-pure, fine silicon carbide powder. Mater. Res. Bull. **40**, 1871–1882 (2005)

Schaaf, P., Kahle, M., Carpene, E.: Reactive laser synthesis of carbides and nitrides. Appl. Surf. Sci. **247**, 607–615 (2005)

Sivasankaran, S: Synthesis of palladium based metal oxides by sonication, Publication no. WO 2012172396 A1/US patent US20130004412 (2012)

Sivasankaran, S., Sankaranarayanan, S., Ramakrishnan, S.: A novel sonochemical synthesis of metal oxides based Bhasmas. Mater. Sci. Forum **754**, 89–97 (2013)

Suslick, K.S.: Sonochemistry. Science **247**, 1439 (1990)

Suslick, K.S, Doktycz, S.J.: In: Mason, T.J. (ed.) Advances in Sonochemistry, vol. 1, pp. 197. JAI Press, New York, (1990)

Tong, L., Ramana, G.R.: Synthesis of titanium carbide nano-powders by thermal plasma. Scripta Mater. **52**(12), 1253–1258 (2005)

Yu, J.C., Yu, J., Ho, W., Zhang, L.: Preparation of highly photocatalytic active nano-sized TiO2 particles via ultrasonic irradiation. Chem. Communication (Camb). **19**, 1942–3 (2001)

Part II
Biosorption and Degradation

Denitrification Under Aerobic Condition in Draft Tube Spouted Bed Reactor

Keshava Joshi, N. Lokeshwari, G. Srinikethan and M.B. Saidutta

1 Introduction

Worldwide rise in population and the industrialization in recent years have resulted in degradation of the natural resources (Satish et al. 2012). Water and water resources has been the worst victim of population explosion and growing industrialization. Nitrogen in various forms is a major pollutant from industries viz, fertilizers production, food processing, metallurgical, coking, petrochemicals, manufacturing of alcohol, explosives, organic chemicals etc. Biological treatment methods for nitrogen removal are more efficient and cost effective for the waste water treatment process.

Denitrification is accomplished by chemoheterotrophic bacteria that obtain energy from chemical reactions involving carbon source at very low dissolved oxygen concentration (anoxic conditions). Recently, researchers have reported that, denitrification can also be accomplished even under aerobic conditions by the specific species that are isolated from various locations like ponds, canals and soil etc. These species possess the capacity to utilize oxygen and nitrate simultaneously as electron acceptors. These species include *Thiosphaera pantotropha* (*Paracoccus denitrificans*) (Su et al. 2001), *Alcaligenesfaecalis* (Joo et al. 2005), *Citrobacter diversus* (Huang et al. 2001), *Pseudomonas stutzeri* (Su et al. 2001) and *Pseudomonas aeruginosa* (Chen et al. 2006), *Pseudomonas putida AD21* (Kim et al. 2008). Aerobic denitrification has some advantages over anoxic denitrification: (i) aerobic denitrifiers can be easily controlled during the operation (ii) it can

K. Joshi (✉) · N. Lokeshwari
Department of Chemical Engineering,
S. D. M. College of Engineering and Technology, Dharwad, India
e-mail: joshikeshava@gmail.com

G. Srinikethan · M.B. Saidutta
Department of Chemical Engineering,
National Institute of Technology Karnataka, Surathkal, India

R. Mohan B. et al. (eds.), *Materials, Energy and Environment Engineering*,
DOI 10.1007/978-981-10-2675-1_10

accelerate nitrification and denitrification by directly utilizing nitrite-N and nitrate-N, as the denitrifying reactant and (iii) Nitrification and denitrification can be carried out simultaneously in a single reactor to reduce the cost of construction and operational problems (Kim et al. 2008).

Biodenitrification studies have been conducted using different continuous bioreactors, viz. activated sludge bioreactor, stirred tank bioreactor, airlift bioreactor, packed bed bioreactor, trickling filter bioreactor, oxidation ponds, fluidized bed bioreactor (Gonzalaz et al. 1997), spouted bed reactor and membrane bioreactor (MBR). In airlift reactor system, maximum carbon is utilized in draft tube as oxygen level is high in this region (Dhamole et al. 2007). At the annular part of the reactor, more nitrate is utilized by microorganism since oxygen level is low (Dhamole et al. 2007). In the current study, the evaluation of performance of the draft tube spouted bed reactors of two different sizes for nitrate removal using immobilized particles under different process parameters was carried out.

2 Materials and Methodology

An isolate *Pseudomonas syringe* kjdr2 was used for all experiments. It was isolated from waste water treatment plant of a fertilizer industry (Joshi et al. 2014). The isolate was sub cultured once in 15 days periodically on nutrient agar slants and was stored at 4 °C.

2.1 Growth Media Composition

Media composition for growth of microorganism used is as follows.

NH_4Cl-0.3 g/L, KH_2PO_4-1.5 g/L, $Na_2HPO_4.7H_2O$-7.9 g/L, KNO_3-2 g/L, $C_4H_4O_4Na_2.6H_2O$-27 g/L $MgSO_4.7H_2O$ Solution (5 mL/L): 20 g/L

Trace element solution (5 mL/L):

$ZnSO_4$-2.2 g/L; EDTA-50.0 g/L; $MnCl_2.4H_2O$-5.06 g/L; $CaCl_2$-5.5; $(NH_4)_6Mo_7O_{24}.4H_2O$-1.1 g/L; g/L$FeSO_4.7H_2O$-5.0 g/L; $CuSO_4.5H_2O$-1.57 g/L and $CoCl_2.6H_2O$-1.61 g/L.

All the reagents used were of analytical grade.

2.2 Analytical Method

Samples were collected from the draft tube spouted bed reactor at regular intervals and analysed for various characteristics. The culture samples were initially centrifuged at 10,000 rpm for 10 min (REMI, India) and supernatant was analysed for nitrate and nitrite concentration. Nitrate and nitrite concentrations were analysed using standard method using UV spectrophotometer. Attached biomass was analysed by weight difference before and after washing of solids with 0.25 N NaOH solution and drying it at 105 °C for 24 h.

3 Experimentation

Table 1 presents the specifications of reactor I and II. The experimental setup consists of cylindrical column which has been provided with a conical bottom made of brass where the angle of the cone is 60°. The central body of the column has been connected to top portion which contains the outlet and with conical portion at the bottom by flanges. Two guides have been connected each at the top and bottom of the column to hold the draft tube in its position. Air and water were introduced into the column from the bottom of the cone. Two screens of size less than 0.5 mm have been fixed at the inlet and outlet to avoid carryover of immobilized particles.

3.1 Experimental Procedure

Figure 1 presents the schematic representation of draft tube spouted reactor used for denitrification. Initially the feed bottle, medium and all connecting tubes were sterilized. Synthetic media of different nitrate concentrations (carbon to nitrogen ratio of 3:1 in distilled, reverse osmosis processed water) were pumped using a peristaltic pump as shown in Fig. 1 In all runs, the reactor was loaded with known mass of granular activated carbon immobilized with *P. syringae*, kjdr2. Compressed air was introduced through filter at the controlled rate just to spout the solids in the

Table 1 Specifications of reactor I and II

Parameter	Reactor-I	Reactor-II
Diameter (D), cm	5.0	7.6
Height (L), cm	57	60
Draft tube diameter, cm	1.2	2.5
Draft tube length	45	53
Volume of reactor excluding conical section, cm^3	1139	2720
Volume of reactor, L	1.2	2.8
Volume of draft tube, cm^3	51	260
Cone angle	60°	60°
Water inlet diameter, cm	0.3	0.3
Air inlet diameter, cm	0.4	0.4

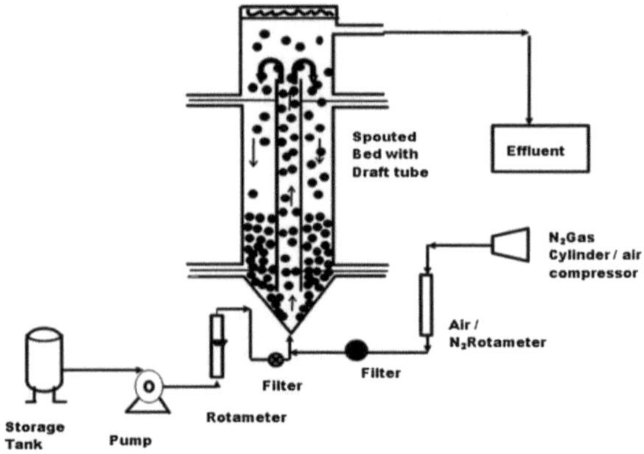

Fig. 1 Schematic diagram of experimental setup

draft tube. Effluent samples were collected at appropriate time intervals and centrifuged at 10,000 rpm for 10 min. Supernatant was analyzed for the nitrate and nitrite and residue was analyzed for biomass. Reaction was carried till the steady state was attained. Then, contents were emptied and attached biomass and biofilm thickness were measured.

4 Results and Discussion

4.1 Effect of Influent Nitrate Concentrations and Dilution Rates on Time to Attain Steady State

Figures 2 and 3 present the results of the effect of influent nitrate concentration and dilution rate on time taken to reach steady state during start up of reactor-I and II respectively. It is observed in Figs. 2 and 3 that, more time is required to attain a steady state for higher dilution rates due to less contact time between the liquid and the solid phase (substrate in liquid media and biofilm on GAC particle) for both reactors. Increase in velocity led to more attrition and shear. This has resulted in wash out of biomass from the reactor and hence, it took longer time in formation of a stable biofilm. With increase in influent nitrate concentration the biofilm became thicker and denser and hence, there would be increase in resistance to mass transfer (Stewart 2003).

For a given dilution rate and influent nitrate concentration, as the GAC loading is increased, there is an increase in the attached biomass quantity. The increase in volume of GAC with increased loading might have resulted in decrease in DO level in the reactor as volume of GAC occupies more space. Therefore it is observed that faster nitrate removal and early steady state attainment for higher GAC loading.

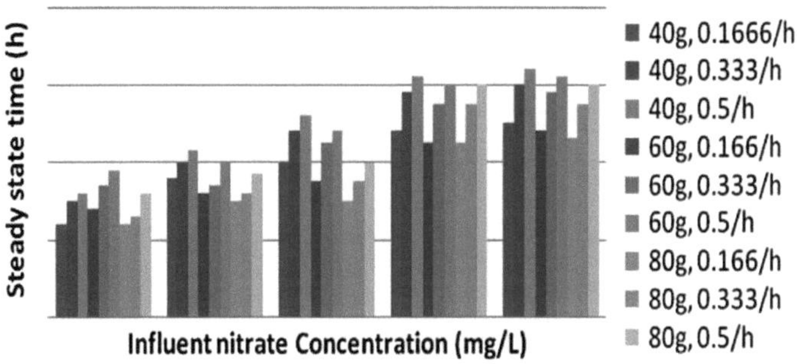

Fig. 2 Effect of influent nitrate concentration on time taken to reach steady state at different dilution rate and GAC loading in reactor-I

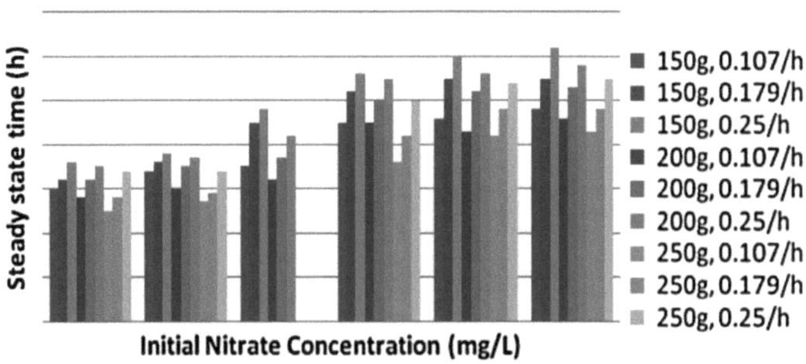

Fig. 3 Effect of influent nitrate concentration on time taken to reach steady state at different dilution rate and GAC loading in reactor-II

4.2 Effect of Nitrate Loading Rate on Removal Rate at Different GAC Loading

Figures 4 and 5 present the results on effect of nitrate loading rate on removal rate at different GAC Loading for reactor-I and reactor-II respectively. It can be seen from Figs. 4 and 5 that, for reactor-I and II in the range of nitrate loading studied, removal rate has increased linearly with nitrate loading for all GAC loadings. In the lower range of loading rate up to 200 mg/L-h, (Fig. 4) the observed performance could be attributed to an increase in biomass with a relatively very little influence of particle attrition and shear. In the middle range of volumetric nitrate loading rate of 200–400 mg/L-h, the nitrate removal rate is the same, in spite of the increase in the nitrate loading rate. In this range of loading rate, the effects of influent nitrate concentration and dilution rate, seem to neutralize each other's effect, leading to the nitrate removal rate remaining more or less the same. Beyond volumetric nitrate loading rate of 400 mg/L-h, the availability of nitrate is more, leading to thicker attached biomass at lower dilution rates or thinner biomass at higher dilution rate, due to particle attrition and shear (Nicolella et al. 2000).

Fig. 4 Effect of volumetric nitrate loading rate on volumetric nitrate removal rate at different GAC loading in reactor-I

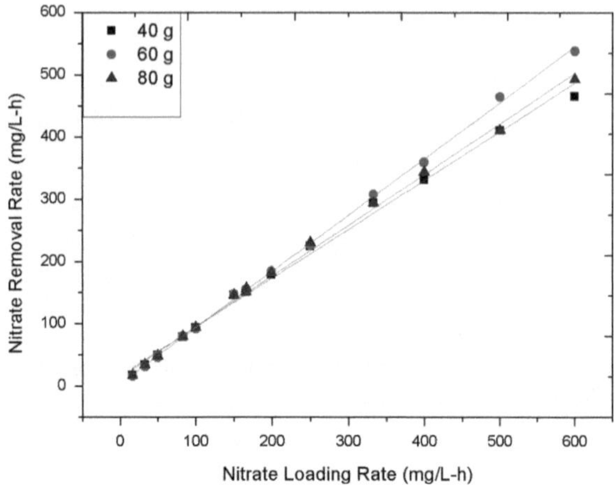

Fig. 5 Effect of volumetric nitrate loading rate on volumetric nitrate removal rate at different GAC loading in reactor-II

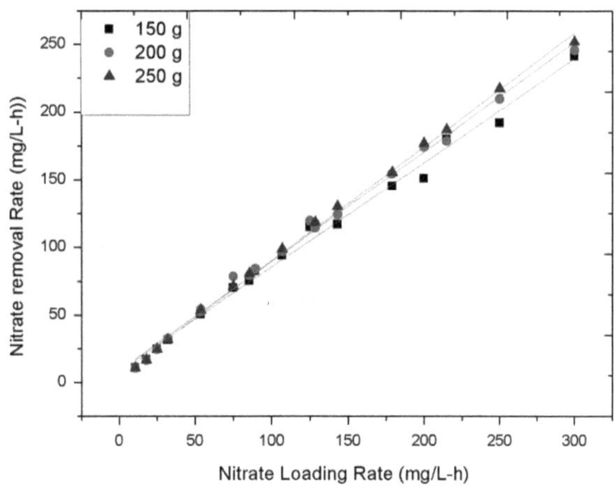

4.3 *Effect of Ratio of Nitrate Loading Rate to Attached Biomass Weight on Percentage Nitrate Removal*

Figures 6 and 7 present the results of the effect of ratio of influent nitrate loading rate to attached biomass on percentage denitrification in Reactors-I and II respectively. With increase in amount of nitrate loaded per unit time per unit dry biomass weight, the percentage nitrate removal decreases. Beyond the ratio of 0.5 g/h-g for reactor-I and 0.42 g/h-g for reactor-II, nitrate removal efficiency decreased to below 90 %. In reactor-I the solids circulation rate was higher compared to reactor II. Higher circulation rate reduces the external mass transfer resistance and also internal diffusional resistances. The external resistance was reduced because of the higher superficial velocities. But higher superficial velocities also lead to higher

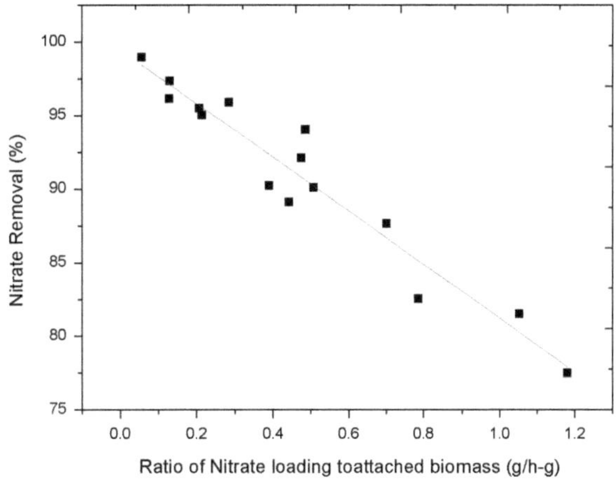

Fig. 6 Effect of ratio of nitrate loading to quantity of attached biomass on percentage denitrification (reactor-I)

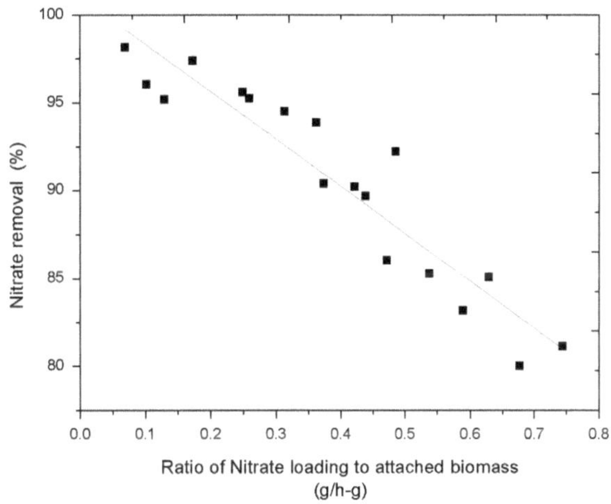

Fig. 7 Effect of ratio of nitrate loading to quantity of attached biomass on percentage denitrification (reactor-II)

attrition and shear there by thinner biofilm on GAC (Hare et al. 2011). Thinner biofilm offer lesser differential resistance which is benificial in nitrate removal.

5 Conclusion

The biodenitrification studies in draft tube spouted bed reactor show that the reactor performance depends upon the dilution rate, GAC loading and influent nitrate concentration. The start up time increased with increase in dilution rates and influent nitrate concentration while the start up time has decreased with increase of GAC loading. An increased dilution rate and influent nitrate concentration leads to decreased percentage denitrification but increase in GAC loading resulted in increase in percentage denitrification. In the present study, it was found that beyond 0.333/h, dilution rate the percentage nitrate removal reduces. Maximum removal was possible in reactor-I, II up to a nitrate loading rate of 400 mg/L-h, 200 mg/L-h respectively. The higher removal rate in draft tube spouted bed reactor would be due the systematic recirculation of biomass inside the reactor, maintaining active film of biomass on solid. The effect of GAC loading was found to be predominant in the start-up time.

Acknowledgments The authors wish to thank, Department of Chemical Engineering, National Institute of Technology Surathkal, Karnataka for providing research facilities. One of the authors, Keshava Joshi is thankful to the Management and Principal, SDM College of Engineering and Technology Dharwad for their support and encouragement.

References

Chen, F., Xia, Q., Ju, L.K.: Competition between oxygen and nitrate respirations in continuous culture of *Pseudomonas aeruginosa* performing aerobic denitrification. Biotechnol. Bioeng. **93**, 1069–1078 (2006)

Dhamole, P.B., Nair, R.R., D'Souza, S.F., Lele, S.S.: Denitrification of high strength nitrate waste. Bioresour. Technol. **98**, 247–252 (2007)

Gonzalez, L.E., Canizares, R.O., Baena, S.: Efficiency of ammonia and phosphorus removal from Colombian agro industrial wastewater by the microalgae *Chlorella vulgaris* and *Scenedesmusdimorphus*. Biores. Technol. **60**, 259–262 (1997)

Hare, C., Ghadiri, M., Dennehy, R.: Prediction of attrition in agitated beds. Chem. Eng. Sci. **66** (20), 4757–4770 (2011)

Huang, H.K., Tseng, S.K.: Nitrate reduction by Citrobacter diversus under aerobic environment. Appl. Microbiol. Biotechnol. **55**, 90–94 (2001)

Joo, H.S., Hirai, M., Shoda, M.: Characteristics of ammonium removal by heterotrophic nitrification-aerobic denitrification by Alcaligenesfaecalis No. 4. J. Biosci. Bioeng. **100**, 184–191 (2005)

Joshi, K., Joseph, J., Srinikethan, G., Saidutta, M.B.: Isolation and characterization of Pseudomonas syringae for nitrate removal under aerobic conditions. J. Biochem Tech. **5**(2), 693–697 (2014)

Kim, M., Jeong, S.Y., Yoon, S.J., Cho, S.J., Kim, Y.H., Kim, M.J., Ryu, E.Y.: Aerobic denitrification of Pseudomonas putida AD-21 at different C/N ratios. J. Biosci. Bioeng. **106**(5), 498–502 (2008)

Nicolella, C., Loosdrecht, M.C., Heijnen, J.J.: Waste water treatment with particulate biofilm reactors. J. Biotechnol. **80**, 1–33 (2000)

Satish, S., Chandra, H., Santosh, K.S., AshishKumar, B.: Environmental sinks of heavy metals: investigations on the effect of steel industry effluent in the urbanised location. Int. J. Adv. Eng. Res. Stud. **1**(2), 235–239 (2012)

Stewart, P.: Diffusion in Biofilms. J. Bacteriol. **185**(5), 1485–1491 (2003)

Su, J.J., Liu, B.Y., Liu, C.Y.: Comparison of aerobic denitrification under high oxygen atmosphere by Thiosphaerapantotropha ATCC 35512 and Pseudomonas stutzeri SU2 newly isolated from the activated sludge of a piggery wastewater treatment system. J. Appl. Microbiol. **90**, 457–462 (2001)

Feasibility of Anaerobic Ammonium Oxidation in the Presence of Bicarbonate

S.S. Ramratan, G. Anjali, P.C. Sabumon and S.M. Malliyekkal

1 Introduction

Ammonia present in wastewaters and industrial effluents is an important pollutant and requires attention for its removal. The discharge of ammonia bearing wastewaters to water bodies causes high dissolved oxygen (DO) depletion, eutrophication and toxicity to certain species of algae and fishes (Zhang et al. 2011). A number of new processes and configurations have been developed for ammonia removal from wastewaters with a potential to meet stringent discharge norms. Ahn (2006) reported a comprehensive review and descriptions of new ammonia removal processes. Among the new processes, anammox process is a potential process for removal of ammonia from wastewaters of higher concentrations.

Compared to conventional nitrification and denitrification, anammox consumes 50 % less oxygen, 100 % less organic carbon and saves 90 % of operating costs in sludge disposal (Chamchoi and Nitisoravut 2007; Molinuevo et al. 2009). The bacteria responsible for ANAMMOX reaction are anaerobic and chemolitho-autotrophic and are classified under deep members of phylum Planctomycetes. ANAMMOX bacteria show diverse metabolism using various electron acceptors like NO_3^-, NO_2^-, SO_4^-, and Propionate (Strous et al. 1998; Fdz-Polanco et al. 2001; Kartal et al. 2007; Sabumon 2009). In this work the feasibility of anammox using HCO_3^- as electron acceptor as per Eq. (1) is explored and such work has

S.S. Ramratan · G. Anjali (✉) · P.C. Sabumon · S.M. Malliyekkal
Environmental Engineering Division, School of Mechanical and Building Sciences, VIT University, Chennai Campus, Vandalur-Kelambakkam Road, Chennai 600 127, India
e-mail: anjaligopakumar@vit.ac.in

R. Mohan B. et al. (eds.), *Materials, Energy and Environment Engineering*, DOI 10.1007/978-981-10-2675-1_11 never been reported. The thermodynamic feasibility of Eq. (1) and/or modified chemical equation is yet to be proven.

$$2NH_4^+ + HCO_3^- = N_2 + CO_2 + H_2O + 7H^+ \tag{1}$$

2 Materials and Methods

2.1 Nutrient Media for Anaerobic Ammonia Oxidation

The nutrient media were prepared as per the composition reported in an anaerobic ammonia removal study by Sabumon (2007, 2009). Predetermined amount of ammonia (using NH_4Cl) and bicarbonate (using $KHCO_3$) were dosed as per requirement of each study. Analytical reagent (AR) grade chemicals was used and the chemicals were supplied by 'Thomas Bakers' (India). Reagent preparation and volume measurements were done by using clean Borosil (India) glassware.

2.2 Biomass

Mixed culture, enriched in simple sequencing batch reactor (SBR) operation over 400 days and dominant in anammox species (Anjali and Sabumon 2014), was used. The biomass sourced from SBR was washed in nutrient media several times to remove the residual NH_4–N and COD present in the aqueous phase.

2.3 Batch Reactor Studies

Batch studies were done to check the feasibility of anaerobic ammonium oxidation in the presence of bicarbonate. Later, Sequencing Batch Reactor (SBR) operation was done to confirm the process. The studies were conducted under anoxic condition in four serum bottles (Abiotic control + three biomass samples) of 70 mL capacity containing 15 mL of biomass of known MLSS (0.7316 g/L) each. Nitrogen purged (5 min) nutrient media containing 250 mg/L NH_4–N and 4 g/L of HCO_3^- was added till the brim of the serum bottle. The reactors were closed tightly with butyl rubber septum and aluminum cover. Studies were conducted for 3 days and the performance of each reactor was monitored every day. Studies were conducted in a temperature controlled shaker incubator (Scigenics Biotech, India) at a liquid temperature of 30–32 °C and mixed at 130 RPM. Analyses of pH, alkalinity, ORP, NH_4–N, NO_2–N, NO_3–N and COD were carried out on each sample drawn. After confirming the feasibility of the process, the specific batch study was carried out to know the rate of reaction and specific substrate utilization rate.

2.4 Kinetic Studies

Kinetic studies were performed using 7 serum bottles (70 mL capacity) as reactors. The biomass used in feasibility and confirmation studies was mixed together and washed several times to remove the residual NH_4–N and COD present in the aqueous phase. Each reactor contained 10 mL of biomass to which nitrogen purged nutrient media containing 350 mg/L of NH_4–N and 4300 mg/L HCO_3^- (@ HCO_3^-/NH_4–N ~ 12) was added till the brim. The reactors were closed using butyl rubber septum and aluminium cover to ensure anoxic condition. The reactors were kept in a temperature controlled (30–32 °C) shaker incubator and mixed at 130 RPM. The samples were drawn at 2, 4, 6, 8, 10, 12 and 24 h from the respective serum bottles. The parameters like pH, ORP, alkalinity, NH_4–N, NO_2–N, NO_3–N, COD, MLSS and MLVSS were analyzed in each sample. An abiotic sample was kept and monitored in the same experimental condition.

A second kinetic study was done using the settled biomass from the respective serum bottles using nitrogen purged nutrient media containing 130 mg/L of NH4–N and 1600 mg/L HCO_3^- (@ HCO_3^-/NH_4–N ~ 12). All other experimental conditions, sampling and analysis procedure were same as described above.

2.5 Analytical Techniques

All the physico-chemical analyses were carried out as per the procedure given in the Standard Methods (APHA 1998). The instrument used to measure pH and ORP was pH meter (WTW inoLab pH 720, Germany). A double junction platinum ORP electrode connected to a calibrated pH meter in mV mode was used for measuring ORP. ORP electrode (Pt–Ag/AgCl), was calibrated using RH 28 supplied by WTW, Germany. Ion Chromatography (Basic IC plus 883, Metrohm) was used for the analyses of aqueous nitrogen species with conductivity detector and the results were processed using integrated magic net software. Closed reflux method was used for COD analysis by using the COD digester (WTW-CR 3200, Germany).

3 Results and Discussion

3.1 Feasibility of Anaerobic Ammonium Oxidation Using HCO_3^- as Electron Acceptor

The results of the feasibility and the confirmatory studies (Table 1) show that that removal of ammonia is possible in anoxic condition. The ratio of HCO_3^-/NH_4–N was changed as the reaction progressed and the removal was maximum (35.8 %)

Table 1 Results of batch reactor studies

Batch reactor		pH	ORP mV	NH_4–N mg/L	NO_2–N mg/L	NO_3–N mg/L	HCO_3^- mg/L as $CaCO_3$	COD mg/L	Ratio of HCO_3^-/NH_4-N	% removal of NH_4–N	% removal of total nitrogen
F	R1	7.33	−26.3	160.4	2.6	3.9	1860	0	11.6	35.8	33.2
	R2	7.54	−38.4	180.4	0.0	0.0	1900	0	10.5	27.8	27.8
	R3	7.59	−47.4	189.5	0.0	0.0	1840	≤32	9.7	24.2	24.2
C	R1	7.78	−30.8	176.7	2.7	4.0	2220	32	12.5	29.3	26.6
	R2	7.91	−38.3	181.2	0.0	0.0	2120	32	11.7	27.5	27.5
	R3	7.84	−55.5	182.6	0.0	0.0	2120	32	11.6	27.0	27.0
Abiotic		7.88	81.5	240	1.1	2.3	3920	0.0	16.33	04.0	02.6

F feasibility study, *C* confirmation Study; *R1* monitored after day 1, *R2* monitored after day 2 and so on
Initial values pH = 7.90; NH_4–N = 250 mg/L and HCO_3^- = 4000 mg/L as $CaCO_3$

within 24 h. As the nutrient media did not contain NO_2–N; it is not possible to have conventional anammox as per the Eq. (2) (Strous et al. 1998).

$$\begin{aligned} & NH_4^+ + 1.31NO_2^- + 0.066HCO_3^- + 0.13H^+ \longrightarrow 1.02N_2 \\ & \quad + 0.26NO_3^- + 0.066CH_2O_{0.5}N_{0.15} + 2.03H_2O; \\ & \quad \Delta G^\circ = -357\ \mathrm{kJ/M}\ NH_4^+ \end{aligned} \tag{2}$$

It is hypothesized that in the absence of NO_2–N or any other electron acceptor, the enriched biomass used in this study is capable of using HCO_3^- as an alternative electron acceptor. Also, the NH_4^+ is first oxidised to NO_2–N with H^+ and H_2O as by-products. Once these products are formed, it is easy for conventional anammox to take place as per Eq. (1). The results of the batch studies after 24 h indicate the possibility of the proposed hypothesis due to the presence of both NO_2–N and NO_3–N in anoxic environment.

After day 1, it is observed that NH_4–N was increased. This could be due to sludge hydrolysis of the stressed biomass. As, the mixed culture used in this study was enriched for autotrophic anammox as per Eq. (2), once the suitable substrate (NO_2–N) becomes unavailable, some biomass can be vulnerable to death and lyses. The increase of NH_4–N along with disappearance of both NO_2–N and NO_3–N indicate the sludge lyses. The disappearance of both NO_2–N and NO_3–N could be as a result of heterotrophic denitrification as per Eq. (3). The mixed culture used in this study was having competing anammox and denitrification potential (Anjali and Sabumon 2014). Therefore, in the presence of organic carbon (generated as a result of lyses of cells), denitrification (Eq. 3) is thermodynamically more powerful than anammox (Eq. 2). The presence of COD as the batch progresses indicates lyses of biomass. The increase of pH and alkalinity observed is another indication of competing denitrification.

$$\begin{aligned} & 1.25CH_3COOH + 2NO_3^- \xrightarrow{\text{Heterotrophic denitrificans}} N_2 \\ & \quad + 2.5CO_2 + 1.5H_2O + 2OH^-; \Delta G^o = -527.4\,\mathrm{kJ/M} \end{aligned} \tag{3}$$

The results of the confirmatory studies were also in the same trend as that of the feasibility studies. The ORP values in all batch reactors were negative, indicating the anoxic environment maintained inside the reactors. The interesting results of this study are that removals of both NH_4–N and total nitrogen were almost the same without any external addition of energy and organic carbon. Therefore, this study indicates the feasibility of economic removal of ammonia in anoxic condition in the presence of HCO_3^- and may find future applications. For example, any ammonia bearing wastewater could be treated economically by scrubbing CO_2 emitted from an industrial stack. Also, the results showed the versatility of anammox species to use HCO_3^- as an alternative electron acceptor which has not been reported. Such versatility of anammox species to work in the presence of various inorganic electron acceptors (Strous et al. 1998; Polanco et al. 2001; Sabumon 2009) will help in the economic autotrophic anoxic ammonia removal from various wastewaters.

3.2 *Kinetic Studies*

As the feasibility and the confirmatory studies showed higher removal of ammonia within a day, kinetic studies were performed at two different initial concentrations of NH_4–N by keeping the ratio of HCO_3^-/NH_4–N ~ 12. Figure 1 shows the results of the kinetics of ammonia removal at an initial concentration of 350 mg/L. It is interesting to note that the major NH_4–N removal had occurred within 2 h and is ~26 % with a corresponding consumption of HCO_3^-. While, NH_4–N removal in the abiotic sample was only ~2.2 % within 2 h. Also, the MLSS/MLVSS values increased corresponding to the decrease in both NH_4–N and HCO_3^-. The specific substrate utilization rate for ammonia oxidation was 31 mg/g MLVSS/h considering the initial 2 h phase. The reason for a slowing of the reaction after 2 h is not clear.

The NO_2–N and NO_3–N concentrations were not very high, resulting in effective nitrogen removal. There were generation and oxidation of NH_4–N as the time progressed indicating occurrence of sludge lyses and anoxic ammonia oxidation. The sludge lyses were confirmed by the presence of COD in the reactor. The MLVSS values also were fluctuating as the reaction progressed. As observed in the earlier batch studies, pH and ORP were decreased as the reaction progressed. The profiles of NH_4–N and HCO_3^- indicate that a correlation exists between the two parameters in ammonia removal. However, at this point of time, the role of HCO_3^- as an electron acceptor is not proved and needs further research.

The results of kinetic study of ammonia removal at an initial concentration of 130 mg/L (Fig. not shown) also showed that the rate of ammonia removal was highest within 2 h. The ammonia removal was 36 % with a corresponding higher rate of consumption of HCO_3^-. The specific substrate utilization rate for ammonia oxidation was found to be 16 mg/g MLVSS/h. Therefore the two different sets of kinetic studies indicate that the rate of reaction is concentration dependent and if higher the initial concentration, the higher is the rate of removal. Such kinetics are,

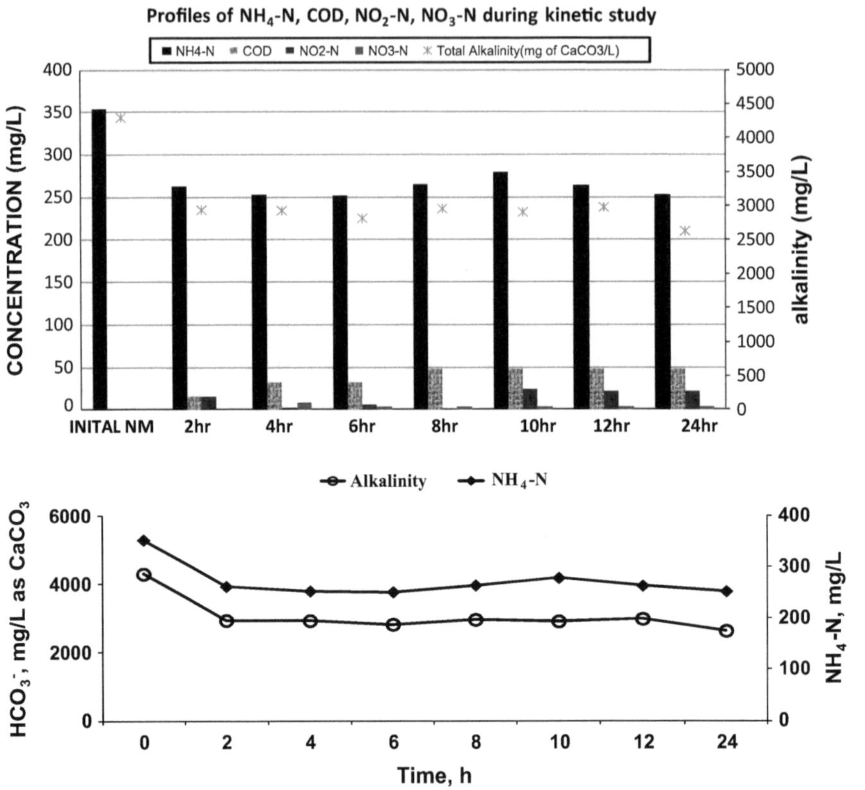

Fig. 1 Kinetics of ammonia removal at an initial concentration of 350 mg/L (HCO_3^-/NH_4–N ~ 12)

in general, an identity of a biochemical process. Therefore, the results of kinetics study also indicate the anoxic ammonia oxidation is feasible by using HCO_3^- as an alternative electron acceptor.

4 Conclusions

Following conclusions can be drawn from the results of this study:

1. Enriched mixed culture containing anammox species is capable of anaerobic ammonia oxidation using HCO_3^- as an electron acceptor.
2. The highest rate of ammonia removal was achieved within 2 h.
3. The specific substrate utilization rate of the biomass for ammonia oxidation was 31 mg/g MLVSS/h.

4. Feasibility of bio-chemical anaerobic ammonia oxidation using HCO_3^- as an electron acceptor was proved and the oxidation rate is dependent on the initial concentrations of substrates.

References

Ahn, Y.H.: Sustainable nitrogen elimination biotechnologies: a review, process Biochem. **41**, 1709–1721 (2006)

Anjali, G., Sabumon, P.C.: Unprecedented development of anammox in presence of organic carbon using seed biomass from a tannery common effluent treatment plant (CETP). Bioresour. Technol. **153**, 30–38 (2014)

APHA-AWWA-WPCF: Standard Methods for the Examination of Water and Wastewater. (1998)

Chamchoi, N., Nitisoravut, S.: Anammox enrichment from different conventional sludges. Chemosphere **66**, 2225–2232 (2007)

Fdz-Polanco, F., Fdz-Polanco, M., Garcia, P.A., Villaverde, S., Uruen, M.A.: New process for simultaneous removal of nitrogen and sulphur under anaerobic conditions. Water Res. **35**, 1111–1114 (2001)

Kartal, B., Rattray, J., van Niftrik, L.A., van de Vossenberg, J., Schmid, M.C., Webb, R.I., Schouten, S., Fuerst, J.A., Damst, J.S., Jetten, M.S.M., Strous, M.: Candidatus "Anammoxoglobus propionicus" a new propionate oxidizing species of anaerobic ammonium oxidizing bacteria. Syst. Appl. Microbiol. **30**, 39–49 (2007)

Molinuevo, B., Garcia, C.M., Karakashev, D., Angelidaki, I.: Anammox for ammonia removal from pig manure effluents: effect of organic matter content on process performance. Bioresour. Technol. **100**, 2171–2175 (2009)

Sabumon, P.C.: Anaerobic ammonia removal in presence of organic matter: a novel route. J. Hazard. Mater. **149**, 49–59 (2007)

Sabumon, P.C.: Effect of potential electron acceptors on anoxic ammonia oxidation in the presence of organic carbon. J. Hazard. Mater. **172**, 280–288 (2009)

Strous, M., Heijnen, J.J., Kuenen, J.G., Jetten, M.S.M.: The sequencing batch reactor, as a powerful tool for the study of slowly growing anaerobic ammonium-oxidizing microorganisms. Appl. Microbiol. Biotechnol. **50**, 589–596 (1998)

Zhang, J., Wu, P., Hao, B., Yu, Z.: Heterotrophic nitrification and aerobic denitrification by the bacterium Pseudomonas stutzeri YZN-001. Bioresour. Technol. **102**, 9866–9869 (2011)

Denitration of High Nitrate Bearing Alkaline Waste Using Two Stage Chemical and Biological Process

Sayali Titre, Akshay Jakhete, Avinash Sahu, Tessy Vincent, Mahendra L. Bari and Ajaygiri K. Goswami

1 Introduction

The most ubiquitous contaminant present in surface water and ground water has to be nitrates and is known to be a cause of concern across many locations worldwide (Pintar 2003). Increased urbanization and rampant use of resources by humans along with excessive mineralization of land, livestock exploitation, use of septic tank systems and wastes from industries viz. fertilizer, explosives and nuclear along with human wastes have resulted in the excess nitrate contamination of water (Xie et al. 2011; Mousavi et al. 2011; Berndt et al. 2001; Glass and Silverstein 1999). In gastrointestinal track, nitrate is converted to nitrite which reacts with hemoglobin to inhibit the transfer of oxygen, which causes methemoglobinemia or the blue baby syndrome (Salome et al. 2009; Canter 1996). Studies have linked consumption of nitrate infested water to increase the chance of gastric cancer, non-Hodgkin's lymphoma, coronary cardiac diseases and hypertrophy of the thyroid (Karanasios et al. 2010). Hence it is imperative to remediate nitrates. Indian standard for nitrate in water has been set at 100 mg/L (ppm), which is the maximum permissible limit. The World Health Organization has set a limit of 10 ppm NO_3 for human consumption and 100 ppm NO_3^- for animals (Naik and Setty 2011). Alkaline wastes having high nitrate content from 100,000 to 150,000 ppm are generated during back end operations of nuclear fuel cycle. The current technologies available for

S. Titre · A. Jakhete (✉) · M.L. Bari · A.K. Goswami
University Institute of Chemical Technology, North Maharashtra University, Jalgaon 425001, Maharashtra, India
e-mail: akshay.jakhete@gmail.com

S. Titre
e-mail: sayali0903@gmail.com

A. Sahu · T. Vincent
Bhabha Atomic Research Centre, Trombay, Mumbai 400085, Maharashtra, India
e-mail: tessyv@barc.gov.in

R. Mohan B. et al. (eds.), *Materials, Energy and Environment Engineering*,
DOI 10.1007/978-981-10-2675-1_12

nitrate reduction can be generally divided into three: physiochemical methods, biological denitrification and chemical denitration. For biological denitrification processes, organic carbon acts as a limiting substrate and thus extra carbon is added by many wastewater treatment plants to achieve denitrification (Naik and Setty 2012). Chemical denitration method refers to nitrate reduction to elemental nitrogen in presence of a suitable solid catalyst using reducing agent and is the most suitable technique that can be used for nitrate reduction of higher concentrations.

In this paper, we aim to develop a new nitrate remediation technique for the waste streams containing highly alkaline salts. The process consists of two stage nitrate reduction—chemical denitration of alkaline waste using formaldehyde as reducing agent in the presence of a Pd-Cu/Al_2O_3 bimetallic catalyst followed by biological denitrification using suitable denitrifying bacteria immobilized on sand. The chemical denitration helps in reducing nitrates up to 5000 ppm approx. levels, after which the nitrate reduction is continued using biological denitrification. The paper discusses development and application of chemical denitration and process development of biological denitrification.

2 Materials and Method

2.1 Chemical Denitration

Chemicals of AR grade were used for lab scale studies—sodium nitrate, formaldehyde (37–41 %), formic acid (90 %) and sodium hydroxide (all manufactured by Thomas and Baker). The commercially available catalyst was used, manufactured by M/s Catalysts & Technologies, Mumbai, which has given the metal content as 2 wt% palladium and 0.6 wt% copper with the support of γ-alumina. After verifying the feasibility process with sodium nitrate, the process was applied for simulated waste consisted of 1.63 M sodium nitrate, 0.12 M sodium nitrite, 0.7 M sodium hydroxide and 0.12 M sodium aluminate. The pH of the prepared solution was approximately 13. Selection of catalyst and reducing agent was based on literature (Hirano et al. 2006; Prüsse and Vorlop 2001).

The study of various parameters and process development was done for synthetic sodium nitrate waste and the setup consisted of a acrylic column (fixed bed catalytic reactor with re-circulating mode) with a volume of one liter and bed height of 20 cm. Peristaltic pump was used to transfer the nitrate waste from the flask to the column. Liquid percolates through the catalyst bed from bottom to top. Constant space velocity of 6.626/h was used and provisions were made for the constant agitation and heating of flask via magnetic stirrer cum heater. The experiment starts only after the desired temperature of 45 °C is achieved. After achieving this temperature, calculated amount of formaldehyde was added to the flask (in periodic batch sequence) following which only, the peristaltic pump was turned on. It is imperative to maintain the pH above 12 and was done by adding NaOH pellets at regular intervals. This entire process was carried out for 5–6 h. Samples were taken at regular intervals and analyzed for nitrate and nitrite content using UV-Vis spectrophotometer at 275 and 360 nm respectively. To remediate nitrites, another step needs to be carried out by the addition of formic acid reductant in presence of a monometallic palladium catalyst. The nitrite reduction method was adopted for a 500 ml scale and consisted of a glass column of 100 cm height with an internal diameter of 1.6 cm and was filled with monometallic palladium catalyst with an approx. bed height of 70 cm. The experiment was carried in a batch recirculation mode and peristaltic pump was used to transfer the nitrite rich waste to Pd catalyst bed at a space velocity of 6.626/h. The experiment was carried out at 45 °C temperature and amount of formic acid to be added was calculated as per desired stoichiometry. Sampling was done periodically for 180 min and analyzed using UV-Visible spectrophotometer. The final mixture thus obtained had nitrites in trace amounts along with nitrates, unreacted formaldehyde and sodium hydroxide. The same procedures were applied for nitrate and nitrite reduction in simulated waste. For all the experiments that were carried out using catalytic reduction method, the extent of denitration achieved was determined by analyzing the samples for residual nitrates and nitrites present using UV-Vis spectrophotometer. A Labindia UV-Visible spectrophotometer (UV 3000) with 1 cm matching quartz cell was used for the absorbance measurements.

2.2 *Biological Denitrification*

Nutrient broth has been originally designed for use in the Standard Method for Examination of Water and Wastewater and thus can be used for maintenance of microorganism growth as basic culture media. The denitrifying organisms were obtained from Effluent Treatment Plant, BARC, Trombay (Sandhya et al. 2012). Nutrient broth (13 g/L) was prepared to be used as growth media for revival of microbes.

Unlike nutrient broth, which is general growth media, chemically defined media was used to encourage the growth of particular microbes inhibiting the growth of others. The composition (per liter of demineralized water) of chemically defined media (Lee et al. 2008) selected for growth were: sodium nitrate ($NaNO_3$) (6.85 g); magnesium sulfate ($MgSO_4$) (0.967 g); mono-potassium phosphate (KH_2PO_4) (0.853 g); ferrous sulfate ($FeSO_4$) (2.667 mg) and copper sulfate ($CuSO_4$) (0.133 mg). Micronutrients: (prepared per 100 ml of demineralized water) consisting of boric acid (H_3BO_3) (0.1 g); manganese sulphate ($MnSO_4$) (0.1 g); zinc sulphate ($ZnSO_4$) (0.1 g); ammonium molybdate ($Mo_7O_{24}(NH_4)_6$) (1 g) and cobalt nitrate ($Co(NO_3)_2$) (0.1 g). 1 mL of micronutrients is added to 1 L of chemically defined media.

Experiments have been performed to select the carbon source as food for the required process. Six different carbon sources, namely methanol, ethanol, sodium acetate, succinic acid, acetic acid and citric acid were added to the flasks. The C/N ratio for all the carbon sources was kept 3. The growth analysis was done by checking the growth of biomass by checking optical density at 600 nm in the mixture using UV Spectrophotometer. Maximum growth was observed in the flask with sodium acetate at the end of the fifth day. Thus, sodium acetate was selected as the carbon source for the process.

Immobilization of microbes was done on sand as a carrier. Sand was distributed in five different flasks. A solution of 5000 ppm nitrate along with chemically defined media at pH 3.5–4 and Sodium acetate was added in each flask. Autoclaving was done at 121 °C for 15 min. The microbes were allowed to grow on sand for 5 days.

The packed bed reactor with recirculation mode was used in order to increase the adherence between biomass and sand particles. The waste containing the inoculums was sent in an upward direction with a very low velocity so as to increase the contact time between inoculum and the sand particles. This resulted in an intense organism growth on the carrier particle leading to increased consumption of biodegradable wastes present in the mixture. The reaction flow rate was affected by the velocity of substrate solution and thus recycling method is advantageous.

Experiments were performed in a packed bed reactor with recirculation mode. The reactor was constructed from Plexiglas with 5.08 cm of internal diameter and height 50 cm. Feed containing carbon source, nitrate and growth medium, stored in a reservoir was pumped from the bottom of the bioreactor using a peristaltic pump with a velocity of 10 ml/min. The liquid was introduced to the bed through a distributor, which is used to pump the liquid evenly and to support the bed. For gas venting and the outlet of the column the vent was provided at the top of the column. The bed height was kept 6 cm. Feed recirculation through the reservoir was done via the side outlet provided at the top of the column. Ball valve was used on the inlet and the outlet to control the flow. The analysis of nitrate and nitrite was carried out using Ion Chromatograph (Dionex, ICS-1000).

3 Result and Discussion

3.1 Chemical Denitration

The process development was carried out in three stages—parametric studies and determination of most optimum value; process development using synthetic nitrate waste prepared using sodium nitrate salt and its application on simulated waste

Table 1 Nitrate reduction parameters

Parameter	Determined value
Scale	1 L
Formaldehyde: nitrate molar ratio	3
Volume of formaldehyde added	360 mL
pH	13.00
Catalyst to nitrate ratio	1:10
Space velocity	6.626/h
Temperature	45 °C

(which is similar to the one obtained during reprocessing operations). The experimental work for process development was carried out at an inactive stage for 1 L scale. Initially nitrate reduction experiments were carried out on 1 L scale using sodium nitrate salt containing nitrate concentrations of ~100,000 ppm. Process parameters like the molar ratio of formaldehyde to nitrate, reaction temperature,

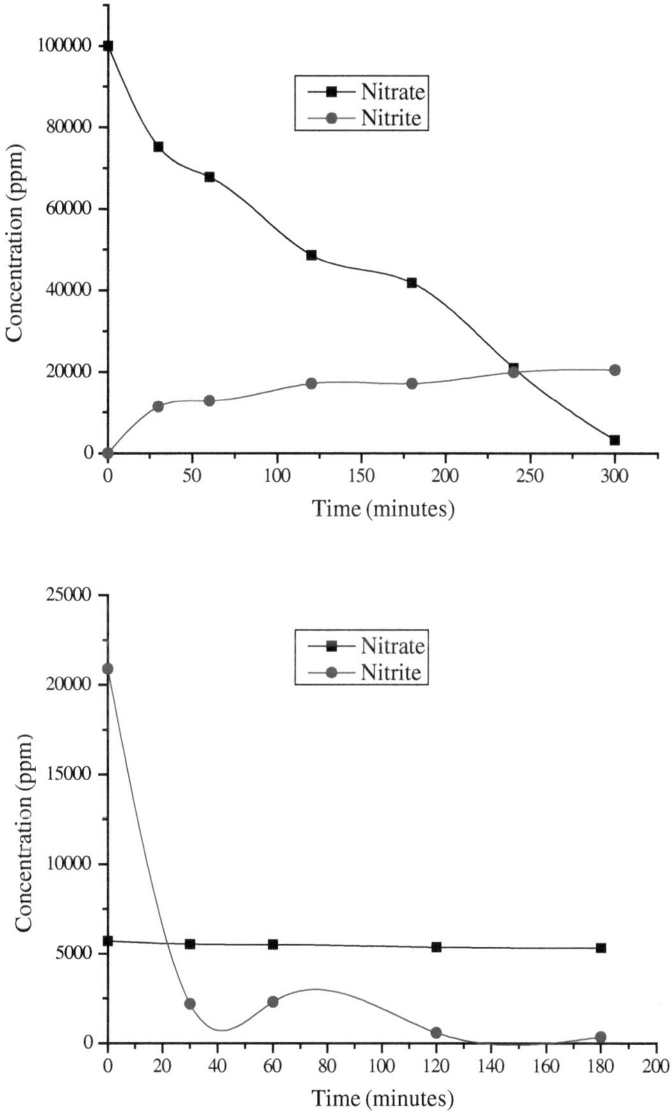

Fig. 1 Nitrate and nitrite reduction for denitration of synthetic sodium nitrate waste

catalyst quantity, space velocity, pH and reaction time were chosen based on trial and error. Details of the parameters are given in Table 1.

Once the parametric values were determined, process development was done using synthetic waste of 100,000 ppm concentration which was prepared by sodium nitrate salt, made alkaline by adding few pellets of sodium hydroxide. The process and its subsequent analysis were carried out as per directions given in Sect. 2.1 and was repeated several times. It was observed that the nitrate mixture thus prepared was reduced to a level of approximately 4000 ppm showing an average reduction close to 97 % as shown in Fig. 1. An important observation was that nitrite concentration increased from 0 to approx. 20,500 ppm. The rise in concentration was studied extensively and it was found that as pH increases the concentration of OH^- ions also increases which inadvertently blocks the monometallic palladium sites. These sites are important for nitrites to adsorb, over which these can be converted to elemental nitrogen hence we can observe a steep increase in nitrite concentration. Its remediation can be done by treating the nitrite rich mixture with formic acid at pH below 3.5 (to avoid precipitation of dissolved salts) over monometallic palladium sites. Similar studies were done at 500 mL scale to determine the most suitable parametric value, details of which are given in Table 2.

When the process was developed it was to remediate nitrites obtained from the preceding experiment and it was found that the nitrite concentration decreased from 20,500 to approx. 300 ppm, accounting for near 99 % nitrite reduction. The nitrite decrease did not have any effect on the nitrate concentration.

With the successful development of the process, it was applied subsequently to the simulated waste that was prepared and is similar to the waste streams generated during the back end operation of nuclear fuel cycle. The same procedure was followed whilst doing experimental studies, and it was found that nitrate concentration decreased from 100,000 to approx. 4000 ppm resulting in close to 96 % reduction with subsequent increase in nitrites from 5000 to 40,000 ppm. The nitrites then were remediated using formic acid which resulted in a final concentration of 2000 ppm accounting for a 95 % reduction. The graphs are shown in Fig. 2.

Table 2 Nitrite reduction parameters

Parameters	Determined value
Scale	500 mL
Molar ratio	2
Volume of formic acid added	58 mL
pH	3.5
Catalyst to nitrate ratio	1:10
Space velocity	6.5/h
Temperature	45 °C

3.2 Biological Denitrification

The final mixture obtained after chemical denitration consisted of approx. 3000–4000 ppm of nitrates with some trace amounts of nitrites, formaldehyde and formic acid in small quantities, traces of residual palladium, copper and unreacted sodium

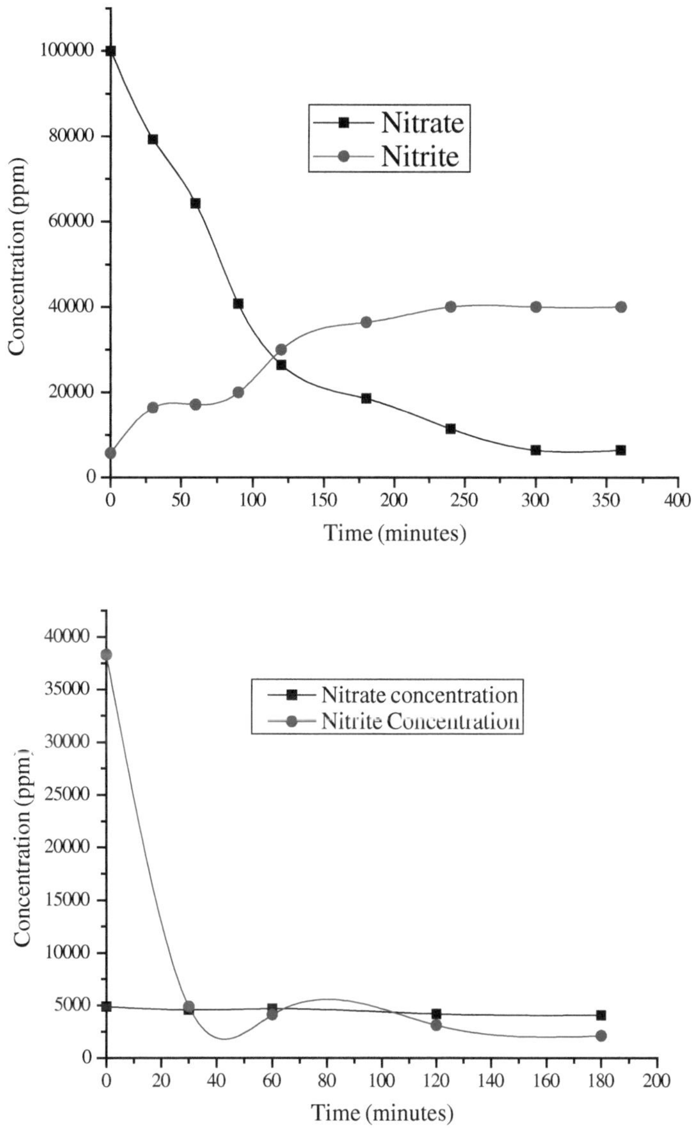

Fig. 2 Nitrate and nitrite reduction for denitration of simulated nitrate waste

hydroxide. The biological denitrification process was incorporated so that the nitrates can be further treated to levels considered optimal by various environmental agencies. Another purpose of using biological denitrification was to remove the residual formaldehyde, which in even in trace amounts is hazardous to all living beings. The studies that were undertaken during process development of biological denitrification were—anaerobic denitrification of 5000 ppm nitrate in the presence of nutrient broth; anaerobic denitrification of 5000 ppm nitrate in the presence of formaldehyde.

The first study performed was the biological denitrification in the presence of 5000 ppm nitrate and nutrient broth at pH 6. The analysis was done using Ion Chromatograph (Dionex, ICS—1000) and the results were noted for nearly 50 h and is as shown in Fig. 3. The graph shows the decreasing concentration of nitrate and simultaneously increasing and decreasing concentration of nitrite. As the curve shows the trends it was observed that the graphs were following the trend of denitrification. And the motive of growth and immobilization of bacteria was achieved. Reduction achieved was about 99.4 % for nitrate and around 99.2 % for nitrite. The final pH of the mixture was 9.

A second study to be performed was denitrification in the presence of 5000 ppm nitrate and formaldehyde. With nutrient broth as the carbon source and pH of 6.22, reduction mechanism of nitrates by denitrifying bacteria was observed for 15–16 days. The results are shown in Fig. 3. It was observed that within a period of 15–16 days the nitrate concentration was about 45 ppm. It is seen that the reduction in the earlier stage was very slow. This was because the microorganisms at earlier stage try to adjust to the environment following which the growth starts and subsequent reduction occurs.

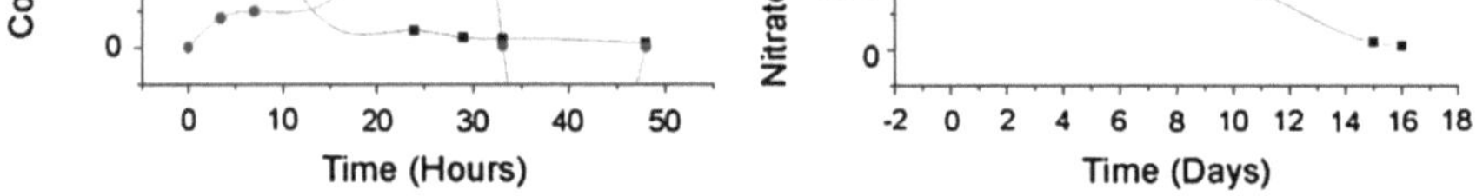

Fig. 3 Nitrate reduction graph for biological denitrification in presence of nutrient broth and formaldehyde

4 Conclusion

The aim of this experimental study was to develop and apply process for the remediation of alkaline nitrate waste originating from radio-chemical plant having concentration up to 100,000 ppm and having a pH greater than 12. It is evident that both chemical and biological process are capable enough for nitrate remediation, but they have their own drawbacks—mainly, chemical denitration is applicable for larger nitrate reduction aspect and is unsuitable for remediation at lower concentration whereas the vice versa is true for biological denitrification. This experimental study combines these two processes for achieving quantitative nitrate removal having initial concentration of 100,000 ppm. The chemical denitration has been developed successfully and applied on simulated waste with nitrate reduction exceeding 96 % for both cases. Process development for biological denitrification has shown excellent results in nitrate remediation in the presence of formaldehyde with reduction up to 99 %.

References

Berndt, H., Monnich, I., Lucke, B., Menzel, M.: Tin promoted palladium catalyst for nitrate removal from drinking water. Appl. Catal. B. Environ. **30**, 111–122 (2001)

Canter, L.W.: Nitrates in Groundwater. CRC Press, Boca Raston (1996)

Glass, C., Silverstein, J.: Denitrification of high nitrate, high salinity wastewater. Water Res. **33**, 223–229 (1999)

Hirano, M., Kato, T., Mihara, S., Ito, H.: Method of treating waste water containing nitrate-nitrogen. European Patent Application, Patent No. EP 1826184 A1, 1–12 (2006)

Karanasios, K.A., Vasiliadou, I.A., Pavlou, S., Vayenas, D.V.: Hydrogenotrophic denitrification of potable water: a review. J. Hazard. Mater. **180**, 20–37 (2010)

Lee, O.M., Oh, J.H., Libert, M., Hwang, D.S., Choi, Y.D., Park, J.H., Chung, U.S., Jo, B.K., Kim, M.J., Lee, S.J.: Bio-denitrification of the nitrate waste solution from the lagoon sludge in a continuous bio-reduction process. Korean J. Chem. Eng. **25**, 787–792 (2008)

Mousavi, S.A.R., Ibrahim, S., Aroua, M.K., Ghafari, S.: Bio-electrochemical denitrification—a review. Int. J. Chem. Environ. Eng. **2**(2), 140–146 (2011)

Naik, S.S., Setty, Y.P.: Effect of carbon sources on biological denitrification of wastewater by immobilized *Pseudomonas stutzeri* bacteria in a fluidized bed bio reactor (FBBR). In: 2nd International Conference on Chemical Engineering and Applications IPCBEE, vol. 23, 114–118 (2011)

Naik, S.S., Setty, Y.P.: Biological denitrification of wastewater—a mini review on carbon source. In: International Conference on Chemical, Environmental Science and Engineering (ICEEBS'2012), Pattaya (Thailand) (2012)

Pintar, A.: Catalytic processes for the purification of drinking water and industrial effluents. Catal. Today **77**, 451–465 (2003)

Prüsse, U., Vorlop, K.D.: Supported bimetallic palladium catalysts for water-phase nitrate reduction. J. Mol. Catal. A Chem. **173**, 313–328 (2001)

Salome, O., Soares, G.P., Orfao, J.J.M., Pereira, M.F.R.: Bimetallic catalysts supported on activated carbon for the nitrate reduction in water: optimization of catalysts composition. Appl. Catal. B Environ. **91**, 441–448 (2009)

Sandhya, U., Banerjee, D., Singh, I.J., Wattal, P.K.: Denitrification of high sodium nitrate bearing effluents using flow through bioreactor. Desalin. Water Treat. **38**, 52–58 (2012)

Xie, Y., Cao, H., Li, Y., Zhang, Y., Crittenden, J.: Highly selective Pd-Cu/amorphous silica-alumina (ASA) catalyst for groundwater denitration. Environ. Sci. Technol. **45**, 4066–4072 (2011)

Optimization Study of Cadmium Biosorption on Sea Urchin Test: Application of Response Surface Methodology

D. John Babu, Y. Prasanna Kumar, Pulipati King and K. Vidya Prabhakar

1 Introduction

Rapid industrialization has led to increased discarding of wastewater into the environment which caused the deposition of hazardous metal ions in water bodies. These metal ions present in the waste waters became precarious pollutants due to their nondestructive nature, high toxicity, accumulation through food chain and subsequent biomagnifications (Iynengar and Venkobachar 2000). Cadmium (Cd (II)) is one of the metals that are extremely toxic to organisms even at low concentration. Exposures to Cd cause severe health problems such as renal failure, liver diseases, bone degradation and hypertension (Nordberg et al. 1992). Due to this, Department of Environment, UK has included Cd in red list of priority pollutants (UK Red List Substances: 1991) and in List I (the "black list") of Directive 76/464/EEC (Council Directive 1976). USEPA has also classified Cd as group B1 carcinogen (US Environmental Protection Agency 1999).

D. John Babu (✉)
School of Biotechnology, Vignan University, Vadlamudi, India
e-mail: johnbabud77@gmail.com

Y. Prasanna Kumar
Sankaitika Vidya Parishad Engineering College, Visakhapatnam, India
e-mail: prasanna_ky1@yahoo.co.in

P. King
Department of Chemical Engineering, Andhra University College of Engineering, Visakhapatnam, India
e-mail: pulipatiking@yahoo.co.in

K. Vidya Prabhakar
Department of Biotechnology, Vikrama Simhapuri University, Nellore, India
e-mail: kodalividyaprabhakar@gmail.com

R. Mohan B. et al. (eds.), *Materials, Energy and Environment Engineering*,
DOI 10.1007/978-981-10-2675-1_13

Biosorption is a simple and effective technology for elimination of heavy metals from waste water (Holan et al. 1993). The major advantages of biosorption are low cost, high efficiency of metal removal, regeneration of biomass, elevated selectivity for certain metals, short process time and no generation of additional sludge in the form of secondary compounds which might be hazardous (Spinti et al. 1995; Srinath et al. 2002). However earlier studies suggest that unlike other approaches, biomass derived from calcium carbonate can be an efficient and low cost biosorbent for the removal of metal ions from waste effluents (Du et al. 2011; Pena-Rodriguez et al. 2010). Biomass prepared from shell of sea urchins can be considered as an inexpensive source of calcium carbonate. The sea urchins are found across the ocean floors worldwide, and sea urchin tests amass as waste at seashore. Thus sea urchins test can be considered as biosorbent. The present study was carried out to asses this hypothesis and to evaluate optimum metal biosorption capacity of the shell dust of the sea urchins for removal of Cd(II) ions form synthetic metal ion solutions.

2 Materials and Methods

2.1 Biosorbent Preparation

Purple sea urchins were collected from the nearby beach area and cleaned well with fresh water. The spines and the flesh were removed and each test was broken into two halves and dried to about a day in the sunlight. The remaining flesh and organic matter if any adhered to the test surface was also removed. The test pieces were again cleaned with tap water first thoroughly and again with distilled water in order to neutralize the traces of salinity if any in the laboratory. These test pieces were allowed to dry in open air at room temperature and powdered using ball mill. The test powder was graded by using BSS sieve set to get the required average size for using as an adsorbent.

2.2 Preparation Synthetic Cd(II) Stock Solution

Synthetic aqueous solution of Cd(II) was prepared by dissolving appropriate amount cadmium nitrate salt in distilled water. Stock solution of concentration 1000 ppm was prepared and working solutions were prepared by appropriate dilution. HNO_3 (0.1 N) and NaOH (0.1 N) were used for adjusting the pH of the solution. All the analytical grade chemicals used for experiments were purchased from Merck India Ltd.

2.3 Biosorption Experiments (Batch Mode)

Cd removal capacity of the sea urchin test was estimated by varying pH, temperature, initial Cd ion concentration and biomass dosage while keeping the volume of reaction mixture as 30 ml. Twenty nine experiments designed by Box-Behnken design were conducted. The mixture was agitated in orbital shaker at 120 rpm. Samples were collected after 1 h contact time and filtered using Whatman filter paper (No. 1). Residual metal ions concentration of was find out using atomic absorption spectroscopy (GBC Avanta Ver 1.32, Australia). Each experiment was repeated four times and the average values have been reported. To make sure that no adsorption has taken place on the walls of the conical flasks, blank experiments were also conducted.

2.4 Design Variables for Biosorption Study

The ranges of the process parameters for Box-Behnken design for biosorption include: biosorbent loading (0.1–0.5 g/l), Cd(II) metal ion concentration (20–100 ppm), temperature (20–40 °C) and pH (2–6). The three levels (coded) for each parameter in Box-Behnken design were shown in Table 1 and the experimental design was shown in Table 2.

2.5 Process Optimization

RSM is a systematic methodology designed using mathematical and statistical techniques to develop a mathematical model to relate the process parameters and responses to understand the process with minimal number of experiment runs. The fundamental objective of RSM is to find out optimum values of operational conditions to maximize the process (Myers and Montgomery 2002). A standard experimental design of RSM called Box-Behnken design for four variables and each with three levels (Box and Behnken 1960) was used to find out the optimum pH, temperature, metal ion concentration and biomass dosage. The design was

Table 1 Three levels (coded) of process parameters for experimental design

Factors	Name	Units	Low actual	High actual	Low coded	Middle coded	High coded
A	pH		2	6	−1	0	1
B	Temperature	°C	20	40	−1	0	1
C	Initial concentration	mg/L	20	100	−1	0	1
D	Adsorbent dosage	g/L	0.1	0.5	−1	0	1

Table 2 Experimental design in terms of coded factors of the Box–Behnken model

Run	pH	Temperature (°C)	Initial concentration (mg/l)	Biomass dosage, (g/l)	Response (% removal)
1	0	−1	−1	0	85.087
2	0	−1	1	0	72.48
3	0	0	−1	1	85.482
4	0	0	1	1	72.875
5	0	0	0	0	76.375
6	0	0	0	0	76.375
7	0	0	1	−1	69.81
8	−1	−1	0	0	70.325
9	0	−1	0	−1	75.96
10	0	0	0	0	76.375
11	0	0	−1	−1	83.167
12	0	1	0	−1	74.1
13	0	0	0	0	76.375
14	1	−1	0	0	81.327
15	1	1	0	0	79.467
16	0	1	−1	0	83.227
17	1	0	−1	0	88.535
18	−1	0	1	0	64.927
19	0	1	0	1	76.415
20	−1	0	0	−1	68.407
21	1	0	0	1	81.722
22	−1	0	0	1	70.722
23	1	0	1	0	75.927
24	−1	1	0	0	68.467
25	0	0	0	0	76.375
26	0	−1	0	1	78.275
27	0	1	1	0	70.617
28	−1	0	−1	0	77.532
29	1	0	0	−1	79.407

taken as it fulfills most of the requirement for optimization of the biosorption study. For better accuracy the second-order model was used. The mathematical equation relating the four independent process variables and the removal percentage of metal ions has been expressed by the following quadratic model

$$Y = \beta_0 + \beta_1 a + \beta_2 b + \beta_3 c + \beta_4 d + \beta_{11} a^2 + \beta_{22} b^2 + \beta_{33} c^2 + \beta_{44} d^2 \\ + \beta_{12} ab + \beta_{13} ac + \beta_{14} ad + \beta_{23} bc + \beta_{24} bd + \beta_{34} cd \tag{1}$$

where Y is the output of the process, β_n is the coefficient associated with factor n, and the letters *a*, *b*, *c*, *d* are the process variables in the model

3 Results and Discussion

3.1 Competency of the Model for Cd(II) Removal

In this study two way interaction analysis was carried out to study individual and double interactive effect of various parameters on biosorption and also to develop appropriate regression equation for analysis of experimental data. To check the competency of models three different tests, such as the sequential model, sum of squares, lack of fit test were carried out and the results were shown in Table 3. *P* value <0.01 for the quadratic model indicated that the model is significant. Fitness of the model was evaluated using determination coefficient (R^2). For the present quadratic model the values of determination coefficient, Predicted R^2 and Adjusted R^2 are 0.9985 and 0.9995 (Table 4). These values approximately equal and also equal to 1.0. This indicates the better fitness of experimental data to quadratic model. The minimum value of standard error design 0.479 around the centroid and maximum predication variance 0.536 at the design points also

Table 3 Selection of a satisfactory model for Cd(II) removal: sequential model sum of squares

Source	Sum of squares	df	Mean square	F value	*p*-value Prob > F	
Mean versus total	1.694E+005	1	1.694E+005			
Linear versus mean	877.63	4	219.41	155.90	<0.0001	
2FI versus linear	0.14	6	0.023	0.013	1.0000	
Quadratic versus 2FI	33.40	4	8.35	499.3	<0.0001	Suggested
Cubic versus quadratic	0.19	8	0.023	3.00	0.0991	Aliased
Residual	0.047	6	7.813E−003			

Table 4 Model summary statistics

Source	SD	R^2	Adjusted R^2	Predicted R^2
Linear	1.19	0.9629	0.9568	0.9419
2FI	1.37	0.9631	0.9426	0.8775
Quadratic	0.13	0.9997	0.9995	0.9985
Cubic	0.088	0.9999	0.9998	0.9926

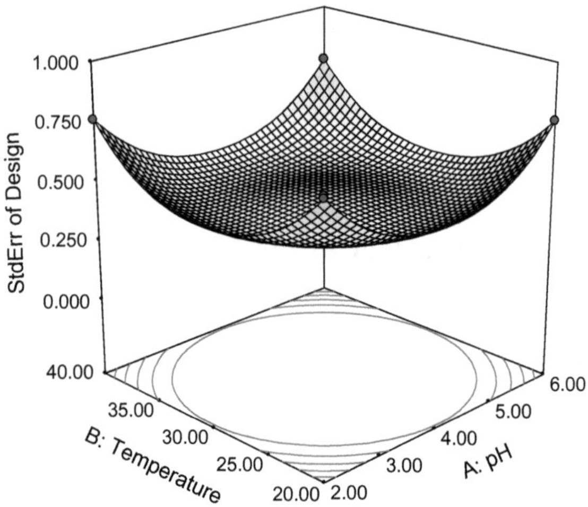

Fig. 1 Standard error design of the model temperature versus pH holding initial metal ion concentration and adsorbent loading at central value

indicated that present model can be used to navigate the design space in the present study (Fig. 1).

3.2 *Regression Analysis*

Models developed in regression analysis represent removal percentage of Cd(II) as functions of pH (A), temperature (B), initial metal concentration (C) and biosorbent dosage(D). The final equation developed for the metal removal capacity of sea urchin test as function of four parameters was expressed in coded terms by the quadratic model equation:

$$\begin{aligned}\%\text{Removal} = & +76.38 + 5.5 \times a - 0.93 \times b - 6.37 \times c + 1.22 \times d \\ & - 5.000\text{E}{-}004 \times a \times b - 7.500\text{E}{-}004 \times a \times c \\ & + 0.000 \times a \times d - 7.500\text{E}{-}004 \times b \times c + 0.000 \times b \times d \\ & + 0.19 \times c \times d - 1.27 \times a^2 - 0.15 \times b^2 + 1.59 \times c^2 - 0.072 \times d^2\end{aligned}$$

The coefficient values of the above equation reveals the affects of individual factors and double interaction factors on Cd(II) biosorption from the aqueous metal ions solution on sea urchin test. Negative coefficient values indicate that individual or interaction factors negatively affect Cd(II) biosorption (i.e., biosorption capacity decreases), whereas positive coefficient values mean that factors increase Cd(II) removal in the tested range. From the equation it can be seen that factors A and D had positive effect, while factors B and C had negative effect on Cd(II) removal.

3.3 ANOVA for Response Surface Quadratic Model

The ANOVA for biosorption of Cd(II) ions on sea urchin test was used to develop an equation relating the removal percentage of the cadmium and the four process parameters and to check the actual relationship between the response and the significant variables represented by the equation. The larger the value of F and the smaller the value of p, the more significant is the corresponding coefficient term (Kalavathy et al. 2005). The value of p was lower than 0.05, indicating that the model may be considered to be statistically significant. For the present study, the ANOVA results (Table 5) indicated that the F value for the model was 3891.72, implying that the model is significant. In this study, a, b, c, d, cd, a^2, b^2, c^2 were significant factors. The other model terms whose p values are listed as being greater than 0.1000 in Table 5 are not significant factors.

Table 5 Two way analysis of variance (ANOVA) for Cd(II) ions removal

Source	Sum of squares	DF	Mean square	F Value	*p*-value Prob > F	
Model	911.18	14	65.08	3891.72	<0.0001	S
a	363.06	1	363.06	21709.02	<0.0001	S
b	10.38	1	10.38	620.72	<0.0001	S
c	486.34	1	486.34	29080.71	<0.0001	S
d	17.86	1	17.86	1067.99	<0.0001	S
ab	1.000E−006	1	1.000E −006	5.980E −005	0.9939	NS
ac	2.250E−006	1	2.250E −006	1.345E −004	0.9909	NS
ad	0.000	1	0.000	0.000	1.0000	NS
bc	2.250E−006	1	2.250E −006	1.345E −004	0.9909	NS
bd	0.000	1	0.000	0.000	1.0000	NS
cd	0.14	1	0.14	8.41	0.0116	S
a^2	10.45	1	10.45	624.97	<0.0001	S
b^2	0.14	1	0.14	8.34	0.0119	S
c^2	16.47	1	16.47	984.57	<0.0001	S
d^2	0.034	1	0.034	2.02	0.1767	NS
Residual	0.23	14	0.017			
Lack of fit	0.23	10	0.023			
Pure error	0.000	4	0.000			
Cor total	911.41	28				

S Significant; *NS* Nonsignificant

3.4 Optimization and Confirmation

The main objective of this study was to find out the optimum values of the process parameters for maximum removal of Cd(II). Optimum process conditions suggested by the soft ware were pH 5.97, temperature 31.98 °C, initial concentration of Cd, 20.56 mg/l and biosorbent loading of 0.5 g/l. A new experiment was conducted at predicted optimum process parameters from model such as pH of 6, temperature of 32 °C, initial Cd(II) concentration of 20 mg/l and biosorbent dosage of 0.5 g/l. the obtained value of removal percentage is 89.624, which approximately equal to the predicted value and validated the design.

4 Conclusion

Box-Behnken Design was used for RSM to study individual and interactive effects of the adsorption variables, namely pH, temperature, initial Cd(II) ion concentration and biosorbent dosage on the removal of cadmium. Quadratic model was developed to connect the variables to response. Regression analysis showed that, the parameters, pH and initial cadmium ion concentration were found to have significant effect on the removal of cadmium from. Optimum values of process parameters were determined as pH 5.97, temperature 31.98 °C, initial concentration of Cd, 20.56 mg/l and biomass loading 0.5 g/l. at these conditions the maximum cadmium removal was found to be 89.642. Hence, adsorbent prepared from sea urchin would be a valuable alternative in effluent treatment.

References

Box, G.E., Behnken, D.W.: Some new three level design for the study of quantitative variables. Technometrics **2**, 455–475 (1960)

Council Directive 76/464/EEC of 4 May 1976 on pollution caused by certain dangerous substances discharged into the aquatic environment of the Community

Du, Y., Lian, F., Zhu, L.: Biosorption of divalent Pb, Cd and Zn on aragonite and calcite mollusk shells. Environ. Pollut. **159**, 1763–1768 (2011)

Holan, Z.R., Volesky, B., Prasetyo, I.: Biosorption of lead and nickel by biomass of marine algae. Biotechnol. Bioeng. **41**, 819–825 (1993)

Iynengar, P.L., Venkobachar, C.: Biosorption of U, La, Pr, Nd, Eu and Dy by *Pseudomonas aeruginosa*. J. Ind. Microbiol. Biotechnol. **25**, 1–7 (2000)

Kalavathy, M.H., Karthikeyan, S., Rajgopal, S., Miranda, L.R.: Kinetcs and isothrem studies of Cu (II) adsorption onto H_3PO_4-activated rubber wood sawdust. J. Colloid Interface Sci. **292**, 354–362 (2005)

Myers, R.H., Montgomery, D.C.: Response Surface Methodology: Process and Product Optimization Using Design Experiments, 2nd edn. Wiley, New York (2002)

Nordberg, G.F., Herber, R.F.M., Alessio, L.: Cadmium in the Human Environment: Toxicity and Carcinogenicity (IARC Scientific Publications No. 118). Lyon, IARC (1992)

Pena-Rodriguez, S., Fernandez-Calvino, D., Novoa- Munoz, J.C., Nunez-Delgado, A., Fernandez-Sanjurjo, M.J., Alvarez Rodriguez, A.: Kinetics of Hg(II) adsorption and desorption in calcined mussel shells. J. Hazardus Mater. **180**, 622–627 (2010)

Spinti, M., Zhuang, H., Trujillo, E.M.: Evaluation of immobilized biomass beads for removing heavy metals from waste waters. Water Environ. Res. **67**, 943–952 (1995)

Srinath, T., Verma, T., Ramteke, P.W., Garg, S.K.: Chromium (VI) biosorption and bioaccumulation by chromate resistant bacteria. Chemosphere **48**, 427–435 (2002)

UK Red List Substances: Environmental Protection (Prescribed Processes and Substances) Regulations, 1991 (SI 1991/472)

US Environmental Protection Agency: Integrated Risk Information System (IRIS) on Cadmium, National Centre for Environmental Assessment, Office of Research and Development, Washington, DC (1999)

Optimization of Nickel (II) and Cadmium (II) Biosorption on Brewery Sludge Using Response Surface Methodology

Rajeswari M. Kulkarni, K. Vidya Shetty and G. Srinikethan

1 Introduction

The present study is focused on optimizing the factors pH and biosorbent dosage for biosorption of nickel (II) and cadmium (II) ions in aqueous solution onto brewery sludge. Optimum conditions for the biosorption of Ni (II) and Cd (II) by brewery sludge were determined by response surface methodology using the results of set of experiments designed as per central composite design.

Of the numerous optimization methods, response surface methodology (RSM) has been used by various researchers for the biosorption studies (Can et al. 2006; Preetha et al. 2007; Garg et al. 2008; Ghorbani and Younesi 2008; Amini et al. 2009). The screening and optimization of process parameters is achieved by the application of statistical design. This allows quick identification of significant factors and their interactions (Myers et al. 2009; Edris et al. 2014). The combined effect of the process parameters can be predicted and optimization can be achieved with the help of experimental design tool, which is practically difficult in conventional experimentation. Ghorbani and Younesi (2008) applied RSM for optimization of cadmium biosorption by *Saccharomyces cerevisiae*. Amini et al. (2009) used RSM for optimization of nickel removal from aqueous solution by *Aspergillus niger*. Aytar et al. (2014) used RSM for lead and nickel biosorption with a fungal biomass isolated from metal mine drainage. Optimization of nickel and cadmium biosorption process was performed by Rajeswari et al. (2014) using RSM.

Cd (II) and Ni (II) metals are considered in the present work due to their toxicity and high disposal rate (Aksu and Dönmez 2006; Kulkarni et al. 2014). Cadmium,

R.M. Kulkarni · K. Vidya Shetty · G. Srinikethan
Department of Chemical Engineering, NITK, Surathkal, India

R.M. Kulkarni (✉)
Department of Chemical Engineering, M. S. Ramaiah Institute of Technology, Bangalore, India
e-mail: rajeswari.kulkarni@yahoo.in

R. Mohan B. et al. (eds.), *Materials, Energy and Environment Engineering*,
DOI 10.1007/978-981-10-2675-1_14

Cd (II) enters the aquatic systems through the effluents discharged from industries like smelting, electroplating, alloy manufacturing, plastic, pigments and dyes, cadmium-nickel batteries, pesticides and fertilizers, mining, textile operations and refining industries (Grayson and Othumer 1978; Zouboulis et al. 2004; Förstner and Wittmann 2012; Edris et al. 2014). Nickel, Ni (II) is another metal that is widely used in industries like mining, batteries manufacturing, electroplating, metal finishing, paint formulations and porcelain enameling (Sittings 1976; Padmavathy et al. 2003; Işik 2008; Kulkarni et al. 2013). Biosorption offers a very encouraging approach to tackle these toxic metals from waste water.

2 Materials and Methods

Brewery sludge was obtained from United Breweries Limited, India. The wet biomass was sun-dried and then oven dried. The biomass was powdered and used for biosorption studies. Aqueous solutions of cadmium (II) [Cd (II)] and nickel (II) [Ni (II)] ions were prepared using cadmium nitrate [$Cd(NO_3)_2 \cdot 4H_2O$] and nickel sulphate [$NiSO_4 \cdot 6H_2O$] salts in distilled water. The residual concentration of cadmium (II) and nickel (II) was determined by atomic absorption spectrometer (GBC 932 plus).

pH and biosorbent dosage are the two important factors that affect the biosorption process for a fixed metal ion concentration. As pH in the range of 4.0–6.0 for Ni (II) and 5.0–7.0 for Cd (II) by brewery sludge were found to be favorable through initial experiments, these ranges were chosen for central composite design (CCD). Similarly, biosorbent dosage in the range of 5–15 g/L (average particle size 150 μm) was fixed as it was favorable for biosorption of Ni (II) and Cd (II) using brewery sludge. It is important to obtain the optimum values of these factors for maximum biosorption.

Batch biosorption experiments were conducted with initial Ni (II) and Cd (II) concentrations of 50 mg/L, shaker speed 150 rpm, temperature of 30 ± 2 °C and with pH and biosorbent dosage conditions as per the CCD matrix for a biosorption time 120 min. Percentage biosorption of the metals after a contact time of 120 min was taken as the response variable. The Statistica 7 software was used for analyzing the obtained data and for optimization using response surface methodology. Optimization studies were carried out for biosorption of Ni (II) and Cd (II) using brewery sludge by response surface methodology using the results of set of experiments designed as per CCD.

3 Results and Discussion

Biosorption experiments were conducted as per experimental design with two factors and five levels for optimization of process variables pH and brewery sludge dosage for biosorption of Ni (II) and Cd (II) on brewery sludge using response surface methodology. The responses to be maximized were percentage biosorption of Ni (II) and Cd (II). The Statistica 7 software was used for analyzing the obtained experimental data. The experimental values of percentage biosorption of Ni (II) and Cd (II) by brewery sludge as responses are given in Table 1. The input-output data presented in Table 1 were fitted onto a second order polynomial equation relating pH and biosorbent dosage with response (% Biosorption) using Statistica 7 software. By applying multiple regression analysis on the experimental data, the experimental results of the design were fitted on the quadratic equation which relates percentage biosorption (Y) with main effects (X_1, X_2), square effects (X_1^2, X_2^2) and interaction effects (X_1X_2) of process variables. X_1 represents pH and X_2 represents brewery sludge dosage (g/L). The regression coefficients were calculated from the analysis. The model predicted values of the responses are also presented in Table 1. The data were analyzed by analysis of variance (ANOVA) using Statistica 7 software in order to ensure the statistical significance of the model equations for biosorption study of Ni (II) and Cd (II) on brewery sludge.

The results of ANOVA are shown in Table 2. The observed R^2 and adjusted R^2 values for removal of Ni (II) using brewery sludge biosorbent were found to be 0.967 and 0.926 and for Cd (II) as 0.931 and 0.845 indicating the fitness of model for the experimental data. The regression equation obtained from ANOVA which gave percentage biosorption as a function of pH and brewery sludge dosage with the coefficient values are given in Table 3. The significance of each coefficient was determined by F-values and *P*-values. The *p*-values less than 0.05 indicated significant model terms. It was observed from Table 2, the coefficients for the main effects were highly significant as compared to square and interaction effects.

The three dimensional surface plots (Figs. 1 and 2) show the simultaneous effect of initial pH and biosorbent dosage on Ni (II) and Cd (II) biosorption on brewery sludge at constant initial metal ion concentration of 50 mg/L, average particle size 150 μm, shaker speed 150 rpm, temperature 30 ± 2 °C and biosorption time

Table 1 Central composite design along with observed and predicted responses for biosorption of Ni (II) and Cd (II) by brewery sludge at temperature 30 ± 2 °C, 2 h biosorption time, 150 rpm shaker speed, 150 μm average particle size, 50 mg/L initial metal ion concentration

Run order	X_1	X_2	% Ni (II) biosorption		X_1	X_2	% Cd (II) biosorption	
			Experimental	Predicted			Experimental	Predicted
1	4.0	5	9.944	12.961	5.0	5	61.975	64.136
2	4.0	15	28.164	29.270	5.0	15	70.464	70.935
3	6.0	5	32.712	34.156	7.0	5	68.962	71.135
4	6.0	15	35.275	34.808	7.0	15	75.539	76.022
5	3.6	10	17.137	14.739	4.6	10	64.012	62.699
6	6.4	10	33.754	33.587	7.4	10	72.575	71.245
7	5.0	2.93	28.145	25.518	6.0	2.93	72.527	70.010
8	5.0	17.07	37.433	37.509	6.0	17.07	78.400	78.273
9	5.0	10	37.130	37.138	6.0	10	77.310	77.310
10	5.0	10	37.130	37.138	6.0	10	77.310	77.310

Table 2 ANOVA for the quadratic model for biosorption of Ni (II) and Cd (II) by brewery sludge

Model term	Ni (II) biosorption			Cd (II) biosorption		
	Standard error	F-value	*p*-value	Standard error	F-value	*p*-value
X_1	1.781	56.307	0.0016	1.569	14.837	0.0182
X_2	1.779	22.727	0.0088	1.569	13.871	0.0203
X_1^2	2.363	30.500	0.0052	2.075	24.815	0.0075
X_2^2	2.352	5.721	0.0750	2.075	2.330	0.2015
X_1X_2	2.515	9.685	0.0358	2.219	0.185	0.6887

Table 3 Second order quadratic equations relating response and process variables

Biosorbent	Metal ion	Quadratic equations
Brewery sludge	Ni (II)	$Y = -218.313 + 79.77X_1 - 6.52X_1^2 + 7.01X_2 - 0.112X_2^2 - 0.782X_1X_2$
	Cd (II)	$Y = -144.82 + 66.00X_1 - 5.16X_1^2 + 2.42X_2 - 0.063X_2^2 - 0.095X_1X_2$

Y, X_1, X_2 represents % biosorption, pH and biosorbent dosage

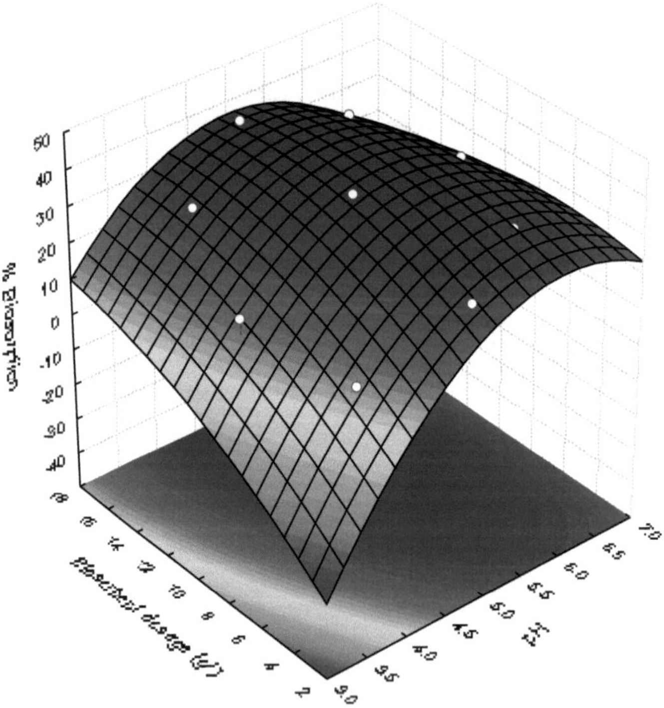

Fig. 1 Surface plot for simultaneous effect of pH and biosorbent dosage on percentage biosorption of Ni (II) on brewery sludge

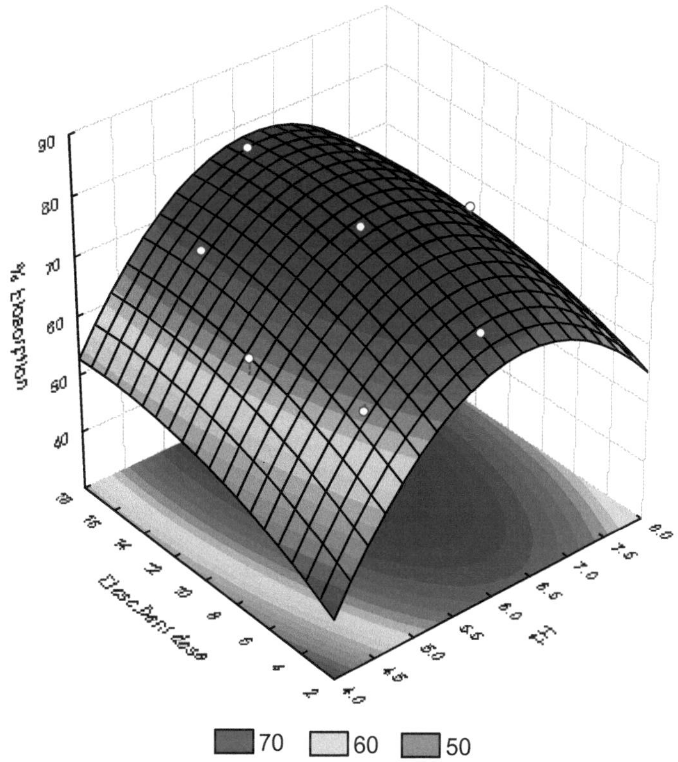

Fig. 2 Surface plot for simultaneous effect of pH and biosorbent dosage on percentage biosorption of Cd (II) on brewery sludge

120 min. The surface plot shows removal efficiency as function of pH and biosorbent dosage while other process variables were constant. pH is a significant parameter in biosorption of Ni (II) and Cd (II) on brewery sludge. Increase in initial pH resulted in increase in percentage biosorption up to a certain pH value and then further increase in pH decreased the percentage biosorption. It can be seen that Ni (II) and Cd (II) biosorption on brewery sludge were increased with increase in dosage due to increase in active sites for biosorption. The optimum levels of selected factors pH and biosorbent dosage were obtained using RSM with Statistica software from the results of experimental design. The optimum conditions for Ni (II) removal was obtained at initial pH of metal solution 5.3 and biosorbent dosage 12.6 g/L which resulted in maximal Ni (II) removal efficiency of 39.41 %. Similarly, a maximal Cd (II) removal efficiency of 78.98 % was obtained at optimum conditions of initial pH of solution 6.2 and biosorbent dosage of 14.42 g/L (Table 4). Experiments were further conducted under optimum conditions. The model predicted values of percentage biosorption at optimum conditions matched well with the experimental values as shown in Table 4. It indicates the effectiveness of the optimization strategy.

Table 4 Optimum values of process variables

Metal ion	Optimum values		% biosorption	
	pH	Biosorbent dosage (g/L)	Predicted	Experimental
Ni (II)	5.3	12.60	39.41	37.20
Cd (II)	6.2	14.42	78.98	78.62

4 Conclusion

Optimum conditions of pH and biosorbent dosage for the biosorption of Ni (II) and Cd (II) on brewery sludge were determined by response surface methodology using experiments designed as per CCD. The optimum conditions for Ni (II) removal was obtained at initial pH of metal solution 5.3 and biosorbent dosage 12.6 g/L. Similarly, optimum conditions for Cd (II) removal was obtained at initial pH of metal solution 6.2 and biosorbent dosage 14.42 g/L. The model predicted values of percentage biosorption at optimum conditions matched well with the experimental values indicating the effectiveness of the optimization strategy.

Acknowledgments Authors thank the Department of Chemical Engineering, M.S. Ramaiah Institute of Technology, Bangalore and Department of Chemical Engineering, NITK, Surathkal for providing the necessary resources to carry out the experiments.

References

Aksu, Z., Dönmez, G.: Binary biosorption of cadmium (II) and nickel (II) onto dried *Chlorella vulgaris*: co-ion effect on mono-component isotherm parameters. Process Biochem. **41**(4), 860–868 (2006)

Amini, M., Younesi, H., Bahramifar, N.: Biosorption of nickel (II) from aqueous solution by *Aspergillus niger*: response surface methodology and isotherm study. Chemosphere **75**(11), 1483–1491 (2009)

Aytar, P.N.A.R., Gedikli, S.E.R.A.P., Buruk, Y.E.L.I.Z., Çabuk, A., Burnak, N.: Lead and nickel biosorption with a fungal biomass isolated from metal mine drainage: box–Behnken experimental design. Int. J. Environ. Sci. Technol. **11**(6), 1631–1640 (2014)

Can, M.Y., Kaya, Y., Algur, O.F.: Response surface optimization of the removal of nickel from aqueous solution by cone biomass of *Pinus sylvestris*. Bioresour. Technol. **97**(14), 1761–1765 (2006)

Edris, G., Alhamed, Y., Alzahrani, A.: Biosorption of cadmium and lead from aqueous solutions by *Chlorella vulgaris* biomass: equilibrium and kinetic study. Arab. J. Sci. Eng. **39**(1), 87–93 (2014)

Förstner, U., Wittmann, G.T.: Metal pollution in the aquatic environment. Springer, Berlin (2012)

Garg, U.K., Kaur, M.P., Garg, V.K., Sud, D.: Removal of Nickel (II) from aqueous solution by adsorption on agricultural waste biomass using a response surface methodological approach. Bioresour. Technol. **99**(5), 1325–1331 (2008)

Ghorbani, F., Younesi, H.: Biosorption of cadmium (II) ions by *Saccharomyces Cerevisiae* biomass from aqueous solutions. J. Water Wastewater **68**(4), 33–39 (2008)

Grayson, M., Othumer, K.: Encyclopedia of Chemical Technology, 43rd edn. Wiley, New York (1978)

Işik, M.: Biosorption of Ni (II) from aqueous solutions by living and non-living ureolytic mixed culture. Colloids Surf. B **62**(1), 97–104 (2008)

Kulkarni, R.M., Srinikethan, G., Shetty, K.V.: Biosorption of Nickel (II) from Aqueous Solution Using *Bacillus laterosporus*. In Prospects in Bioscience: Addressing the Issues, pp. 415–418. Springer, India (2013)

Kulkarni, R.M., Shetty, K.V., Srinikethan, G.: Cadmium (II) and nickel (II) biosorption by *Bacillus laterosporus* (MTCC 1628). J. Taiwan Inst. Chem. Eng. **45**(4), 1628–1635 (2014)

Myers, R.H., Montgomery, D.C., Anderson-Cook, C.M.: Response Surface Methodology: Process and Product Optimization Using Designed Experiments, vol. 705. Wiley, New York (2009)

Padmavathy, V., Vasudevan, P., Dhingra, S.C.: Biosorption of nickel (II) ions on Baker's yeast. Process Biochem. **38**(10), 1389–1395 (2003)

Preetha, B., Viruthagiri, T.: Application of response surface methodology for the biosorption of copper using *Rhizopus arrhizus*. J. Hazard. Mater. **143**, 506–510 (2007)

Rajeswari, M.K., Srinikethan, G., Vidya, Shetty K.: Optimization of factors influencing Ni (II) and Cd (II) biosorption on *Bacillus laterosporus* using response surface methodology. J. Chem. Eng. Res. **2**(1), 181–186 (2014)

Sittings, M.: Environmental Sources and Emission Handbook. Noyes Data Corporation, Park Ridge (1976)

Zouboulis, A.I., Loukidou, M.X., Matis, K.A.: Biosorption of toxic metals from aqueous solutions by bacteria strains isolated from metal-polluted soils. Process Biochem. **39**(8), 909–916 (2004)

Biosorption of Copper from Wastewater Using *Spirulina* Species

B. Prathima, Praphulla Rao and M.R. Mangala Mahalakshmi

1 Introduction

Water is one of the primary requirements of life and advancement in every part of the world. It is assessed that man can live for twenty days without solid food, however begins battling for life without water soon after one day. In man's body 70% is water. Blood, cells and bones contain 18, 75 and 22% water respectively. Water pollution is characterised by changes in physical, chemical and boilogical parameters of water which in turn has detrimental effect on environment and human health. (Rao et al. 2014).

The expulsion of heavy metals from water bodies is an imperative issue confronted by industry releasing waste water to river. Huge amounts of heavy metals are released from mechanical procedures, for example, electroplating, plastics assembling, mining and metallurgical procedures. At present various distinctive advances exist for treating heavy metals containing streams, for example, Ion Exchange, Chemical Precipitation, Solvent Extraction, Adsorption and Membrane Separation. These advances have certain impediments like sludge disposal, high cost and so on. Biosorption is the upcoming innovation to treat the heavy metals from wastewater and is the capacity of natural materials to collect heavy metals from wastewater by metabolizing it or by using physico-chemical mechanism. Microorganisms like algae, bacteria, fungi and yeasts have shown the capability of adsorbing heavy metals (Chopra and Pathak 2010).

B. Prathima · Praphulla Rao (✉) · M.R. Mangala Mahalakshmi
B.M.S. College of Engineering, Bangalore, India
e-mail: praphullarao.bt@bmsce.ac.in

B. Prathima
e-mail: prathimab.civ@bmsce.ac.in

M.R. Mangala Mahalakshmi
e-mail: mangala.ml@gmail.com

R. Mohan B. et al. (eds.), *Materials, Energy and Environment Engineering*,
DOI 10.1007/978-981-10-2675-1_15

In this study, batch experiments were performed to remove Copper from wastewater using dead *Spirulina* sp. The process parameters considered are biosorbent dosage, contact time, pH, and initial metal concentration.

2 Materials and Methods

Dead *Spirulina* sp. powder was procured from Antenna Nutritech Foundation, Madurai. Biosorbent was soaked in 0.12 N HCl for an hour. Then it was washed with distilled water and dried, keeping in oven at 1000 °C for 12 h. The biosorbent was finally crushed into powder.

The influencing parameters such as contact time, biosorbent dosage, pH and initial metal ion concentration were considered for the study. Batch experiments were conducted at room temperature in Erlenmeyer flasks of capacity 250 ml. In every experiment 100 ml of Copper solution with known initial concentration was shaken with a varied amount of biosorbent, contact time and pH at a speed of 150 rpm using rotary flask shaker. After reaching the equilibrium the biosorbent was separated from the metal solution by using Whatman filter paper No. 42. The metal concentration remaining in the solution was analyzed using Atomic Absorption Spectrophotometer for Copper (Kanchana et al. 2011).

Column Studies were carried out to know the behaviour of *Spirulina* sp., in removing heavy metals under optimized condition which were obtained in the batch studies. Immobilised *Spirulina* sp., beads were prepared using following procedure. 1 g of treated dry *Spirulina* biomass was added to 25 ml of distilled water, the equal volume of 2% (w/v) sodium alginate solution was prepared and mixed with biomass solution. A 100 ml of sodium alginate cell suspension was added dropwise to 1000 ml of 2% calcium chloride solution. Precipitate of Sodium alginate obtained by coming in contact with calcium chloride. Algal cells were entrapped in gel beads. These beads were then soaked in calcium chloride solution for 4 h to complete the gelling and then washed with saline containing 0.85% of NaCl to take away the excess calcium and algal cells. Totally 4 g dry *Spirulina* sp., was used to prepare beads required to fill the column depth of 18 cm. The synthetic metal solution was passed through the column packed with immobilized algal beads. The treated sample was collected at regular intervals and analyzed for residual metal ion concentration using Atomic Absorption Spectrophotometer.

3 Results and Discussions

3.1 *Effect of Contact Time (min)*

As *Spirulina* sp., has the ability to absorb Copper has been confirmed, varying contact time through batch studies was conducted. The biosorbent dosage was fixed as 3 g/100 ml.

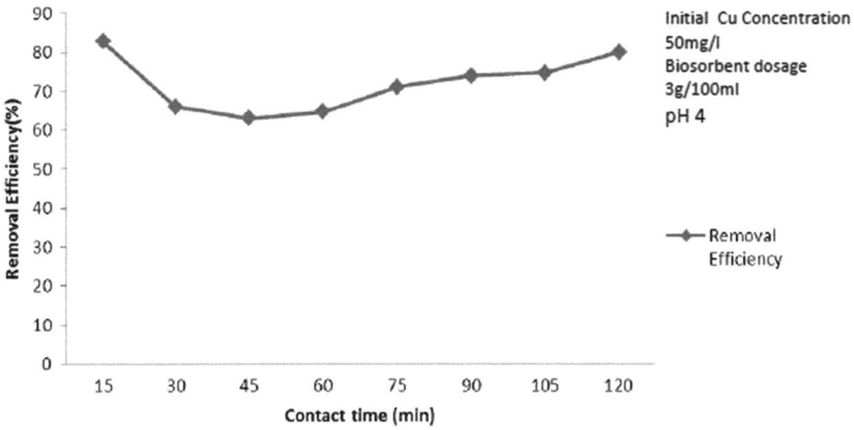

Fig. 1 Efficiency of removal of Cu by varying the contact time

The effect of contact time on Copper removal is shown in Fig. 1. In the starting phase of adsorption of Copper, the concentration gradient between the film and the accessible pore destinations is substantial and consequently the rate of adsorption of Copper is quicker i.e., found in the 15 min, the removal efficiency is 82.7%. The rate of adsorption declines in later phase of the Copper adsorption most likely because of moderate pore dispersion of the solute particle into the main part of the biosorbent.

3.2 *Effect of Biosorbent Dosage*

Metal sorption is highly related to the biosorbent dosage. *Spirulina* sp., dosage changed from 1 g/100 ml to 3.5 g/100 ml to observe the efficiency of Copper removal.

In Fig. 2 it was shown that increase in the biosorbent dosage from 1 to 3.5 g, the percentage removal increased and gave maximum efficiency of 83.4% for 3.5 g for a contact time of 120 min for the Copper concentration of 50 mg/l. Further increase in dosage showed decrease in efficiency.

3.3 *Effect of pH*

The pH is a vital parameter influencing biosorption of heavy metals (Kumar and Goyal 2008).

Figure 3 shows the removal of Copper ions at different pH. It is seen that maximum Copper ion removal by *Spirulina* sp., is obtained at pH 4(85%) for

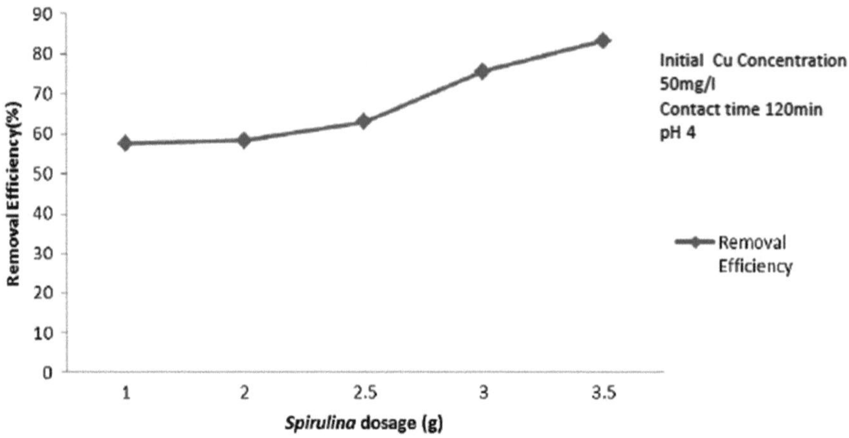

Fig. 2 Efficiency of removal of Cu by varying the biosorbent dosage

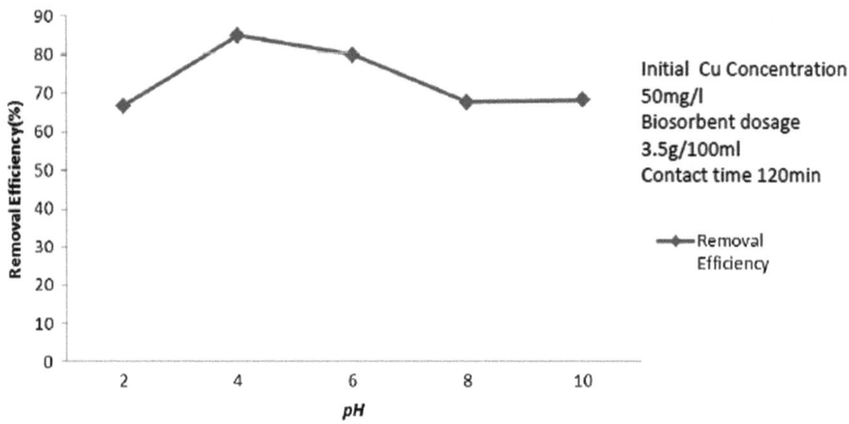

Fig. 3 Efficiency of removal of Cu by varying the pH

3.5 g/100 ml of 50 mg/l concentration of Copper solution. The contact time was 120 min and speed was 150 rpm.

3.4 *Effect of Initial Cu Ion Concentration*

The pH, biosorbent dosage and contact time for batch experiment was fixed as 4, 3.5 g/100 ml and 120 min respectively.

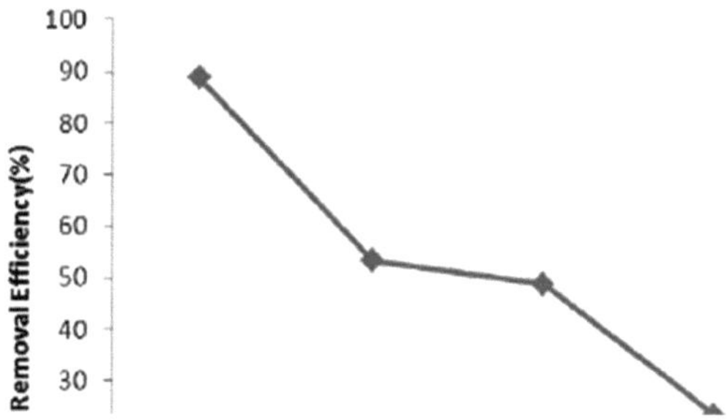

Fig. 4 Efficiency of removal of Cu by varying the initial Cu ion concentration

The influence of initial metal concentration on Copper removal by *Spirulina* species is shown in Fig. 4. Maximum removal efficiency is seen for 50 mg/l. Gradual increase in metal ion concentration of Copper resulted in decrease in percentage of Copper removal because it was evident that the biosorbent was approaching saturation.

3.5 *Adsorption Isotherm Study of Cu Metal*

The adsorption isotherm was evaluated for Copper with initial pH 4 for the contact time of 120 min with varying *Spirulina* sp., dosage from 2.5 to 4.0 g. The various adsorption isotherms are evaluated for Copper.

Langmuir equation was used to linearize the data and isotherm was plotted as $1/qe$ versus $1/Ce$ shown in Fig. 5. The high estimate of regression correlation coefficient ($R^2 = 0.9451$) was obtained which demonstrate a good agreement between the experimental values and isotherm parameters furthermore affirms the single layer adsorption of Copper onto the *Spirulina* sp., surface.

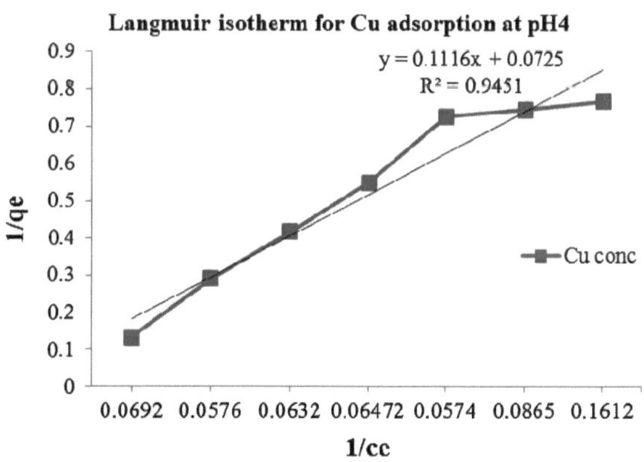

Fig. 5 Langmuir adsorption isotherm of Cu

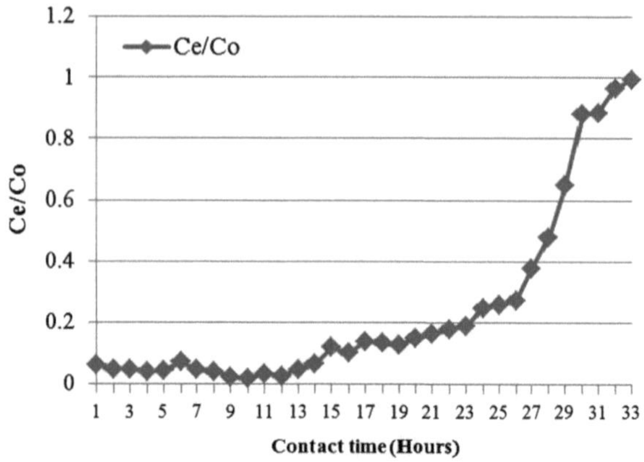

Fig. 6 Breakthrough curve for Cu biosorption on *Spirulina*—sodium-alginate beads

3.6 Column Studies

Column studies were performed in a glass column (2.5 cm inner diameter and 25 cm height), filled with a known quantity of sodium alginate algae beads of height 18 cm. A 50 mg/l concentration of Copper solution was passcd through the column, at flow rate of 0.6 ml/min of pH 4 obtained from optimized condition of batch studies. Effluent samples were collected at every 15 min.

The initial concentration of Copper solution was 48 mg/l before running through the column. To attain saturation Ce/Co should be 1, which reaches after 33 h. The break through curve s shaped formed as shown in Fig. 6.

3.7 Experiment on Industrial Sample

The experiment on industrial sample was carried out using the optimised condition of batch studies. The flow rate was set to 0.6 ml/min. The pH of the industrial

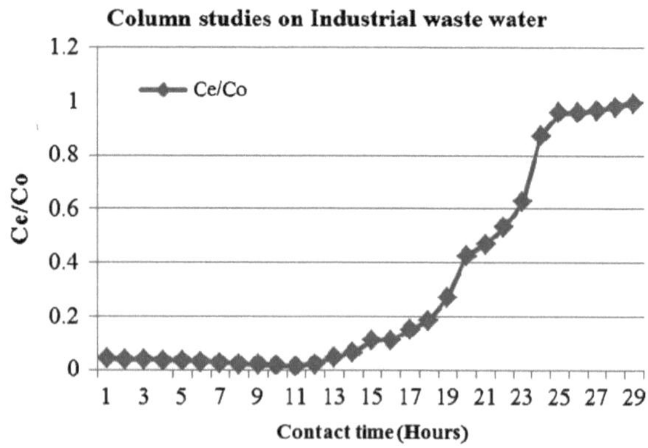

Fig. 7 Breakthrough curve for Cu biosorption from industrial wastewater onto *Spirulina*—sodium-alginate beads

sample was 12. Hence using 1 N HCl acid the pH was reduced to 4. The column was packed with 18 cm height of *Spirulina*—sodium alginate beads. The Concentration of industrial sample was 116 mg/l. The concentration of industrial sample was diluted to 50 mg/l. The effluent before running through the column, the concentration was 49.45 mg/l. After running through the column the sample was collected at every 1 h.

The break through curve versus time is shown in Fig. 7. Till 5 h the concentration of Copper decreased to 1.11 mg/l and left for continuous flow. After 13 h when the concentration of the column effluent was checked the concentration was 9.2 mg/l, means biosorbent is started exhausting. The break through time is attained at 5th hour. The saturation point reached at 29th hour.

4 Summary/Conclusion

From the study it can be concluded micro algae *Spirulina* sp. has the potential to remove the Copper from wastewater. Batch studies conducted for Copper removal showed the optimum parameters as contact time as 120 min, biosorbent dosage as 3.5 g, pH as 4 and metals concentration as 50 mg/l. The removal efficiency of Copper at this condition was 85% . Adsorption isotherm studies were conducted and the correlation coefficient R^2 is determined for Langmuir isotherms for Copper which confirms monolayer adsorption of Copper by *Spirulina* sp. Column studies carried out for synthetic sample of Copper and Electroplating industrial wastewater showed good biosorption capacity of Copper from *Spirulina* sp.

References

APHA: Standard Methods for the Examination of Water and Wastewater. American Public Health Association, AWWA and WPCF, Washington DC (2002)

Chopra, A.K., Pathak, C.: Biosorption technology for removal of metallic pollutants—an overview. J. Appl. Nat. Sci. **2**, 318–329 (2010)

Gokale, S.V., Jyothi, K.K., Lele, S.S.: Modeling of chromium (VI) biosorption by immobilized *Spirulina platensis* in packed column. J. Hazard. Mater. **170**, 735–743 (2009)

Kanchana, S., Jeyanthi, J., Dinesh Kumar, R.R.: Equilibrium and kinetic studies on biosorption of chromium (VI) on to *chlorella* species. Eur. J. Sci. Res. **63**, 255–262 (2011)

Kumar, R., Goyal, D.: Comparative biosorption of Pb^{2+} by live algal consortium and immobilsed dead biomass from aqueous solution. Indian J. Exp. Biol. **46**, 690–694 (2008)

Praphulla Rao, Prathima, B., Mangala.: Removal of chromium (VI) from synthetic solution using dead *Spirulina* species. Int. J. Innovations Eng. Technol. **4**, 244–248 (2014)

A Study on Simultaneous Photocatalytic Removal of Hexavalent Chromium and Pharmaceutical Contaminant from Aqueous Phase

Sarungbam Pipileima, Srimanta Ray and Leichombam Menan Devi

1 Introduction

The pollution of environment due to Chromium (Cr) release is a cause of serious concern in developing countries. Among various sources, the wastewater from leather tanning and mordant dyeing are the major contributor towards Cr pollution, and the wastewater from leather tanning is single largest source of Cr release to the environment. 60–70 % of Cr used in the tanning process reacts with the hides and the remaining 30–40 % of Cr is discharged with the solid and liquid wastes (spent tanning solution in particular) (Patterson 1985; Zayed and Terry 2003). Cr released in the environment occurs either occur as Cr(+III) or Cr(+VI). Cr(+III) plays a vital role in development of living organism. Also, Cr(+III) is readily precipitated or adsorbed on a variety of inorganic and organic substrates at neutral or alkaline pH (Zayed and Terry 2003; Esmaeili et al. 2005). Alternately, Cr(+VI) is highly toxic to most organisms including humans and do not readily precipitate or become bound to components of soil. Toxicity kinetics of Cr(+VI) show higher rate of penetration into biological membranes as compared to Cr(+III). The ulcer formation and carcinogenic reactions are some well documented adversities due to Cr(+VI) in human. Thus, Cr(+VI) contaminate ground water aquifers and surface water sources. These Cr(+VI) contaminated water sources present hazard to living beings

S. Pipileima · L. Menan Devi
Department of Civil Engineering, National Institute of Technology, Agartala, Barjala, Jirania 799055, Tripura (West), India

S. Ray (✉)
Department of Chemical Engineering, National Institute of Technology, Agartala, Barjala, Jirania 799055, Tripura (West), India
e-mail: rays.nita@gmail.com

R. Mohan B. et al. (eds.), *Materials, Energy and Environment Engineering*,
DOI 10.1007/978-981-10-2675-1_16

and health hazard to humans. Hence, is considered as a priority pollutant and the removal of Cr(+VI) is highly desirable to minimize the health hazard and adverse environmental effects (Esmaeili et al. 2005; Wittbrodt and Palmer 1995; Johnson et al. 1992).

Several methods have been investigated for removing of Cr(+VI) from aqueous solutions, which include chemical precipitation, ion exchange, membrane processes, solvent extraction, adsorption, etc. (Kotas and Stasicka 2000; Madhavi et al. 2013; Barrera-Díaz et al. 2012). However, these treatment methods are essentially phase separation process and does not effect reduction of Cr(+VI) to less toxic of Cr (+III). The reduction of Cr(+VI) in the natural aqueous environment is an extremely slow process, lasting from several days to months water environment (Feng et al. 2012). Accelerated Cr(+VI) reduction has demonstrated in chemical reduction and electrochemical reduction. Like natural aquatic systems biological reduction of Cr (+VI) is also a slow process (Barrera-Díaz et al. 2012; Kato and Nagai 1991; Durai and Rajasimman 2011). The chemical process is expensive due to cost of reducing chemicals and not environmentally friendly. The electrochemical process is energy intensive, depends on consumption of electrical energy and capital intensive. Hence exploring alternative process of accelerated Cr(+VI) reduction has been of great deal of research attention. Photocatalysis on a photo-illuminated catalyst surface has emerged as a potential technology for treating variety of industrial effluents (Ray et al. 2009). Accordingly, in recent years, the photocatalytic removal of Cr(+VI) have gained significant research focus (Wang et al. 2004; Chakarbarti et al. 2009; Qmar et al. 2011). Among the reported photocatalysts titanium dioxide (TiO_2) and zinc oxide (ZnO) had been widely studied due to high bandgap energy (Kanki et al. 2004; Wang et al. 2008). However, the potential of ZnO as photocatalyst in the removal of Cr(+VI) is less explored. Nonetheless, a recent report exploring the properties of ZnO stated that presence of iron (Fe) ions in ZnO structure shift the bandgap to lower energies favouring enhanced reactivity compared to that of pure ZnO on irradiance with radiation visible or solar radiation (Chand et al. 2014). On the other hand a study investigating the photocatalytic reduction of Cr(+VI) with (TiO_2) reported that presence of sacrificial organic electron donor greatly influence the photocatalytic activity (Chakarbarti et al. 2009). An obvious research need is to investigate the similar effect on ZnO. The present study is addressed to the research need and examined the effect of sacrificial organic electron donor on photocatalytic reduction of Cr(+VI) with ZnO. In addition, the study also investigated the effect of incorporating Fe on the efficacy of ZnO.

2 Materials and Methods

ZnO (molecular weight—81.408 gm/mol) was procured commercially (Nice Chemicals, India). Ferric Chloride ($FeCl_3$) (Nice Chemicals, India, molecular weight—162.206 gm/mol) was used as source of Fe atoms for preparation of Fe incorporated ZnO. ZnO was used as-is for all experiments. Solar radiation at a site

23.80N and 91.50E (National Institute of Technology, Agartala) and an artificial light source [Cool day light, 6500 K, 1010 lm (Philips, India)] were used as energy source for photocatalysis. Solar photochemical reactions were conducted during the month of August–October and March–April when average irradiance was 4.7 kW/m^2/day. The solar energy was exploited to makes the process economic and sustainable. The photochemical experiments were conducted in batches. The deionised (DI) water of 18.2 M ohm resistivity was used as medium. Fe-incorporation in ZnO was performed by a solid solution method, wherein aqueous suspension of ZnO was mixed with different mol percent of Fe^{3+} for 4 h, the mixture was dried and subsequently calcined at 350 °C for 4 h. The mol percent of Fe was varied at three level, 0 (no Fe), 3 % and 5 %. A paste was prepared with the calcined photocatalysts in ethanol (Nice Chemicals, India), applied on pre-cleaned glass slides by doctor's knife technique, sintered at 90 °C for 2 h, cooled, washed in DI water, soaked in DI water overnight and thereafter used for photochemical reaction after characterization. The bandgap energy of the catalyst was quantified from analysis of absorbance in UV-visible spectrophotometer (Lamba 25, Perkin-Elmer, Singapore). The Fe-incorporation or doping in ZnO is confirmed from IR spectroscopy (Nicolet, iS10, Thermo Fisher, USA) after fourier transformation and X-ray diffraction (D8 Advance, Bruker, USA). The synthetic Cr (+VI) contaminated wastewater was prepared by dissolution of potassium dichromate (Nice Chemicals, India). The concentration of Cr(+VI) was varied at three levels, 15, 30 and 60 mg/l. The residual Cr(+VI) concentration was monitored for each photochemical reaction from the absorbance at 372 nm and in-house calibration curve. Pharmaceutical water containing a model pharmaceutical compound was chosen as the organic electron donor. Pharmaceutical contamination in water bodies has greatly increased in recent years due to the increased and widespread manufacture together with extensive use of pharmaceutical compounds. In addition, humans and ecosystems undergo significant risk upon chronic exposure to these pharmaceuticals and cumulative build-up of these contaminants in the various environmental compartments. Moreover, conventional water treatment methods often cannot degrade these pharmaceuticals. The organic electron donor is oxidized when Cr(+VI) is reduced. Accordingly, use of pharmaceuticals wastewater as electron donor in can demonstrate simultaneous utilization of recalcitrant pharmaceutical wastewater with photocatalytic reduction of Cr(+VI). Methylene blue (MB) was chosen as model pharmaceutical due to large scale application and ease of monitoring. The residual MB was determined from the absorbance at 664 nm and in-house calibration (concentration versus absorbance) curve. The residual concentration of Cr(+VI) and MB were computed for different irradiation time to determine the reaction kinetics.

3 Results and Discussions

3.1 Characterization of the Catalyst

The photocatalyst without Fe incorporation (0 %), with 2 mol% Fe (2 %) and with 5 mol% Fe (5 %) were subjected to X-ray diffraction (XRD) studies. The XRD profiles of the catalysts are presented in Fig. 1a. The XRD profile showed that the diffraction peaks exhibit a single phase wurtzite structure (JCPDS #36-1451) and exclude the existence of secondary or impure phases. The profiles clearly infer that Fe ions exist in the lattice sites rather than interstitial sites. The peak position at 34.44° corresponds to (002) plane of ZnO and confirm hexagonal wurtzite structure (Chand et al. 2014). As the mole percent of Fe changes in ZnO the peak (002) changes. The profile of peak (002) for different Fe incorporated ZnO is shown in Fig. 1b.

In comparison with pure ZnO, the peak (002) shifted towards higher angle and there was decrease in intensity with increase in Fe mole percent. The shifting in peak position and decrease in intensity reflect that Fe is successfully replaced Zn in ZnO matrix as diffraction of X-ray changes. Since the ionic radius of the Fe^{3+} (0.64 Å) is very close to the Zn^{2+} radius (0.60 Å), Fe^{3+} can be assumed to penetrate into the ZnO lattice, however, the increase in Fe mole percent declined the crystalline quality of ZnO.

The bandgap profile of ZnO photocatalyst with various level of iron incorporation is shown in the Fig. 2a. The bandgap profile showed that the bandgap decrease with the increase of Fe content in ZnO. The decrease in bandgap energy corresponds to shift in absorbance to higher wavelength. The absorbance shifted from 368 nm for pure ZnO (0 mol% Fe) to 377 nm for Fe incorporated ZnO with 5 mol% ZnO. The shift in absorbance towards visible region suggests that the Fe incorporated ZnO requires less energy for photoactivity. Also the Fe incorporated

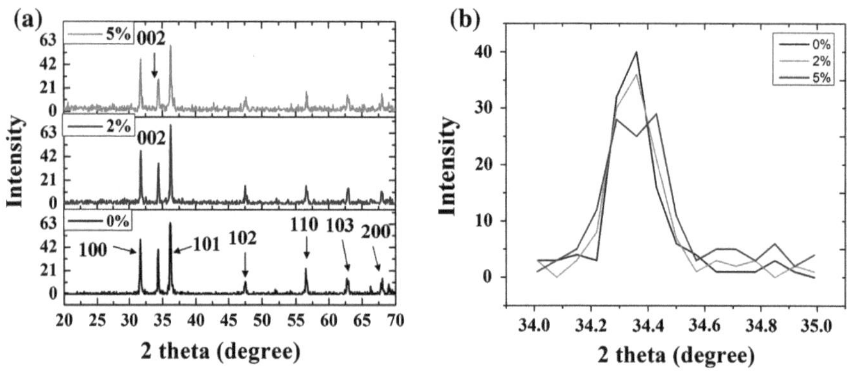

Fig. 1 The X-ray diffraction profile of the various Fe incorporated ZnO **a** complete profile, **b** profile of the peak corresponding to 002 plane

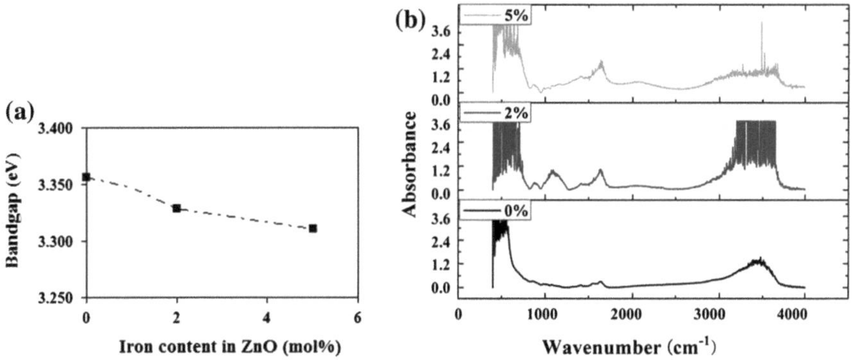

Fig. 2 **a** Bandgap profile **b** FTIR spectroscopic profile of ZnO photocatalyst with various level of iron incorporation

ZnO is expected to be more efficient under solar radiation since visible light accounts for 45 % energy in the solar radiation while UV is less than 10 % (Yang et al. 2009).

The bonding configuration of the various Fe incorporated ZnO were investigated by absorbance measurements in infrared spectroscopy through fourier transformation (FTIR) in the spectral range from 400 to 4000 cm^{-1}. The FTIR spectra of various Fe incorporated ZnO is presented as Fig. 2b. A change in absorbance intensity was noted in the range of 400–700 cm^{-1}. This spectral region is credited to stretching of ZnO molecules. Additional peaks were observed in spectral range from 1000 to 1500 cm^{-1}. The changes in the FTIR profiles can be attributed to incorporation of Fe in the lattice structure of ZnO.

3.2 Reduction of Hexavalent Chromium

The reduction of Cr(+VI) was monitored for 4 h and the degradation rate was computed. ZnO with 2 mol% Fe was chosen for photocatalytic experiments in order to compare the efficacy of Fe incorporated ZnO against pure ZnO in presence and absence of organic electron donor, i.e. methylene blue. A constant decrease of absorbance at λ_{max} was observed in the photocatalytic experiments (Fig. 3a). The hexavalent chromium reduction percent was noted to vary linearly against time (Fig. 3b). The photochemical reduction was also performed under dark condition to confirm the photocatalytic. No significant reduction of Cr(+VI) was noted under dark condition (Fig. 4a). The efficacy of reduction of Cr(+VI) by various catalysts under different radiation source (solar and artificial lamp) is presented in Fig. 4a. Photolytic experiment (without catalyst) were also performed to confirm the photocatalytic effect of ZnO and Fe–ZnO. Increase in initial Cr(+VI) concentration was observed to decrease the reduction percent. Reduction in higher concentration of Cr

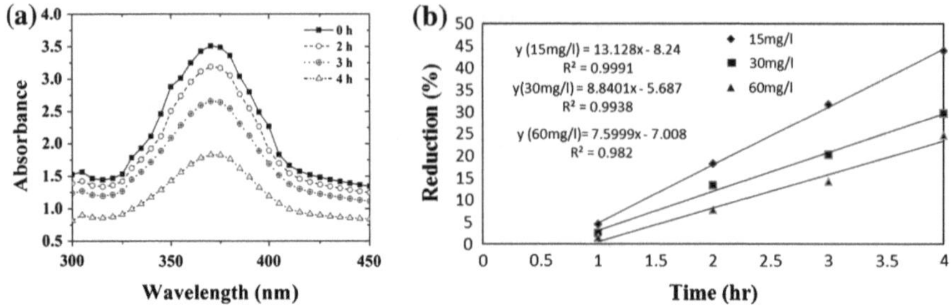

Fig. 3 **a** Absorbance profile of Cr(+VI) versus time, **b** reduction percent of Cr(+VI) versus time, in photochemical experiment with ZnO^{-2} mol% Fe catalyst for 15 mg/l Cr(+VI) under solar radiation

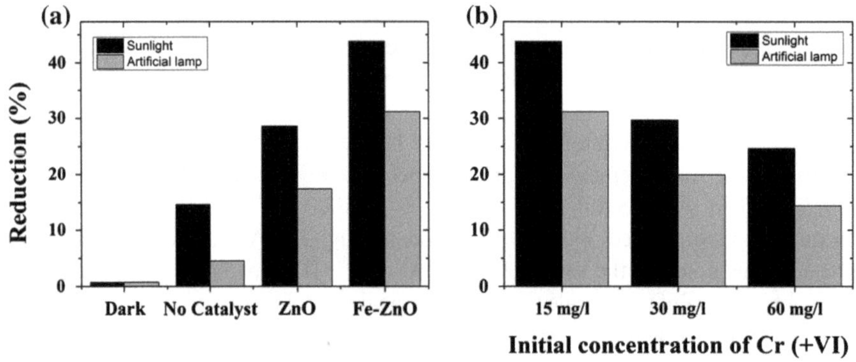

Fig. 4 Reduction percent of Cr(+VI) under different radiation for—**a** various photochemical condition with 15 mg/l Cr(+VI), **b** different initial concentration with ZnO^{-2} mol% Fe catalyst

(+VI) is limited by the available photo-electrons on catalyst surface. In general it was noted that Fe incorporated ZnO faster reduces Cr(+VI) in comparison to ZnO under both radiation source (Fig. 4b). Experiments were performed on 20 mg/l MB with different catalyst to investigate decolorization profile. Around 98 % decolorization was achieved in 4 h with Fe–ZnO catalyst under solar radiation. Increase in catalyst loading enhanced the photo-reduction percent due to increase in photocatalytic surface area (Fig. 5).

Thus, the present study reported successful synthesis of Fe incorporated ZnO in simple steps. Fe–ZnO had higher photocatalytic efficacy compared to ZnO, due to absorbance in the solar or visible light spectra. The photocatalytic reduction of Cr (+VI) is enhanced by the presence of a model pharmaceutical contaminant, as organic electron donor. Similar study on reduction of Cr(+VI) in presence of pharmaceutical contaminant by Fe incorporated ZnO is till date not reported in the literature.

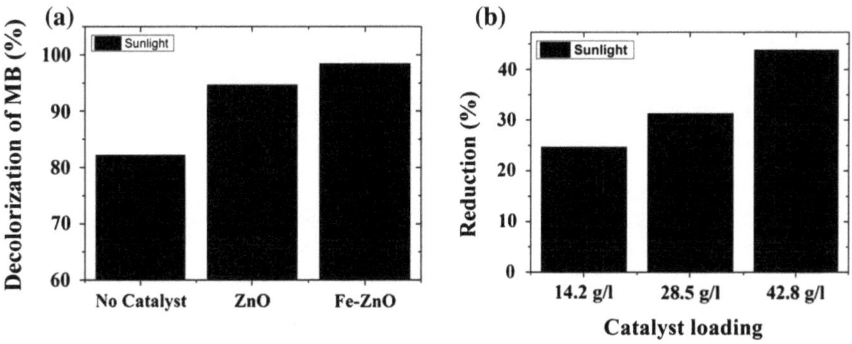

Fig. 5 **a** Decolorization of MB [with Cr (+VI) with various catalyst (20 mg/l MB)], **b** reduction percent of Cr(+VI) at different catalyst loading [ZnO^{-2} mol% Fe and 15 mg/l Cr(+VI)]

4 Conclusions

The study reported fabrication of Fe incorporated ZnO in simple steps. The characterization of the catalyst in XRD and FTIR showed that Fe penetrated into lattice structure of ZnO, resulting in a shift in the bandgap energy towards visible range of the electromagnetic spectra. The shift in bandgap energy allowed better utilization of solar radiation. Higher catalyst loading favored and higher initial Cr(+VI) hindered the photocatalysis. A maximum of 48 % reduction of Cr(+VI) has been reported in the study in 4 h with 15 mg/l initial Cr(+VI) concentration. The presence of MB, pharmaceutical contaminant, as organic electron donor greatly favored the Cr(+VI) reduction in photocatalysis. The present work demonstrates an alternative approach wherein toxic and problematic hexavalent chromium is reduced using another problematic waste and pharmaceutical residues in a sustainable manner.

Acknowledgments The authors also like to express their gratitude to the National Institute of Technology, Agartala (NITA), India for the infrastructure and facilities and TEQIP-II, NITA program for the financial support.

References

Barrera-Díaz, C.E., Lugo-Lugo, V., Bilyeu, B.: A review of chemical, electrochemical and biological methods for aqueous Cr(VI) reduction. J. Haz. Mat. **223**, 1–12 (2012)

Chakarbarti, S., Chaudhuri, B., Bhattacharjee, S., Ray, A.K., Dutta, B.K.: Photo-reduction of Hexavalent chromium in aqueous solution in the presences of ZnO as semiconductor catalyst. Chem. Eng. J. **153**, 86–93 (2009)

Chand, P., Gaur, A., Kumar, A.: Effect of Cr and Fe doping on the structural and optical properties of ZnO nanostructures. Int. J. Chem. Mol. Nucl. Mat. Metall. Eng. **8**, 1286–1289 (2014)

Durai, G., Rajasimman, M.: Biological treatment of tannery wastewater. Environ. Sci. Technol. **4**, 1–17 (2011)

Esmaeili, A., Mesdaghi, A., Vazirinejad, R.: Chromium(III) removal and recovery from tannery wastewater by precipitation. Am. J. Appl. Sci. **2**, 1471–1473 (2005)

Feng, X., Ding, S., Zhang, L.: Photocatalytic reduction of hexavalent chromium induced by photolysis of ferric/tartrate complex. Bull. Korean Chem. Soc. **33**, 3691–3695 (2012)

Johnson, C.A., Sigg, L., Lindauer, U.: The chromium cycle in a seasonally anoxic lake. Limno Oceanogr. **37**, 315–321 (1992)

Kanki, T., Yoneda, H., Sano, N., Toyoda, A., Nagai, C.: Photo-catalytic reduction and deposition of metallic ions in aqueous phase. Chem. Eng. J. **97**, 77–81 (2004)

Kato, I., Nagai, S.: Treatment of Chromium Containing Wastewater. Japan Kokai Tokkyo Koho, pp. 224–692 (1991)

Kotas, J., Stasicka, Z.: Chromium occurrence in the environment and methods of its specification. Environ. Pollut. **107**, 263–283 (2000)

Madhavi, V., Reddy, A.B.V., Reddy, K.G., Madhavi, G., Vara Prasad, T.N.V.K.: An overview on research trends in remediation of chromium. Res. J. Recent Sci. **2**, 71–83 (2013)

Patterson, J.W.: Industrial Wastewater Treatment Technology. Butterworth, Stoneham (1985)

Qmar, M., Gondal, M.A., Yamani, Z.H.: Laser-induced efficient reduction of Cr(VI) catalysed by ZnO nanoparticles. J. Haz. Mat. **187**, 258–263 (2011)

Ray, S., Lalman, J.A., Biswas, N.: Using the Box-Benkhen technique to statistically model phenol photocatalytic degradation by titanium dioxide nanoparticles. Chem. Eng. J. **150**, 15–24 (2009)

Wang, X., Penkonen, S.O., Ray, A.K.: Removal of aqueous Cr(VI) by a combination of photocatalytic reduction and Co-precipitation. Ind. Eng. Chem. Res. **43**, 1665–1672 (2004)

Wang, L., Wang, N., Zhu, L., Yu, H., Tang, H.: Photcatalytic reduction of Cr(VI) over different TiO_2 photocatalysts and the effects of dissolved organic species. J. Haz. Mat. **152**, 93–99 (2008)

Wittbrodt, P.R., Palmer, C.D.: Reduction of Cr(VI) in the presence of excess soil fulvic acid. Environ. Sci. Technol. **29**, 255–263 (1995)

Yang, G.C., Chan, S.W.: Photocatalytic reduction of chromium(VI) in aqueous solution using dye-sensitized nanoscale ZnO under visible light irradiation. J. Nanopart. Res. **11**, 221–230 (2009)

Zayed, A.M., Terry, N.: Chromium in environment: factors affecting biological remediation. Plant Soil **249**, 139–156 (2003)

Effect of Precursor Salt Solution Concentration on the Size of Silver Nanoparticles Synthesized Using Aqueous Leaf Extracts of *T. catappa* and *T. grandis* Linn f.—A Green Synthesis Route

Aishwarya Devadiga, K. Vidya Shetty and M.B. Saidutta

1 Introduction

Silver nanoparticles (AgNPs) have been widely researched upon as they are known to exhibit a huge array of unique properties ranging from optoelectronics, catalysis and antimicrobial activity. Owing to these properties, AgNPs have found applicability in textiles, lifestyle products, healthcare sector and photocatalytic applications. Conventional and traditional chemical and physical routes of AgNPs synthesis involve usage of hazardous and toxic chemicals along with elevated parameters of temperature and pressure. AgNPs synthesized through these routes contain toxic chemical surface agents (Bar et al. 2009) and may lead to bio-incompatibility in medical (DeLousie 2012) and therapeutical applications (Wang and Wang 2014). "Green" routes for the synthesis of AgNPs involving biological resources are being explored to synthesize AgNPs which can be further used in medical applications. Various plants and their parts have been reported for their use in the biosynthesis of AgNPs (Chandran et al. 2006; Song and Kim 2009; Kumar and Yadav 2009). These processes have been found to be efficient, advantageous and faster (Cruz et al. 2010). However indiscriminate usage of commercially valued crop products for the synthesis of AgNPs should be

A. Devadiga · K. Vidya Shetty (✉) · M.B. Saidutta
Department of Chemical Engineering, National Institute of Technology, Karnataka, Srinivasnagar post, Surathkal 575025, Karnataka, India
e-mail: vidyaks68@yahoo.com; vidyaks95@nitk.ac.in

A. Devadiga
e-mail: aishwarya.d.88@gmail.com

M.B. Saidutta
e-mail: mbs@nitk.ac.in

R. Mohan B. et al. (eds.), *Materials, Energy and Environment Engineering*,
DOI 10.1007/978-981-10-2675-1_17

discouraged and alternate resources that are abundantly available as an agrowaste should be utilized for the synthesis of nanoparticles.

T. catappa (Indian almond) tree is widely grown in the tropical belt for its high nutritional value seeds. Leaves of the tree remain as an unharnessed reservoir of various bioactive compounds (Owolabi et al. 2013). The leaves of this tree that serve as "bioreservoirs" of active components go untapped as the leaves are shed off from the tree naturally, during the fall season. While *T. grandis* Linn f. (Teak tree) leaves form the refuse of timber industry and possess a huge repertoire of bioactive compounds (Devadiga et al. 2015). In the present work, these rich sources of "bioactive compounds" have been successfully utilized in the synthesis of AgNPs. This nature mining process for harnessing bioactive compounds from agrowaste for the synthesis of AgNPs offers to be a simple, efficient, cost effective and a green method. In this study, the pivotal role of precursor salt solution concentration in the synthesis of AgNPs and its effect on the morphological characteristics using the aqueous extracts of *T. catappa* and *T. grandis* Linn f. leaves is evaluated.

2 Materials and Methods

2.1 *Collection of the Plant Material*

The plant material was collected from Surathkal, a town geographically positioned at 12°58′60 N 74°46′60 E in Mangalore, western coastal belt region of India, Karnataka. The washed leaves were shade dried for 10 days and powdered using a domestic motor operated blade mixer.

2.2 *Preparation of the Aqueous Extracts of* T. catappa *(ALE) and* T.Grandis *Linn f (TLE) Leaves*

The extract of the leaves was prepared by boiling a known quantity of leaf powder (5 g for ALE and 10 g for TLE) in 100 mL of deionised water for 5 min through open solvent heating process and was filtered through Whattman No.1 filter paper and the filtrate was stored in air tight bottles for further usage.

2.3 *Biosynthesis of AgNPs*

AgNPs were synthesized using aqueous solution of precursor salt of different molar concentrations such as 1, 5, 10, 20 and 100 mM along with ALE or TLE in ratio 1:1(v/v) at ambient temperature of 28 ± 2 °C to study the effect of precursor salt solution concentration on synthesis of AgNPs. Surface plasmon resonance (SPR) analysis of AgNPs synthesized in the synthesis mixture after 24 h of the synthesis time was analyzed using UV-Vis spectroscopy (Labomed, USA) and the conversion of Ag^+ ions to AgNPs was determined by measuring the concentration of residual Ag^+ ions using atomic absorption spectroscopy (AAS- GBC 932plus, USA) according to the methodology reported earlier by Maria et al. (2015) and Devadiga et al. (2015).

3 Results and Discussion

The SPR spectra of the synthesis mixture prepared with different concentrations of precursor salt solution with ALE and TLE at 24 h of synthesis time duration are presented in Fig. 1a, b respectively. As observed in Fig. 1a, b, the intensity of SPR peak emanated by AgNPs increased with an increase in $AgNO_3$ concentration from 1 to 20 mM, expressive of the result that increased concentration of Ag^+ ions are being converted to Ag^0 by the bioactive components present in the reaction mixture which act as reducing agents, yielding a large quantity of AgNPs (Prarthna et al. 2011) and the same is supported by increase in conversion of Ag^+ ions to AgNPs with increased $AgNO_3$ solution concentration as presented in Table 1. As presented in Table 1, almost complete conversion of Ag^+ ions occurred in 24 h with 100 mM $AgNO_3$ solution by the reducing agents present in ALE and TLE. However, with concentrations lesser than 100 mM, complete conversion could not take place in 24 h, due to lower rate of reduction to Ag^0. As observed in Fig. 1a, b, at lower concentration of $AgNO_3$ of 1 and 5 mM, the SPR peak is less intense than that obtained with 10 and 20 mM precursor salt solution. Formation of less intense peaks at lower concentrations of 1 and 5 mM are indicative of the formation of lesser number AgNPs (Dubey et al. 2010) in 24 h owing to lower rate of formation at lower concentration of $AgNO_3$. Higher intensity of the SPR peaks obtained from higher concentrations of

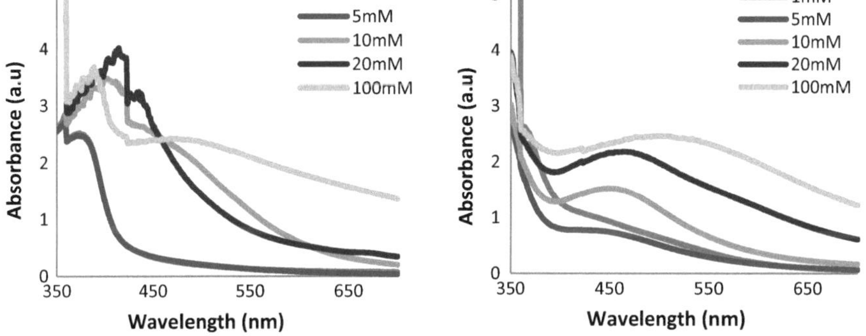

Fig. 1 SPR spectra of AgNPs synthesized from synthesis mixtures containing different concentrations of precursor salt solutions and **a** ALE and **b** TLE

Table 1 Conversion of Ag^+ ions into AgNPs and Approximate size range determination of AgNPs synthesized using ALE and TLE with varying precursor salt concentrations

Precursor salt solution concentration (mM)	ALE		TLE	
	Conversion of Ag^+ ions into AgNPs (%)	Particle size range	Conversion of Ag^+ ions into AgNPs (%)	Particle size range
1	88.87	38–171 nm	97.78	48–102 nm
5	95.03	41–180 nm	97.8	32–97 nm
10	94.94	28–83 nm	97.91	39–92 nm
20	97.31	33–48 nm	98.88	34–41 nm
100	99.91	Polydispersed and aggregated	99.92	Polydispersed and aggregated

$AgNO_3$ signifies large number AgNPs being biosynthesized owing to the larger rate of formation of AgNPs (Prarthna et al. 2011). However, at higher concentrations of 20 mM the number of Ag^+ ions available per unit volume of the reaction mixture for reduction into Ag^0 is larger and thus, the rate of formation of AgNPs is high, leading to the formation of large number of AgNPs, hence the peak is sharper, narrower and intense (Rodriguez-León et al. 2013).

The SEM images of the AgNPs synthesized with different concentrations of precursor salt solution with ALE and TLE are given in Fig. 2a–e and Fig. 3a–e respectively. As seen in Fig. 2e and 3e agglomeration of the nanoparticles forming smaller number of large sized aggregates is observed when 100 mM of $AgNO_3$ salt solution is used for the synthesis of AgNPs. This can be attributed to the presence of very large number of particles in the synthesis solution; the particles collide with each other at higher rate resulting in agglomeration of the nanoparticles forming smaller number of large sized aggregates as indicated by broad SPR band in Fig. 1a, b. The SEM images in Fig. 2a–e and Fig. 3a–e support the analysis and interpretation of SPR characteristics of AgNPs synthesized using ALE and TLE as shown in Fig. 1a, b respectively. The approximate size ranges (determined using Image J software) of the AgNPs obtained with different concentration of precursor salt solution are presented in Table 1. As the concentration of precursor salt solution was increased the size range of AgNPs synthesized using ALE and TLE narrowed down considerably, the narrowest size range being that with 20 mM $AgNO_3$ solution as presented in Table 1. Similar observation of decrease in size of AgNPs with increase in the precursor salt solution in the synthesis solution was made by Bar et al. (2009) using the seed extract of *Jatropha curcus. Spherical particles were formed at precurser salt concentrations of 20 mM and lower.*

The result obtained by the spectral analysis of AgNPs synthesized using ALE and TLE (Fig. 1a); conversion of Ag^+ ions into AgNPs (Table 1) is supported by the morphological study of the AgNPs by SEM analysis (Fig. 2a–e and Fig. 3a–e).

This study concludes that the plant extracts used in the synthesis of AgNPs possess necessary bioactive phytocompounds with reducing efficiency and capping

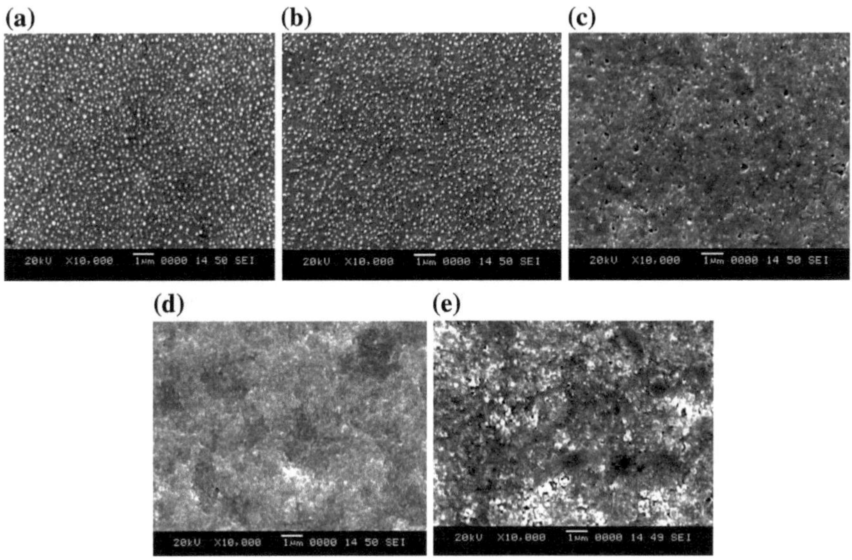

Fig. 2 SEM images of AgNPs synthesized from the synthesis mixture containing ALE and a precursor salt concentration of **a** 1 mM **b** 5 mM **c** 10 mM **d** 20 mM and **e** 100 mM

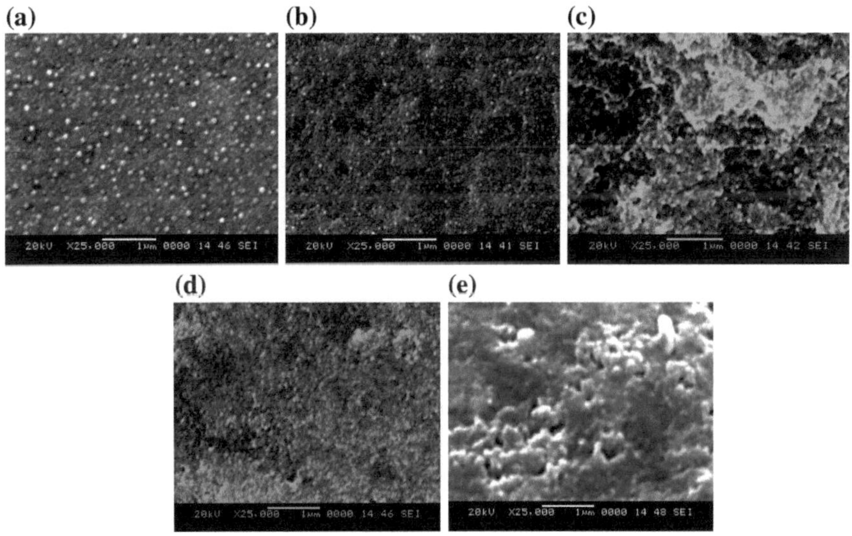

Fig. 3 SEM images of AgNPs synthesized from the synthesis mixture containing TLE and a precursor salt concentration of **a** 1 mM **b** 5 mM **c** 10 mM **d** 20 mM and **e** 100 mM

potential and the concentration of precursor salt solution used in the synthesis mixture affects the size and morphological characteristics of the AgNPs synthesized using ALE and TLE. The narrowest size range of AgNPs was synthesized using 20 mM of $AgNO_3$ salt solution with ALE and TLE.

4 Conclusion

Bio-based routes for the synthesis of AgNPs have been gaining momentum as they overcome the shortcomings of chemical and physical synthesis routes. Nature mining of non nutritional and non commercial plant resources enriched with bioactive phytocomponents for the synthesis of nanoparticles is an efficacious process. The study reveals that the concentration of the precursor salt solution has a pivotal role in the synthesis process as it exhibits a pronounced effect on the morphological characteristics of the AgNPs being synthesized.

References

Bar, H., Bhui, D.K., Sahoo, G.P., Sarkar, P., De, S.P., Misra, A.: Green synthesis of silver nanoparticles using latex of *Jatropha curcas*. Colloids Surf., A **339**, 134–139 (2009)

Chandran, S.P., Chaudhary, M., Pasricha, R., Ahmad, A., Sastry, M.: Synthesis of gold nanotriangles and silver nanoparticles using Aloevera plant extract. Biotechnol. Prog. **22**, 577–583 (2006)

Cruz, D., Falé, P.L., Mourato, A., Vaz, P.D., Serralheiro, M.L., Lino, A.R.L.: Preparation and physicochemical characterization of Ag nanoparticles biosynthesized by *Lippia citriodora* (Lemon Verbena). Colloids Surf., B **81**, 67–73 (2010)

DeLouise, L.A.: Applications of nanotechnology in dermatology. J. Invest. Dermatol. **132**, 964–975 (2012)

Devadiga, A., Shetty, K.V., Saidutta, M.B.: Timber industry waste-teak (*Tectona grandis* Linn.) leaf extract mediated synthesis of antibacterial silver nanoparticles. Int. Nano Lett. **5**, 205–214 (2015)

Dubey, S.P., Lahtinen, M., Särkkä, H., Sillanpää, M.: Bioprospective of *Sorbus aucuparia* leaf extract in development of silver and gold nanocolloids. Colloids Surf., B **80**, 26–33 (2010)

Kumar, V., Yadav, S.K.: Plant-mediated synthesis of silver and gold nanoparticles and their applications. J. Chem. Technol. Biotechnol. **84**, 151–157 (2009)

Maria, B.S., Devadiga, A., Kodialbail, V.S., Saidutta, M.B.: Synthesis of silver nanoparticles using medicinal *Zizyphus xylopyrus* bark extract. Appl. Nanosci. **5**, 755–762 (2015)

Owolabi, M.S., Lawal, O.A., Ogunwande, I.A., Hauser, R.M., Setzer, W.N.: Chemical composition of the leaf essential oil of *Terminalia catappa L.* growing in south-western Nigeria. Am. J. Essent. Oils Nat. Products **1**, 51–54 (2013)

Prathna, T.C., Chandrasekaran, N., Raichur, A.M., Mukherjee, A.: Biomimetic synthesis of silver nanoparticles by Citrus limon (lemon) aqueous extract and theoretical prediction of particle size. Colloids Surf., B **82**, 152–159 (2011)

Rodríguez-León, E., Iñiguez-Palomares, R., Navarro, R.E., Herrera-Urbina, R., Tánori, J., Iñiguez-Palomares, C., Maldonado, A.: Synthesis of silver nanoparticles using reducing agents obtained from natural sources (Rumex hymenosepalus extracts). Nanoscale Res. Lett. **8**, 1–9 (2013)

Song, J.Y., Kim, B.S.: Rapid biological synthesis of silver nanoparticles using plant leaf extracts. Bioprocess Biosyst. Eng. **32**, 79–84 (2009)

Wang, E.C., Wang, A.Z.: Nanoparticles and their applications in cell and molecular biology. Integr. Biol. **6**, 9–26 (2014)

Impact of Hydrochloric Acid on Phase Formation of Titanium Dioxide Nanoparticles

Swati Aggarwal, R.R. Ezhil Venuswaran and P. Balasubramanian

1 Introduction

Titanium dioxide (TiO_2) is a transition metal oxide ceramic material that acquires a prominent position in the wide range of applications (Yin et al. 2013). The field of nanotechnology had pioneered remarkable developments in the physical, mechanical, optical, biological and electronic properties of nanostructured TiO_2 in recent years. TiO_2 is predominantly preferred by most of the researchers in general due to its high refractive index, the capability of higher energy conversion, ease of handling and low toxic nature (Shi et al. 2013). Since titanium metal is the ninth prevalent element in the earth's crust, the preferable focus had been shifted towards the preparation of titanium dioxide nanoparticles in large quantities. Titanium dioxide exists in three different polymorphs—namely anatase, rutile, and brookite. A stable crystalline structure had been observed only with rutile TiO_2. However, the anatase and brookite phases exhibit a metastable crystalline structure (Hanaor and Sorrell 2011). Anatase TiO_2 can be formed under room temperature itself whereas rutile demands for high-temperature treatments for a complete phase formation.

The brookite form of TiO_2 is an extremely complex polymorph that can be synthesized only under specific temperature and optimum pressure conditions. Various investigations had revealed that the properties and efficiency of TiO_2 on a particular application purely depends on the morphology, crystalline phase, grain size and surface area. Thus, the controlled rate of handling the processing parameters

S. Aggarwal · P. Balasubramanian (✉)
Department of Biotechnology and Medical Engineering, NIT Rourkela, Rourkela, India
e-mail: biobala@gmail.com

S. Aggarwal
e-mail: swati.aggarwal9018@gmail.com

R.R. Ezhil Venuswaran
Department of Ceramic Engineering, NIT Rourkela, Rourkela, India
e-mail: ezhilvenus@gmail.com

R. Mohan B. et al. (eds.), *Materials, Energy and Environment Engineering*,
DOI 10.1007/978-981-10-2675-1_18

of a synthesis route is highly important (Li et al. 2008). Various methods can synthesize titanium dioxide nanoparticles—Sol-gel method, chemical vapor deposition method, hydrolysis method, sonochemical method and microwave assisted techniques (Zhou et al. 2012; Su et al. 2006). Each synthesis route underwent different modification in the properties of nanostructured TiO_2 particles. Furthermore, these synthesis methods made a continuous breakthrough in the recent years which gave various forms of TiO_2 as nanotubes, nanowires, nanosheets, etc. (Wu et al. 2002). Among these methods, hydrolysis method was prominently focused by the researchers due to its single step crystallization of TiO_2 nanoparticles.

The hydrolysis method is a simple method that can be carried out easily at room temperature. Under an acid medium, $TiCl_4$ - liquid precursor is hydrolyzed and crystallized for the formation of TiO_2 nanoparticles. The acid used in the synthesis performs a dual role—acts as a catalyst for hydrolysis reaction and solubilizes the liquid precursor. Homogeneity, short range of particle size and purity are the major advantages of this hydrolysis method. The concentration of the acid, volume of $TiCl_4$ and volume of the entire solvent plays a crucial role in influencing the morphology, crystalline phases and surface area of the nanoparticles (Chen and Mao 2007). In this paper, different volumes of solvent had been considered for investigating the significance of the processing parameters on the synthesis of Titanium dioxide nanoparticles.

2 Materials and Methods

Titanium dioxide nanoparticles had been synthesized by simple hydrolysis method in the presence of hydrochloric acid (HCl), 35 % purity ACS Reagent (Zheng et al. 2008). Titanium tetrachloride, 9 %, trace metal basis, Sigma-Aldrich, had been used as a liquid precursor. 1 ml of Titanium tetrachloride had been added to 50, 100, 150 and 200 ml of 1 M HCl and stirred for 12 h at room temperature. The liquid precursor had been hydrolyzed and the prepared solutions were aged by keeping it in a dry air oven for 12–18 h at 100 °C. Then the obtained product had been centrifuged at 7000 rpm and washed with deionized water followed by series of increasing concentrations of ethanol. The obtained phases on titanium dioxide nanoparticles on different volumes of HCl had been investigated by using Philips X-Ray Diffractometer with Ni filter Cu-Kα (λ = 1.5418 Å) with a scanning range of 5°–90°. Based on the following equation, the relative phase percentages of TiO_2 could be expressed as a function of XRD peak areas of different crystallographic forms.

$$\mathbf{W_A = K_A I_A (K_A I_A + I_R)} \quad (1)$$

$$\mathbf{W_R = I_R / (K_A I_A + I_R)} \quad (2)$$

Where, W_R and W_A are the weight fraction of rutile and anatase respectively; I_R and I_A are obtained from characteristic peaks areas of rutile and anatase respectively with $K_A = 0.886$ as constant (Zhang and Banfield 2000). Thermogravimetric analysis of titanium nanoparticles had been carried out by using NetzschSTA/409C, by firing till 650 °C under Argon atmosphere. The particle size and the particle morphology had been determined by using Nova Nanosem 450 FESEM.

3 Results and Discussion

1 ml of $TiCl_4$ had been hydrolyzed against 50, 100, 150 and 200 ml of 1 M HCl solution. After washing with series of concentrations of ethanol and water, the yield of Titanium dioxide nanoparticles had been weighed and reported in Table 1. It had been observed that the yield of Titanium dioxide nanoparticles had been decreasing

Table 1 Yield of titanium dioxide nanoparticles by hydrolysis method

Sample	Volume of HCl solution (mL)	TiO_2 powder yields before washing (g)	TiO_2 powder yields after washing (g)
S1	50	0.775	0.734
S2	100	0.720	0.604
S3	150	0.540	0.490
S4	200	0.310	0.231

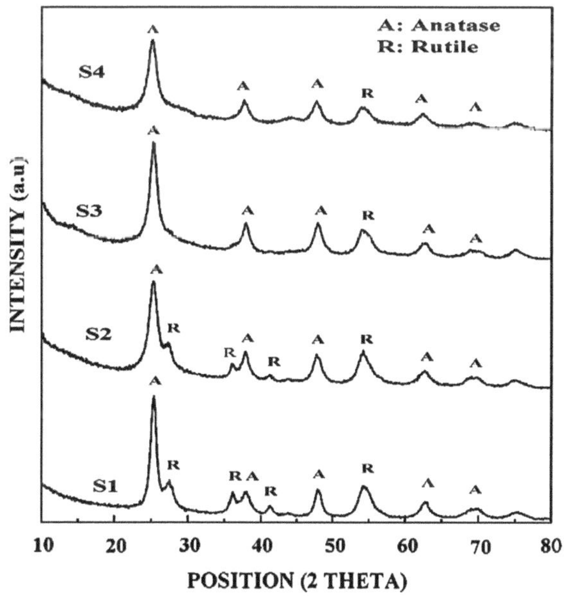

Fig. 1 Phase analysis of titanium dioxide nanoparticles

Table 2 Phase evaluation results of synthesized titanium dioxide nanoparticles

Sample	Volume of HCl solution (ml)	Crystallite size (nm)	JCPDS rutile file	JCPDS anatase file	Anatase (%)	Rutile (%)
S1	50	10.37	01–1292	01–0562	59.1	40.9
S2	100	5.09	01–1292	01–0562	68.6	31.3
S3	150	10.91	02–0494	01–0562	71.9	28.1
S4	200	5.70	02–0494	02–0387	71.9	28.1

gradually with higher volumes of 1 M HCl. Since the nucleation rate will be vigorous, and particle interaction is high in lower volumes of the aqueous solutions, the yield of titanium nanoparticles had been obtained with a higher amount. The crystalline phases obtained from the above mentioned samples S1, S2, S3 and S4 had been determined by XRD technique. The obtained phases on the synthesized Titanium dioxide nanoparticles had been shown in the Fig. 1.

The formation of brookite phase demands for optimum conditions, it had not been observed in these samples. The synthesized titanium dioxide nanoparticles attain neither a pure anatase phase nor a pure rutile phase. The obtained phases are a combination of rutile and anatase phases. When the volume of 1 M HCl increases from 50 ml to 200 ml, the crystallinity of rutile polymorph is reduced. Since the synthesized titanium nanoparticles featured with binary phase of anatase and rutile, the weight fractions of the individual had been evaluated by using Eq. 1 and Eq. 2, which had been reported in the Table 2.

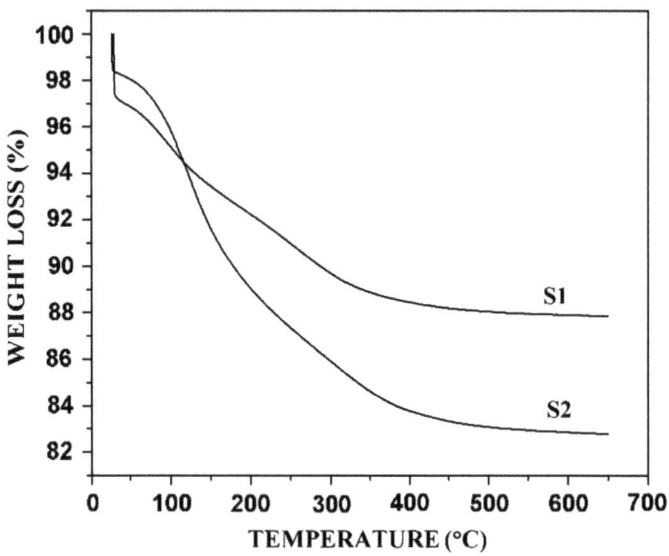

Fig. 2 TGA analysis of titanium dioxide nanoparticles

The thermogravimetric analysis of the samples S1 and S2 had been illustrated in the Fig. 2. For the sample S1, weight loss of around 12 % was noticed when the temperature reaches to 400 °C. Similarly for S2, weight loss of around 18 % had been observed when the temperature reaches to 500 °C. These weight loss might be due to the evaporation of water, HCl and decomposition of unreacted Ti–OH molecules. On further heat treatment, the unreacted Ti–OH molecules decompose to form TiO_2. Thus, if the samples had been heat treated, the weight fractions of anatase and rutile might change due to phase transformation and simultaneous crystallization.

The microstructures of titanium dioxide nanoparticles had been illustrated in the Fig. 3. All the samples acquired particle size in the range of 20–90 nm. It clearly

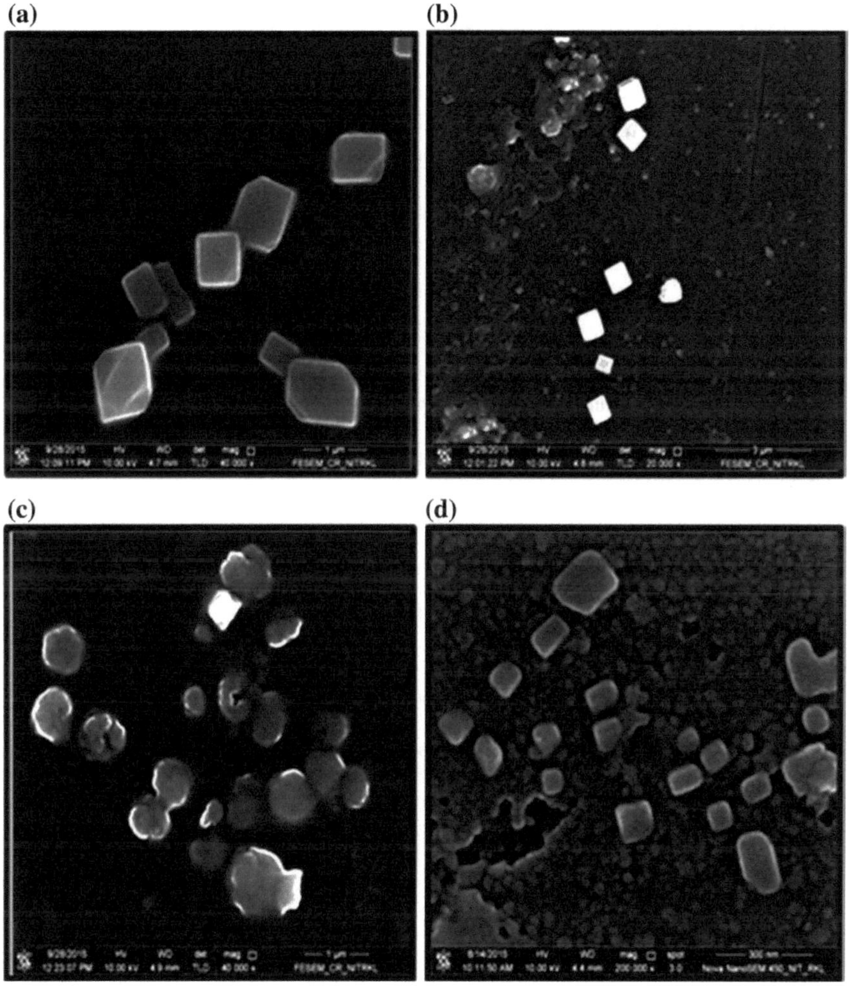

Fig. 3 FESEM images of titanium dioxide nanoparticles **a** S1 **b** S2 **c** S3 **d** S4

shows that the samples S1, S2 and S4 attained a cubical morphology. But the morphology of S3 was partly spherical and partly cubical. This clearly shows an unstable structure that had been attained due to the phase transformation from rutile to anatase. Similar kind of profile had been clearly illustrated in the results of phase analysis.

4 Conclusions

Titanium dioxide nanoparticles had been successfully prepared by hydrolysis method. Instead of pure anatase or pure rutile TiO_2 nanoparticles, anatase-rutile composite nanoparticles had been synthesized. On increasing the volume of HCl, the probability of rutile phase formation is reduced. No drastic change in morphology had been observed due to the changes in the volume of HCl. Hence by varying the processing parameters varieties of titanium dioxides nano-composites could be developed. The knowledge on the impact of processing parameters during synthesis on phase formation of TiO_2 nanoparticles could be implied further on its various environmental applications such as photocatalytic conversion of carbon dioxide to methanol and photodegradation of organic and inorganic compounds.

References

Chen, X., Mao, S.S.: Titanium dioxide nanomaterials: synthesis, properties, modifications, and applications. Chem. Rev. **107**(7), 2891–2959 (2007)

Hanaor, D.A.H., Sorrell, C.C.: Review of the anatase to rutile phase transformation. J. Mater. Sci. **46**(4), 855–874 (2011)

Li, H., Duan, X., Liu, G., Jia, X., Liu, X.: Morphology controllable synthesis of TiO_2 by a facile hydrothermal process. Mater. Lett. **62**(24), 4035–4037 (2008)

Shi, H., Magaye, R., Castranova, V., Zhao, J.: Titanium dioxide nanoparticles: a review of current toxicological data. Part Fibre Toxicol. **10**(1), 15 (2013)

Su, C., Tseng, C.-M., Chen, L.-F., You, B.-H., Hsu, B.-C., Chen, S.-S.: Sol–hydrothermal preparation and photocatalysis of titanium dioxide. Thin Solid Films **498**(1), 259–265 (2006)

Wu, M., Lin, G., Chen, D., Wang, G., He, D., Feng, S., Ruren, X.: Sol-hydrothermal synthesis and hydrothermally structural evolution of nanocrystal titanium dioxide. Chem. Mater. **14**(5), 1974–1980 (2002)

Yin, Z.F., Wu, L., Yang, H.G., Yong, H.S.: Recent progress in biomedical applications of titanium dioxide. Phys. Chem. Chem. Phys. **15**(14), 4844–4858 (2013)

Zhang, H., Banfield, J.F.: Understanding polymorphic phase transformation behavior during growth of nanocrystalline aggregates: insights from TiO_2. J. Phys. Chem. B **104**(15), 3481–3487 (2000)

Zhou, J., Song, B., Zhao, G., Han, G.: Effects of acid on the microstructures and properties of three-dimensional TiO_2 hierarchical structures by solvothermal method. Nanoscale Res. Lett. **7**(1), 1–10 (2012)

Zheng, W., Liu, X., Yan, Z., Zhu, L.: Ionic liquid-assisted synthesis of large-scale TiO_2 nanoparticles with controllable phase by hydrolysis of $TiCl_4$. ACS Nano **3**(1), 115–122 (2008)

Synthesis and Characterization of Mg Doped CuO Nano Particles by Quick Precipitation Method

Rintu Varghese, H. Joy Prabu and I. Johnson

1 Introduction

It is an undeniable fact that, after the invention of carbon nanotubes (CNTs), the fabrication of different morphologies have become much exploring area on nano research (Huang et al. 2001). Among the transition metal oxides, copper oxide (CuO) is an interesting topic of concern due to its diverse applications such as electronics, catalysis, sensors, solar cells and antimicrobial activity (Devi and Singh 2014; Awwad et al. 2015; Luo et al. 2011; Padil and Černík 2013; Song et al. 2011; Seo et al. 2011). Till now, there are several methods well-known for nanoparticles synthesis and some of them brought with many practical limitations. In this perspective, the quick precipitation method is particularly more attractive because of its cost effectiveness and simple operations (Shahmiri et al. 2013). Polyvinyl pyrrolidone (PVP) is the most commonly used polymer in the quick precipitation of metal oxides due to its remarkable properties like distinct shape, dissolving ability and stabilizing capacity (Park et al. 2009; Zhang and Cui 2008; Slistan-Grijalva et al. 2008; Pomogailo and Kestelman 2005). Furthermore, a number of authors have been reported in the literature for the improvement of material properties by the addition of dopants (Das et al. 2009; Kim et al. 2012; Ponnarasan et al. 2014).

In this present investigation, the method used was concentrated on the effects of magnesium (Mg) doping on the structural and optical properties of CuO nanoparticles. In this preparation method, the organic molecule PVP, NaOH, copper acetate and magnesium chloride has been chosen as a stabilizer, reducing agent, precursor material and dopant respectively.

R. Varghese (✉) · H. Joy Prabu · I. Johnson
Centre for Nano Science and Applied Thermodynamics, Department of Physics, St. Joseph's College, Trichy 620002, Tamilnadu, India

R. Mohan B. et al. (eds.), *Materials, Energy and Environment Engineering*,
DOI 10.1007/978-981-10-2675-1_19

2 Materials and Method

All the reagents were used of analytical grade and without further purification. Copper acetate [Cu $(CH_3COO)_2$] and magnesium chloride ($MgCl_2$) were obtained from Sigma Aldrich, Germany and sodium hydroxide (NaOH) were obtained from Merck, Germany.

In this method two separate aqueous solutions, copper acetate (5 M) and NaOH were referred to as solution A and solution B, respectively. Stirring is continued up to the respective metal salts were totally dissolved in distilled water. Then about 1 wt% of PVP is added to solution A. Then the solution B is added to solution A drop by drop and the stirring continued until the formation of a suspension. It was kept at room temperature for 4 h for the completion of reaction. A large amount of black precipitate was formed. The nano structures were collected by centrifugation (4000 rpm) and were washed with absolute Ethanol and distilled water several times to remove the possible residual impurities. The particles were dried in an oven at 60 °C for 3 h. After grinding a dark brown powder was obtained. In order to dope, the dopant solution ($MgCl_2$) was added to the same precursor solution at different concentrations (2, 5 wt %).

3 Results and Discussion

3.1 XRD Analysis

It is to be noted that, in all the Figs. 1a, b and c the peaks observed in the XRD patterns match with those of the CuO (monoclinic) reported in the JCPDS card no. 65-2309. According to literature, these additional peaks in Figs. 1a and b can be attributed to the presence of Cu_2MgO_3.

Based on Scherer equation, the average particle size of undoped CuO was found to be 49 nm while it was 23 and 18 nm for 2 and 5 % Mg doped CuO respectively. It can also be seen that broadening of the XRD peaks indicates the formation of nanoparticles.

3.2 FESEM and EDX Analysis

FESEM has been used to characterize the shape and morphologies of formed nanoparticles. The FESEM result of the Fig. 2a shows the surface morphology of undoped CuO nanoparticles which are almost spherical in shape. The Figs. 2b and c show the Mg doped CuO particles which are having wire like structure.

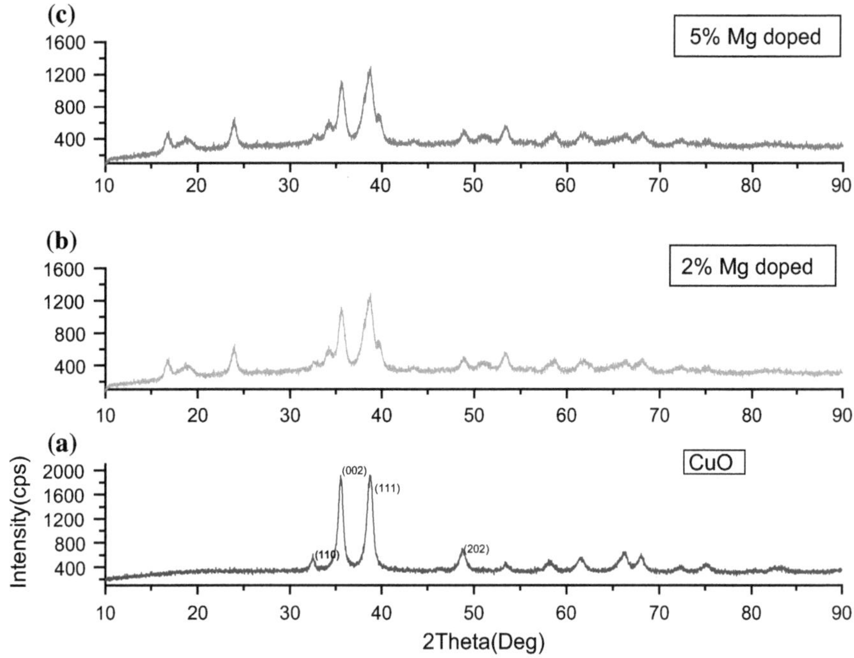

Fig. 1 **a–c** The XRD pattern for undoped and doped CuO nanoparticles

The EDX result of Fig. 2d displays the presence of Cu and O ions in larger amount along with some C. The Fig. 2e and f shows the elemental composition of Cu, O, and Mg. Their weight and atomic percent confirms that Mg was doped on CuO nanowires.

3.3 *UV-Vis Analysis*

The reduction of CuO nanoparticles was Monitored by UV spectrophotometer HITACHI U2300 range of absorbance from 250 to 480 nm. An observed peak at 390 nm which is comparable with the literature value (Fig. 3a). The fig [3b and c] shows the UV excitonic peak at 368 and 365 nm for 2 and 5 % Mg doped CuO nanoparticles respectively.

The direct allowed band gap of CuO, 2 % Mg doped CuO, 5 % Mg doped have been found to be 1.53, 1.67, 1.73 eV respectively. It is observed that the band gap value of samples are higher than bulk CuO (1.2–1.3) (Johan et al. 2011). This blue

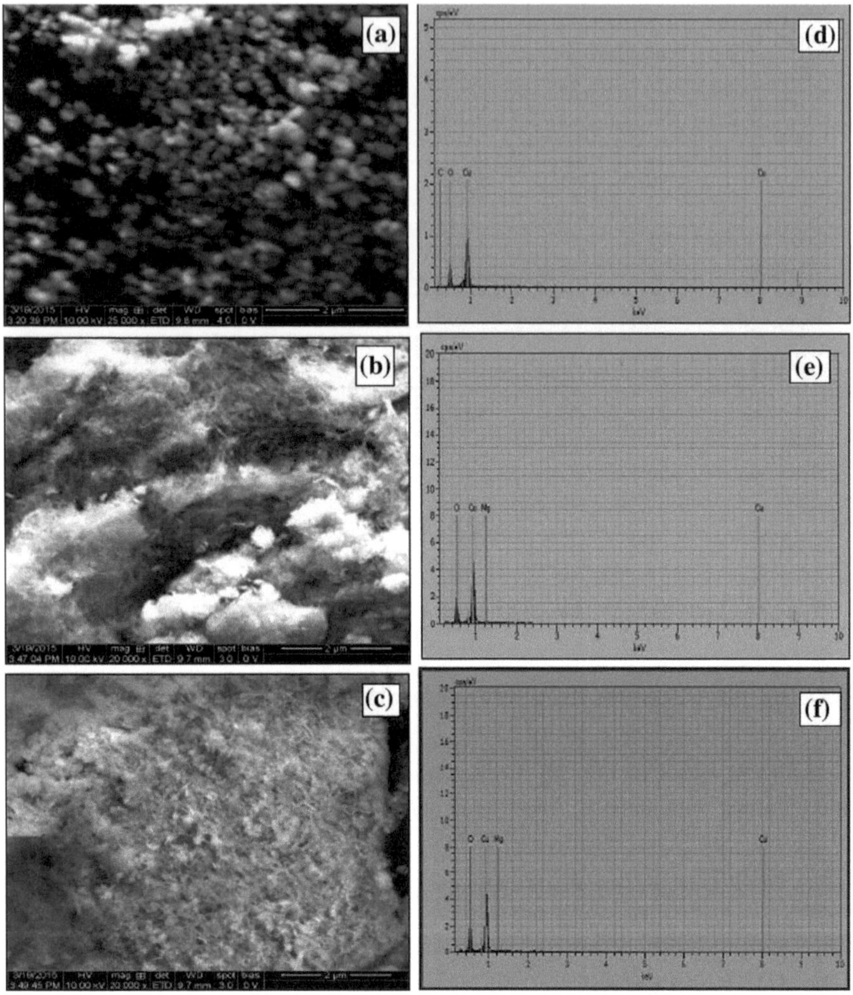

Fig. 2 FESEM image (**a–c**); EDX analysis (**d–f**) of undoped, 2 and 5 % Mg doped CuO NPs respectively

shift in the direct band gap is due to the quantum confinement effect. From the Figs. 4a, b and c it is found that the indirect band gap values are red shifted compared to bulk value due to the formation of surface defects.

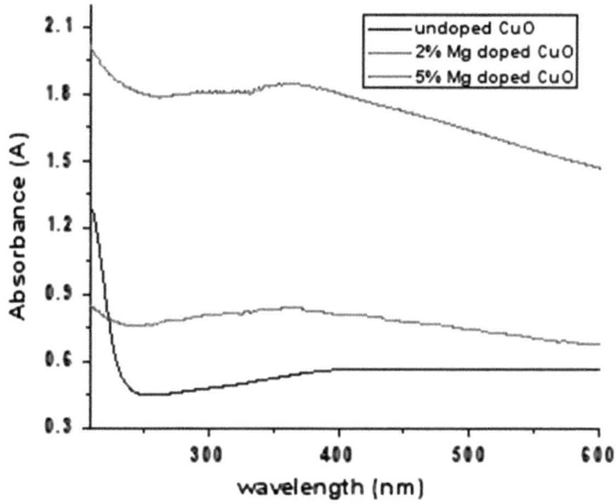

Fig. 3 The UV-Vis absorption spectra of doped and undoped CuO nanoparticles

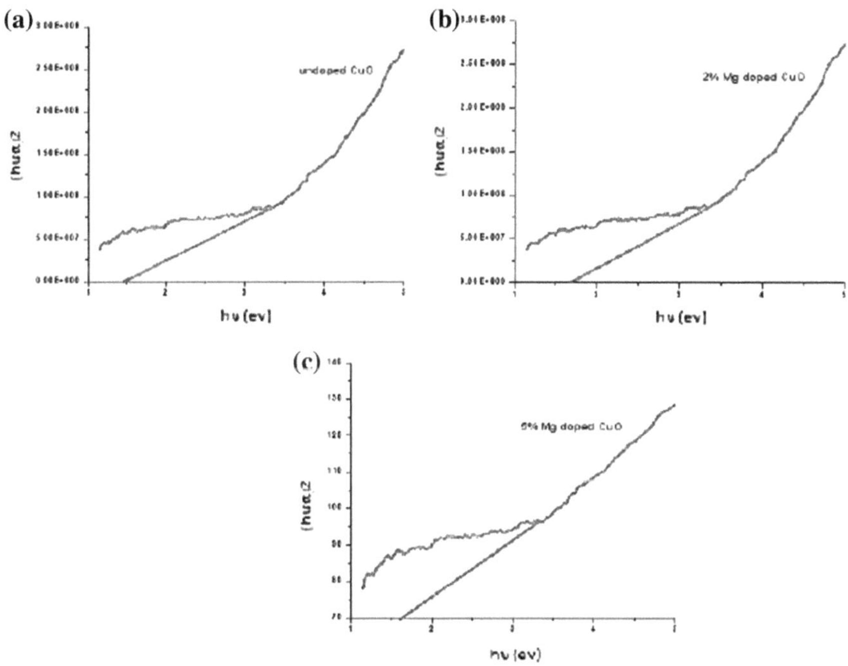

Fig. 4 **a–c** the direct band gap energy of the sample

4 Conclusion

Undoped and Mg doped CuO wire like nano structure were successfully synthesized in PVP by using Quick Precipitation method at a low reaction temperature (60 °C) in the absence of templates and additives. The structural, morphological and optical properties of a CuO sample were investigated. Small amounts of impurity phase were observed in doped XRD pattern. The particle sizes were observed to decrease slightly with the increase in Mg concentration. The blue shift in the direct band gap may be due to the quantum confinement effect. The shapes of the undoped and doped CuO structure were examined. In conclusion, the Mg doping in CuO improves the properties of the CuO nanoparticles and has many interesting applications in gas sensors, solar cell, catalyst etc.

References

Awwad, A., Albiss, B., Salem, N.: Antibacterial activity of synthesized copper oxide nanoparticles using Malva sylvestris leaf extract. SMU Med. J. **2**, 91–101 (2015)

Das, K., Sharma, S.N., Kumar, M., De, S.: Morphology dependent luminescence properties of Co doped TiO_2 nanostructures. J. Phys. Chem. C **113**, 14783–14792 (2009)

Devi, H.S., Singh, T.D.: Synthesis of copper oxide nanoparticles by a novel method and its application in the degradation of methyl orange. Adv. Electron. Electr. Eng. **4**, 83–88 (2014)

Huang, Y., Duan, X., Cui, Y., Lauhon, L.J., Kim, K.-H., Lieber, C.M.: Logic gates and computation from assembled nanowire building blocks. Science **294**, 1313–1317 (2001)

Johan, M.R., Suan, M.S.M., Hawari, N.L., Ching, H.A.: Annealing effects on the properties of copper oxide thin films prepared by chemical deposition. Int. J. Electrochem. Sci. **6**, 6094–6104 (2011)

Kim, K.-M., Jeong, H.-M., Kim, H.-R., Choi, K.-I., Kim, H.-J., Lee, J.-H.: Selective detection of NO_2 using Cr-doped CuO nanorods. Sensors **12**, 8013–8025 (2012)

Luo, S., Su, F., Liu, C., Li, J., Liu, R., Xiao, Y., et al.: A new method for fabricating a CuO/TiO 2 nanotube arrays electrode and its application as a sensitive nonenzymatic glucose sensor. Talanta **86**, 157–163 (2011)

Padil, V.V.T., Černík, M.: Green synthesis of copper oxide nanoparticles using gum karaya as a biotemplate and their antibacterial application. Int. J. Nanomed. **8**, 889–898 (2013)

Park, J.C., Kim, J., Kwon, H., Song, H.: Gram-scale synthesis of Cu_2O nanocubes and subsequent oxidation to CuO hollow nanostructures for lithium-ion battery anode materials. Adv. Mater. **21**, 803–807 (2009)

Ponnarasan, V., Krishnan, A.: Studies on pure and divalent metal doped copper oxide nanoparticles (2014)

Pomogailo, A.D., Kestelman, V.N.: Principles and mechanisms of nanoparticle stabilization by polymers. Metallopolymer Nanocomposites, 65–113

Seo, S.-D., Jin, Y.-H., Lee, S.-H., Shim, H.-W., Kim, D.-W.: Low-temperature synthesis of CuO-interlaced nanodiscs for lithium ion battery electrodes. Nanoscale Res. Lett. **6**, 1–7 (2011)

Shahmiri, M., Ibrahim, N.A., Zainuddin, N., Asim, N., Bakhtyar, B., Zaharim, A., et al.: Effect of pH on the synthesis of CuO nanosheets by quick precipitation method. WSEAS Trans. Environ. Dev. **9**, 137–145 (2013)

Slistan-Grijalva, A., Herrera-Urbina, R., Rivas-Silva, J., Avalos-Borja, M., Castillón-Barraza, F., Posada-Amarillas, A.: Synthesis of silver nanoparticles in a polyvinylpyrrolidone (PVP) paste, and their optical properties in a film and in ethylene glycol. Mater. Res. Bull. **43**, 90–96 (2008)
Song, X., Gao, L., Mathur, S.: Synthesis, characterization, and gas sensing properties of porous nickel oxide nanotubes. J Phys. Chem. C **115**, 21730–21735 (2011)
Zhang, H., Cui, Z.: Solution-phase synthesis of smaller cuprous oxide nanocubes. Mater. Res. Bull. **43**, 1583–1589 (2008)

Studies on Process Parameters of Continuous Production of Nickel Nanoparticles Using Spiral Microreactor

Urvashi V. Bhivgade and Shirish H. Sonawane

1 Introduction

Nano-size metal particles have received the great importance in recent years because of extensive application in diverse technical fields which includes chemical catalyst, magnetic recording media, sensors, capacitor and rechargeable batteries (Gao et al. 2000). Traditional metal particles are typically produced in macro reactors on a large scale with limited scope for finer control and innovative technique. There is a continuous demand for these nanoparticles with very specific properties for above mentioned novel applications. To obtain such properties the synthesis of these metal particles plays important role. There are several preparation technique have been discussed to produce these metal particles such as micro-emulsion synthesis (Zhang et al. 2005), electrochemical method (Reetz and Helbig 1994), microwave method (Xu et al. 2008), hydrothermal reduction method (Zhang et al. 2005; Reetz and Helbig 1994; Xu et al. 2008; Wang et al. 1998). The need of efficient synthesis and rapidly changing techniques made synthesis using hydrazine reduction an ideal option. The continuous production of metal particles using a micro reactor is recently one of the growing components of process intensification in chemical engineering. Due to development of microfabrication technology in last decade, microreactor technology has been introduced in the chemical synthesis engineering. Recent research showed great features of microreactor technology which involve rapid mixing because of reduction in diffusion time and fast heat and mass transfer rates (Zhang et al. 2004). Various metals have been developed so far such as platinum (Baumgard et al. 2013), gold

U.V. Bhivgade · S.H. Sonawane (✉)
Department of Chemical Engineering, National Institute of Technology Warangal, Warangal, India
e-mail: shirishsonawane09@gmail.com

U.V. Bhivgade
e-mail: urvashi364@gmail.com

R. Mohan B. et al. (eds.), *Materials, Energy and Environment Engineering*,
DOI 10.1007/978-981-10-2675-1_20

(Pacławski et al. 2011) and silver (Xue et al. 2004) to increase conversion of production by using microreactors. These study focus on the continuous production of Nickel nanoparticles in U-junction copper spiral microreactor. The effect of temperature, the concentration of reactants and the surfactant over Ni nanoparticles formation is studied.

2 Experimental

2.1 Chemicals

Nickel (II) chloride hexahydrate ($NiCl_2.6H_2O$ purity $\geq$ 97 %) from MERK, hydrazine hydrate ($N_2H_4.H_2O$, 99 %) from SDFCL, sodium hydroxide (NaOH), cetyl trimethyl ammonium bromide (CTAB) and polyvinylpyrrolidone (PVP) as surfactant are received and used without further purification. Ethylene glycol is used as solvent for the preparation of solutions with desired concentrations. Particle size distribution was carried out for stability of solution (ZS90, Malvern Instruments, UK). The particle size and morphology of Ni particles were studied using scanning electron microscopy (SEM) VEGA3TE SCAN.

2.2 Experimental Setup and Synthesis

Setup for the experiment is shown in Fig. 1; the microreactor is made of copper with U-shaped junction. The I.D. of copper spiral microreactor was 0.05 in. and length was 200 cm placed within water bath. The reactor temperature is maintained using water bath. This microreactor junction is made in order to avoid clogging and pressure drop.

The production of nickel particles is performed by chemical reduction method using hydrazine hydrate. First, desired concentration of $NiCl_2$ solution and

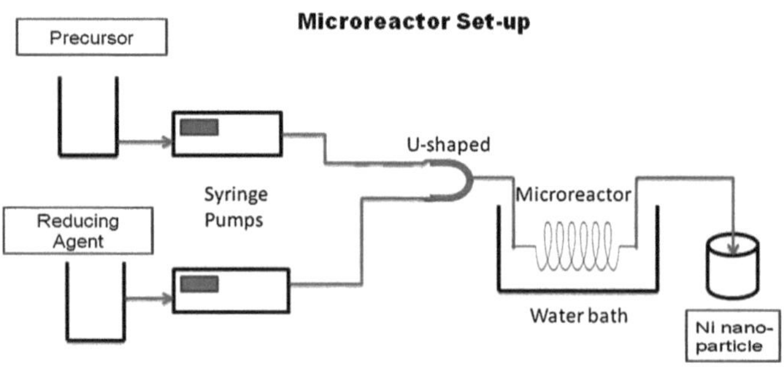

Fig. 1 Experimental setup

Table 1 Experimental condition

Sample name	$NiCl_2$ soln. (M)	N_2H_4 soln. (M)	Surfactant	Temp(°C)
E1	0.1	2	–	80
E2	0.05	0.5	–	80
E3	0.05	1	–	60
E4	0.05	1	–	80
E5	0.05	1	–	100
E6	0.05	1	PVP	80
E7	0.05	1	CTAB	80
E8	0.05	2	–	80

hydrazine hydrate solution were formulated by adding $NiCl_2.6H_2O$ and hydrazine hydrate in ethylene glycol respectively. In the hydrazine hydrate solution, NaOH is added, to maintain pH of the solution of 10–12. In the $NiCl_2$ solution, surfactants cetyl-trimethyl ammonium bromide (CTAB) or poly (vinylpyrrolidone) (PVP) were added, to control the particle size. The precursor and reducing agent solutions were introduced into the U-junction using two different syringes. With the desired flow rate of 90 mL/h, two feeds were fed to spiral copper Microreactor using syringe pumps (Scientech, Hyderabad, India). Reaction of reduction method is carried out at 80 °C in water bath, as bluish color of hydrazine complex changes to dark black color, it confirm the formation of nickel particles from the following reaction

$$2NiCl_2 + N_2H_4 + 4OH^- \rightarrow 2Ni + N_2 + 4H_2O \quad (1)$$

In the experiment the reducing agent was used in excess of stoichiometric requirement as shown in Table 1. Szu-Han and Chen (2003) reported that for the formation of pure Nickel particles the appropriate amount of reducing agent plays important role.

3 Result and Discussion

3.1 *Effect of Temperature*

We firstly carried out reaction by introducing $NiCl_2$ solution and hydrazine hydrate at 80 °C (Eluri and Paul 2012). Then we conducted experiments at 3 different temperatures 60, 80 and 100 with same concentration and flow rates of reactants. Peak size results show improved particle size with increasing temperatures. Nickel nanoparticles particle size distributions indicate the average particle size of them at 60, 80 and 100 °C are 257,173 and 90 nm. This shows that, the increase in reaction temperature increases the nucleation rate, therefore large numbers of nuclei are formed that inhibit the growth of crystal, and favor production of small nanoparticles (Zeng et al. 2012). Figures 2 and 3 show the effect of temperatures on the particle size distribution of nickel nanoparticles.

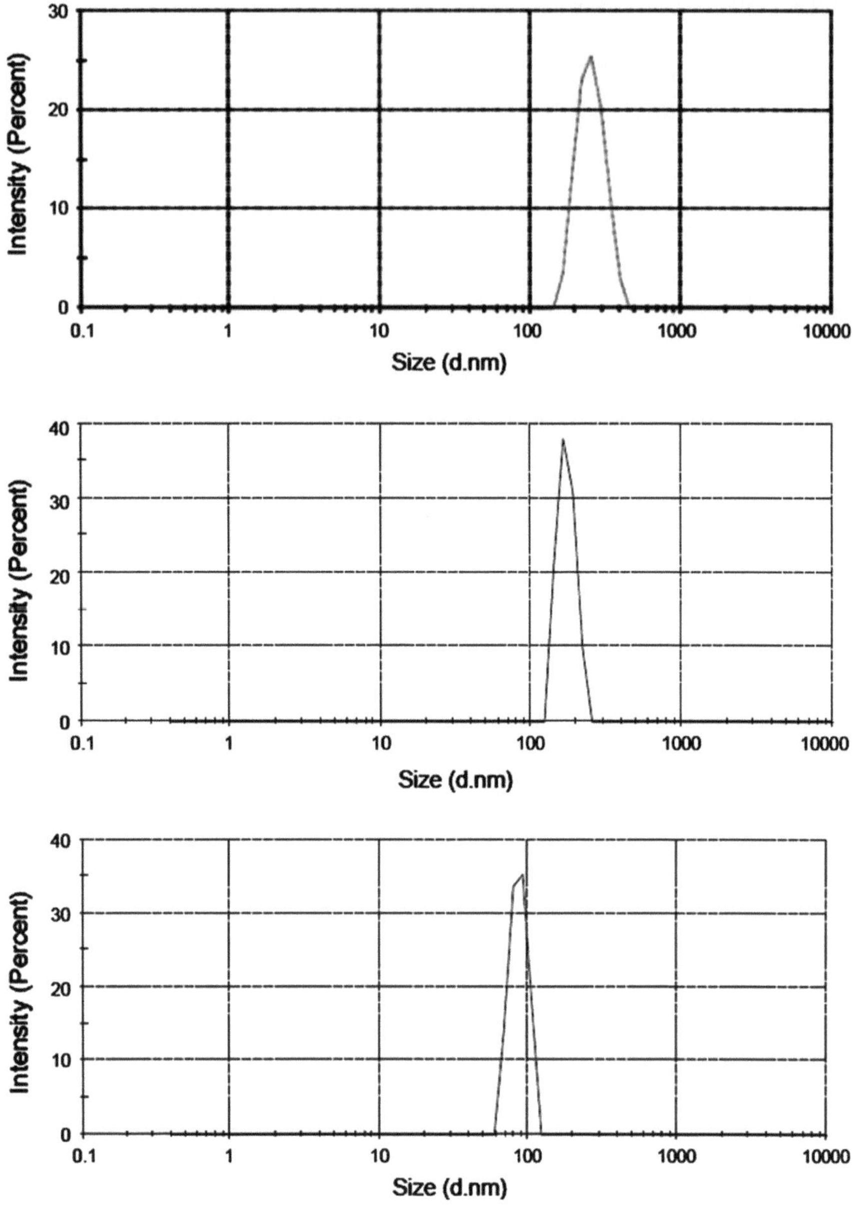

Fig. 2 PSD plots for nickel nanoparticles at different temp

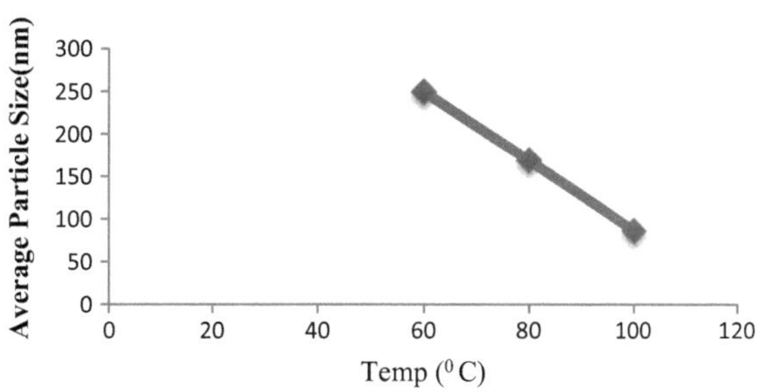

Fig. 3 Effect of temperature on average particle size

3.2 *Effect of Surfactant*

In the experiment conditions two surfactant PVP/CTAB influencing the average particle size of nickel powder, are use to investigate. Surfactant is used to prevent agglomeration of nuclei by sticking on the nickel ion so that particles will not come closer to each other. It decreases the surface tension which leads to restriction of growth of particles (Szu-Han and Chen 2003; Park et al. 2006) (Table 2, Fig. 4).

Table 2 PSD data of nickel nanoparticles without surfactant, with PVP and with CTAB respectively

Sample name	Peak size	Z-average for Ni par ticles (d.nm)	PDI	Intercept	Standard deviation	Intensity (%)
E4	168.8	173.1	0.179	0.865	23.48	100
E6	153.8	159.9	0.324	0.876	26.98	100
E7	155	138.8	0.263	0.836	27.77	100

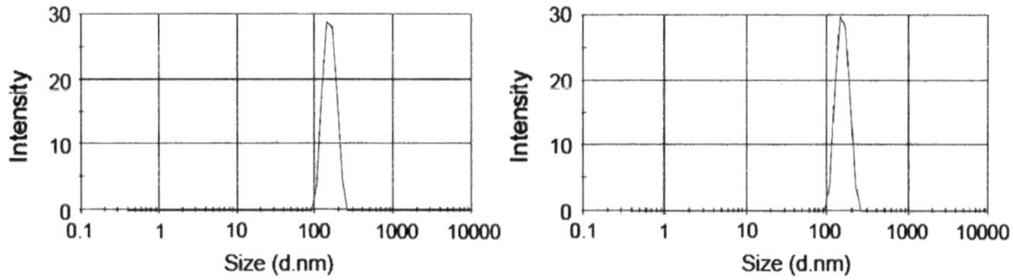

Fig. 4 PSD plots for nickel nanoparticles with PVP and with CTAB respectively

Table 3 PSD data of nickel nanoparticles for different samples

Sample name	Peak size	Z-average for Ni particles (d.nm)	PDI	Intercept	Standard deviation	Intensity (%)
E1	153.9	166.6	0.879	0.920	20.67	100
E2	264.3	257.4	0.186	0.889	58.29	100
E4	168.8	173.1	0.179	0.865	23.48	100
E8	150.8	149.7	0.208	0.859	20.85	100

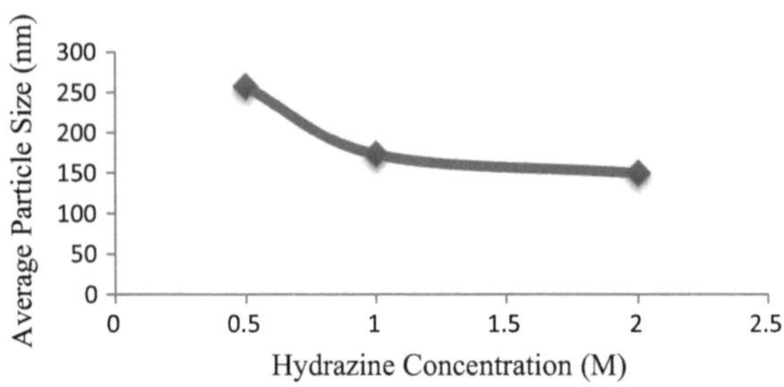

Fig. 5 Effect of hydrazine on average particle size

3.3 *Effect of N_2H_4/Ni^{2+} Molar Ratio*

In the experiment, concentration of hydrazine is varied from 0.5 to 2 M to find the effect of N_2H_4 to Ni^{2+}. It is found that with increase in hydrazine concentration, average particle size is decreased. The rise in hydrazine concentration leads to increase in the reduction rate which favors the production of large number of nuclei and restriction of crystal growth ultimately result into reduction in particle in size (Zhang et al. 2004; Szu-Han and Chen 2003). (Table 3, Fig. 5).

3.4 *SEM Analysis: Nanoparticles Structure*

The typical SEM analysis for the nickel nanoparticles formations are shown in Figure, which indicates that the product consists of nano- sized spheres with spiky surface. Variation in particle size nano sphere is due to temperature variation and surfactant effect (Lei et al. 2015) (Fig. 6).

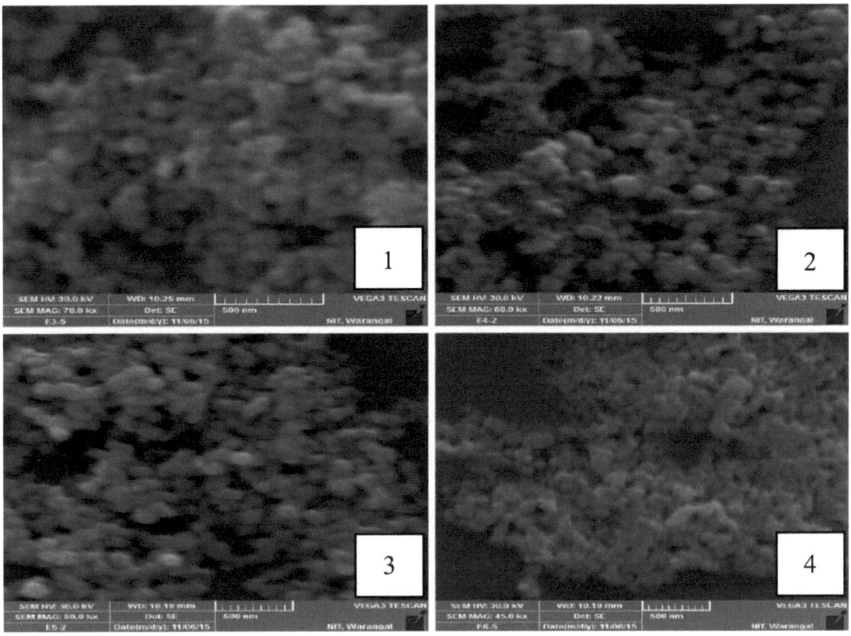

Fig. 6 SEM images of the samples E3, E4, E5 and E6 of Ni particles

4 Conclusion

An attempt was made to produce Ni nanoparticles in continuous spiral microreactor, by reduction method. In this process, the precursor Nickel chloride hexahydrate with solvent ethylene glycol was first injected to capillary along with reducing agent hydrazine monohydrate followed by thermal reduction at temp 80 °C. It was attempted to prepare Nickel in this microreactor of much less nanoparticle size than that produced in conventional reactors and with reduced cost of fabrication of microreactor. As U- junction copper spiral reactor is used, pressure drop reduced and we obtained continuous flow of chemicals. In this work, the effect of type of surfactant, precursor concentration and change in temperature on nanoparticles formation was studied.

References

Baumgard, J., Vogt, A.M., Kragl, U., Jähnisch, K., Steinfeldt, N.: Application of microstructured devices for continuous synthesis of tailored platinum nanoparticles. Chem. Eng. J. **227**, 137–144 (2013)

Eluri, R., Paul, B.: Synthesis of nickel nanoparticles by hydrazine reduction: mechanistic study and continuous flow synthesis. J. Nanopart. Res. **14**, 800–814 (2012)

Gao, H.J., Sohlberg, K., Xue, Z.Q., Chen, H.Y., Hou, S.M., Ma, L.P., Fang, X.W., Pang, S.J., Pennycook, S.J.: Reversible, nanometer-scale conductance transitions in an organic complex. Phys. Rev. Lett. **84**, 1780 (2000)

Lei, X., Srinivasakannan, C., Peng, J., Zhang, D., Chen, G.: Synthesis of nickel nanoparticles by aqueous reduction in continuous flow microreactor. Chem. Eng. Process. **93**, 44–49 (2015)

Pacławski, K., Streszewski, B., Jaworski, W., Luty-Błocho, M., Fitzner, K.: Gold nanoparticles formation via gold(III)chloride complexions reduction with glucose in the batch and in the flow microreactor systems. Colloids Surf A **413**, 208–215 (2011)

Park, J.W., Chae, E.H., Kim, S.H., Lee, J.H., Kim, J.W., Yoon, S.M., Choi, J.-Y.: Preparation of fine Ni powders from nickel hydrazine complex. Mater. Chem. Phys. **97**, 371–378 (2006)

Reetz, M.T., Helbig, W.: Size-selective synthesis of nanostructured transition metal clusters. J. Am. Chem. Soc. **116**, 7401–7402 (1994)

Szu-Han, W., Chen, D.-H.: Synthesis and characterization of nickel nanoparticles by hydrazine reduction in ethylene glycol. J. Colloid Interface Sci. **259**(2), 282–286 (2003)

Wang, C., Zhang, X.M., Qian, X.F., Xie, Y., Wang, W.Z., Qian, Y.T.: Preparation of nanocrystallive nickel powders through hydrothermal reduction method. Mater. Res. Bull. **33** (12), 1747-175 (1998)

Xu, W., Liew, K.Y., Liu, H., Huang, T., Sun, C., Zhao, Y.: Microwave-assisted synthesis of nickel nanoparticles. Mater. Lett. **62**(17–18), 2571–2573 (2008)

Xue, Z.L., Terepka, A.D., Hong, Y.: Synthesis of silver nanoparticles in a continuous flow tubular microreactor. Nano Lett. **4**, 2227–2232 (2004)

Zeng, C., Wang, C., Wang, F., Zhang, Y., Zhang, L.: A novel vapor—liquid segmented flow based on solvent partial vaporization in microstructured reactor for continuous synthesis of nickel nanoparticles. Chem. Eng. J. **204**(206), 48–53 (2012)

Zhang, X., Stefanick, S., Villani, F.J.: Application of microreactor technology in process development. Org. Process Res. Dev. **8**(3), 455–460 (2004)

Zhang, D., Ni, X., Zheng, H., Li, Y., Zhang, X., Yang, Z.: Synthesis of needle-like nickel nanoparticles in water-in-oil microemulsion. Mater. Lett. **59**(16), 2011–2014 (2005)

Optimization of Cassava Pulp Pretreatment by Alkaline Hydrogen Peroxide Using Response Surface Methodology for Bioethanol Production

A. Sudha, V. Sivakumar, V. Sangeetha and K.S. Priyenkadevi

1 Introduction

Environmental impact of fossil fuel usage and the inevitable depletion of reserves have necessitated intense research on the efficient exploitation of agro-based industrial byproducts for the production of bioethanol which is economically viable alternative energy source (Chintala et al. 2013). Cassava pulp, a by-product generated during processing of cassava (*Manihot esculenta* crantz) for starch extraction (Srinivas 2007) has high moisture and organic content. Recently, many studies have examined the potential of utilizing cassava pulp for the production of bioenergy, aroma compounds, mushrooms and organic acids (Pandey et al. 2000). Cassava pulp has starch granules trapped in the fibrous matrix of the non-starch polysaccharides (Kosugi et al. 2009; Hermiati et al. 2012). Acidic and enzymatic hydrolysis has been applied in several studies to enhance the yield of reducing sugar (Euis et al. 2012; Nair et al. 2011), but the fibrous carbohydrates present hinders the extraction of starch without any pretreatment, so pretreatment of cassava pulp before enzymatic hydrolysis can enhance the reducing sugar yield. Alkaline

A. Sudha (✉) · V. Sangeetha · K.S. Priyenkadevi
Department of Food Technology, Kongu Engineering College,
Perundurai, Erode, TN 638052, India
e-mail: sudhaseshu@gmail.com

V. Sangeetha
e-mail: sanpranati@gmail.com

K.S. Priyenkadevi
e-mail: priyenka1989@gmail.com

V. Sivakumar
Department of Chemical Engineering, Alagappa College
of Technology Campus, Anna University, Chennai, India
e-mail: drvsivakumar@yahoo.com

R. Mohan B. et al. (eds.), *Materials, Energy and Environment Engineering*,
DOI 10.1007/978-981-10-2675-1_21

hydrogen peroxide, the pretreatment agent (Amjed et al. 1992; Azzam 1989; Gould and Freer 1984; Hari Krishan et al. 1998) has the advantage of not forming any residues and byproducts, as it breaks down into water and oxygen (Rabelo et al. 2011).

The aim of the study is to examine the effect of alkaline hydrogen peroxide pretreatment (AHP) of cassava pulp on reducing sugar yield upon the enzymatic hydrolysis under microwave energy and develop model to predict reducing sugar yield during AHP pretreatment. Subsequently, fermentation of hydrolysate (with highest reducing sugar yield) with *Saccharomyces cerevisiae* was studied for bioethanol production.

2 Materials and Methods

2.1 Materials

Cassava pulp was obtained from a cassava processing unit located at Rasipuram, Tamilnadu, India. The sample was dehydrated at 60 °C for 12 h, and kept at atmospheric conditions in an air tight container. Alpha amylase (Himedia® Laboratories) was used for saccharification of cassava pulp. The reagents used in the study were of analytical grade.

2.2 Pretreatment

Cassava pulp was pretreated using alkaline hydrogen peroxide. Hydrogen peroxide concentration, pretreatment time and temperature were optimized in the trials. The alkaline peroxide solution was prepared by diluting H_2O_2 in distilled water with pH adjusted to 11 using NaOH. Cassava pulp was treated with the peroxide solution in 0.5 L flask and agitated at 120 rpm in an orbital shaking incubator (Remi CIS-24BL). The pretreated suspension was maintained at 100 °C for 5 min to decompose the residual H_2O_2 before enzymatic hydrolysis.

2.3 Enzyme Hydrolysis

The AHP pretreated suspension of cassava pulp was treated with alpha amylase (5U/g substrate) by adjusting the pH to 6 using 0.1 N HCl and microwave energy at 90 °C (ENERZI Microwave Systems Pvt. Ltd, 2.45 GHz, Max output 800 W) for 1 h and the yield of reducing sugar was analysed in the hydrolysate.

2.4 Fermentation

The hydrolysate from the enzyme hydrolysis of AHP pretreated cassava pulp was subjected to fermentation. The pH of the hydrolysate was adjusted using NaOH until the solution reached 6.5. Commercial baker's yeast *Saccharomyces cerevisiae* was used as an inoculum. *S. cerevisiae* (dried, 5 g) was added into the liquid media consisting of peptone (3.5 g/L), glucose (10 g/L), yeast extract (3 g/L), $MgSO_4{\cdot}7H_2O$ (1.0 g/L), KH_2PO_4 (2.0 g/L), $(NH_4)_2SO_4$ (1.0 g/L) and distilled water before using it as an inoculum in the fermentation process. The culture was propagated aerobically in 500 ml flasks incubated for 15 h at 31 °C in an orbital shaker. For inoculation, 5 ml yeast media was added into flasks containing hydrolysate obtained from AHP pretreated cassava pulp. The fermentation was carried out at 40 °C and pH wasn't controlled. Samples were collected every 4 h, centrifuged (6000 rpm, 10 min) and filtered. The filtrate was used for reducing sugar and alcohol analyses.

2.5 Analytical Methods

The composition analysis of cassava pulp was done according to the standard methods of Association of Official Analytical Chemists (AOAC). Cellulose was estimated by Anthrone method spectrophotometrically. The total reducing sugars released was determined by DNS (3,5-dinitrosalicylic acid) method (Miller 1959). Hemicellulose, lignin and ash contents were estimated according to the methods described by Goering and Van Soest (1970), APHA (2005). The ethanol concentration was determined by dichromate method (Caputi et al. 1968).

2.6 Experimental Design

Response surface methodology (RSM), a mathematical tool was used to develop a multivariate equation influenced by several factors. Box-Behnken design (BBD), a three factor and three level rotatable quadratic design was employed to determine the effects of process variables influencing the reducing sugar yield from pretreatment of cassava pulp on enzymatic hydrolysis under microwave energy. In the study, the influence of three factors namely, H_2O_2 concentration (X_1, 1.0–5.0 %), pretreatment time (X_2, 20–60 min) and temperature (X_3, ambient to 75 °C) were studied. A total run of 15 experiments were carried out with three center points (Kim and Han 2012) and the datas were analysed using Design Expert software (Version 8.0.7.1, Stat-Ease, Inc., USA).

3 Results and Discussion

3.1 *Effect of Solid to Liquid Ratio (SLR)*

The major constituents present in cassava pulp were starch, cellulose and hemicellulose. The composition is same as mentioned elsewhere (Sudha et al. 2015). The pretreatment of cassava pulp using alkaline H_2O_2 was performed at different SLR (g/ml) ranging from 1:15 to 1:30 at 3 % H_2O_2 concentration for 6 h at ambient temperature in order to find the optimum SLR. The pretreated cassava pulp was then subjected to enzymatic hydrolysis under microwave energy. The results obtained are shown in Fig. 1. The results showed that the yield of reducing sugar increases as the solid liquid ratio decreases. The yield of reducing sugar was observed as low when solid liquid ratio was high at SLR of 1:10. It might be due to insufficient volume of solvent for the dispersion of cassava pulp in the medium (Gong et al. 2010). As SLR decreases, the cassava pulp may have better interaction with hydroxyl radicals, which cleaved the aromatic nuclei and propyl side chains of C_α–C_β bonds of complex lignin (Agnemo and Gellerstedt 1979). Experiment with SLR of 1:25 showed maximum yield of reducing sugar, so further study was conducted with this SLR.

3.2 *Model Fitting*

The design matrix for the process parameters and the response, reducing sugar yield were presented in Table 1. The relationship between reducing sugar yield and the process parameters were expressed through the following regression equation in coded units:

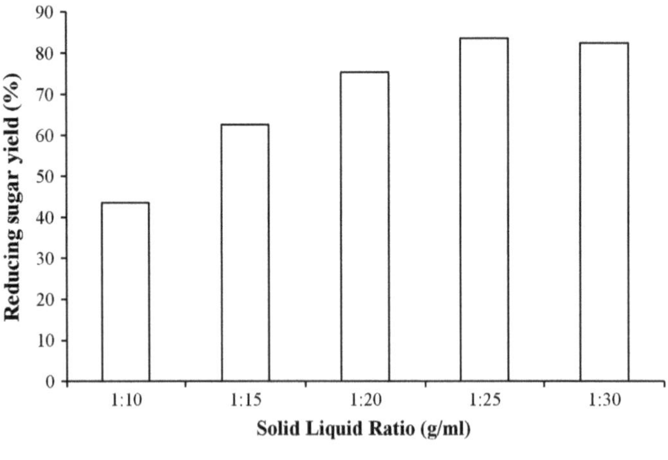

Fig. 1 Effect of SLR on the reducing sugar yield

Table 1 Design of BBD for the independent variables on reducing sugar yield

Run	Experimental factors			Reducing sugar yield (%)	
	H_2O_2 Conc. (%)	Pretreatment time (min)	Temperature (°C)	Experimental	Predicted
1	3	20	30.0	79.3	80.57
2	1	20	52.5	64.5	63.01
3	1	40	30.0	63.9	64.11
4	5	40	75.0	79.3	79.09
5	3	40	52.5	81.4	80.76
6	1	40	75.0	63.2	64.05
7	3	60	75.0	80.8	79.53
8	3	20	75.0	78.3	78.95
9	5	60	52.5	81.8	83.27
10	3	60	30.0	84.9	84.23
11	5	40	30.0	86.2	85.35
12	3	40	52.5	79.7	80.76
13	5	20	52.5	88.3	87.85
14	1	60	52.5	71.4	71.83
15	3	40	52.5	81.2	80.76

$$
\begin{aligned}
Y = 80.76 + 9.07X_1 + 1.06X_2 - 1.58X_3 - 3.35X_1X_2 - 1.55X_1X_3 \\
- 0.77X_2X_3 - 5.97X_1^2 + 1.70X_2^2 - 1.64X_3^2
\end{aligned}
\tag{1}
$$

Analysis of variance (ANOVA) was carried out for the response, reducing sugar yield to check the statistical significance of the mathematical model by F-test (Table 2). From the table, it is observed that model terms are found to be significant with *p*-value of 0.0004 and F value of 41.2, which is below the F critical value.

In general, a regression model is measured to be highly correlated if R^2 value is greater than 0.90 (Qi et al. 2009). Regression coefficient (R^2) for reducing sugar yield was observed to be 0.98 which implies that the model term was statistically significant. The variability of total reducing sugar yield was observed to be around

Table 2 Analysis of variance (ANOVA) for the quadratic model[a]

Source	SS	DF	MS	*F* value	Probability (*P*) > *F*
Model	900.67	9	100.07	41.27	0.0004
Residual (error)	12.12	5	2.42		
Lack of fit	10.39	3	3.46	4.01	0.2058
Pure error	1.72	2	0.86		
Total	912.79	14			

[a]Coefficient of determination (R^2) = 0.98; adjusted R^2 = 0.96; coefficient of variation (CV) 2.0 %; *SS* sum of squares; *MS* mean square; *DF* degree of freedom

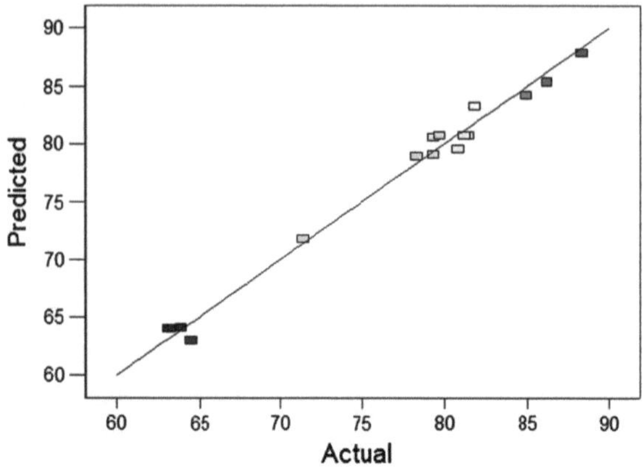

Fig. 2 Observed reducing sugar yield versus the predicted reducing sugar yield

1.4 % of the total variation which might not significantly ascribe the independent variables. The reliability and precision of experiments are analysed by comparing with treatments which shows a relatively low value (2.0 %) of coefficient of variation.

Figure 2 depicts actual reducing sugar yield versus predicted yield for the empirical model. From the figure, the predicted results for the developed model coincides well with the experimental values for the given set of variables considered. The adjusted R^2 value (0.96) (Table 2) reforms the R^2 value for the number of terms and the sample size. A higher significance of the model implies a maximum value of adjusted coefficient of determination (0.96).

Table 3 shows the regression coefficients of the quadratic model of reducing sugar yield for AHP pretreatment of cassava pulp after enzymatic hydrolysis. The coefficient of estimates and the corresponding p values shows that $p < 0.05$ for the independent variables such as H_2O_2 concentration and temperature, the quadratic variable of H_2O_2 concentration and interactive variable of H_2O_2 concentration and

Table 3 Regression coefficients of the regression models of reducing sugar yield

Factor	Coefficient estimate	Standard error	F value	[a]p value
Model	80.76	0.90	41.27	0.0004
X_1	9.07	0.55	271.70	<0.0001
X_2	1.06	0.55	3.72	0.1115
X_3	−1.5875	0.55	8.31	0.0344
$X_1 X_2$	−3.35	0.78	18.51	0.0077
$X_1 X_3$	−1.55	0.78	3.96	0.1031
$X_2 X_3$	−0.77	0.78	0.99	0.3653
X_1^2	−5.97	0.81	54.28	0.0007
X_2^2	1.70	0.81	4.42	0.0894
X_3^2	−1.64	0.81	4.12	0.0980

[a]p value less than 0.05 indicates model terms are significant

pretreatment time, implying that these terms have significant effect on the reducing sugar yield.

3.3 *Effect of Process Variables on Reducing Sugar Yield*

The interaction of the process variables were studied by plotting response surface curves and the optimum level of each factor for maximum response were determined. The response surface of reducing sugar yield versus H_2O_2 concentration and pretreatment time, while temperature is maintained at the central point, is depicted in Fig. 3a. From this figure, it can be seen that reducing sugar yield decreases, when H_2O_2 concentration is at a low level. Significant enhancement in the reducing sugar yield could be obtained by increasing the concentration of H_2O_2. At low concentration of H_2O_2, an insignificant increase in the reducing sugar yield was observed with increase in pretreatment time. The effect might be due to increased oxidative delignification by H_2O_2 at higher concentration and loosening lignocellulosic matrix to release starch granules for improved enzymatic hydrolysis (Silverstein et al. 2007).

The effect of temperature and H_2O_2 concentration on the reducing sugar yield, when pretreatment time is at center point, is shown in Fig. 3b. Figure 3b depicts

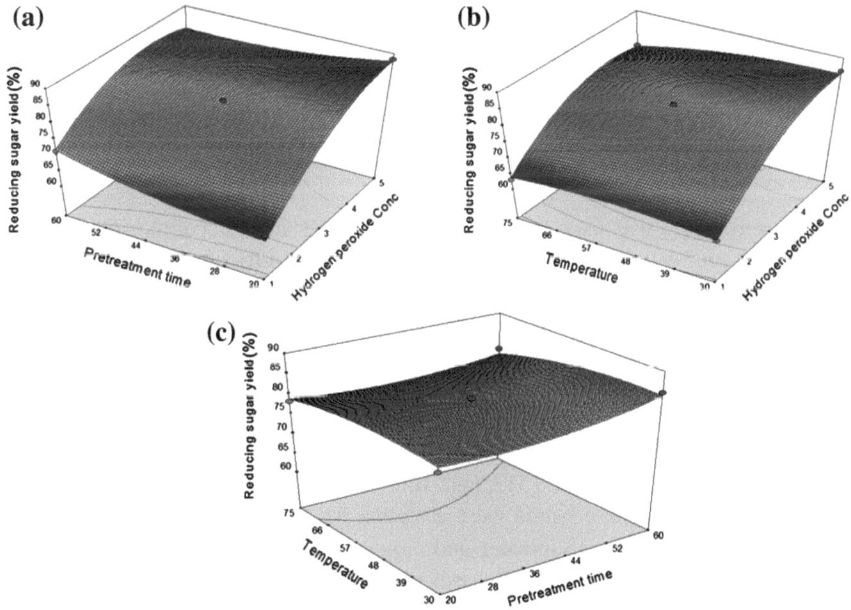

Fig. 3 The surface plots of the effects of H_2O_2 concentration, pretreatment time and temperature on reducing sugar yield: **a** fixed temperature at 52.5 °C; **b** fixed pretreatment time at 40 min; **c** fixed H_2O_2 concentration at 3 %

that, when H_2O_2 concentration is increased keeping the temperature at a low level, then the reducing sugar yield also increases. But when both H_2O_2 concentration and temperature increases, there is a slight decrease in the yield. The oxidative delignification capacity of H_2O_2 decreases at high temperature as it decomposes to water and oxygen (Silverstein et al. 2007).

The effect of pretreatment time and temperature, while H_2O_2 concentration is at center point is shown in Fig. 3c. When the pretreatment time increases with the decrese in temperature, the reducing sugar yield increases (Fig. 3c). But when both pretreatment time and temperature is increased, there is decrease in the reducing sugar yield.

3.4 Confirmation Experiments

The reducing sugar yield at H_2O_2 concentration, pretreatment time and temperature in the range of the experimental design can be predicted using the regression equation. The optimized conditions for maximum reducing sugar yield were $X_1 = 4.9$ %, $X_2 = 20$ min and $X_3 = 42$ °C. To validate the result, experiments were performed at the suggested optimal conditions. Reducing sugar yield of 87.5 % with percentage error of 0.09 % was obtained for the validation experiment. Therefore, the regression model developed was rationally accurate, and the hydrolytic condition of lignocellulosic feedstocks could be predicted and optimized using RSM analysis method.

3.5 Spectral Characterization

FTIR spectroscopy was used to examine the changes in lignocellulosic structures during AHP pretreatment. Figure 4. shows the FTIR spectra of untreated and AHP pretreated cassava pulp. The most illustrative bands can be briefed as follows. The broad absorption at 3340–3412 cm^{-1} is associated to the stretching of H-bonded OH groups, and the absorption at 2900 cm^{-1} to the C–H stretching (Binod et al. 2012). The absorption at 1201 cm^{-1} belonged to the C–O–H in-plane bending at C-6, and the bands at 1237 cm^{-1} was the bending of O–H. The bands at 1080 and 1369 were assigned for unconjugated C=O stretching in acetyl group in the hemicelluloses. The bands at 1512 and 1605 cm^{-1} showed the aromatic skeletal vibrations of lignin. These bands were assigned for C=O stretching in conjugated p-substituted aryl ketones (Subhedar and Gogate 2014).

The native and AHP pretreated cassava pulp had shown different FTIR spectra which specified the structural changes in lignocellulosic material. The high intensity peak at 1156 cm^{-1}, shows the presence of arabinosyl side chains is absent in pretreated sample indicating the partial removal of hemicellulose (Ramadoss and Muthukumar 2015). Lignin, ester groups and xylan shows absorption band at

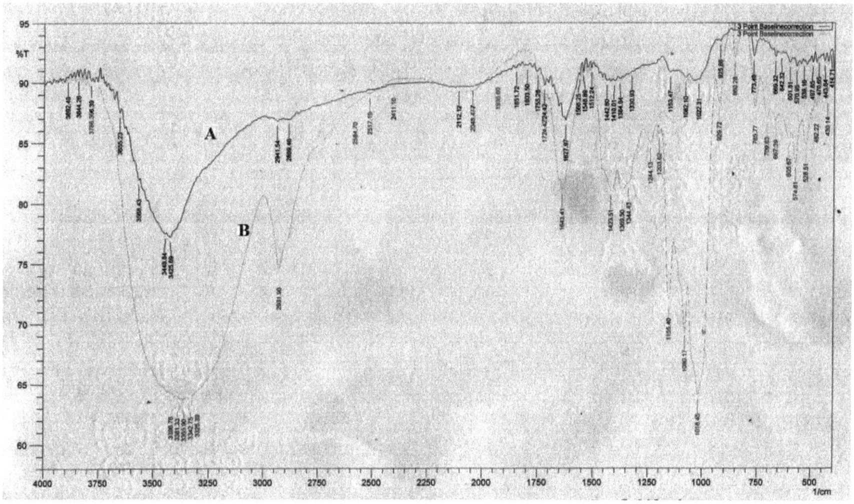

Fig. 4 FTIR spectra of native and AHP pretreated cassava pulp (A-AHP pretreated pulp; B-native pulp)

1238 cm^{-1} due to C=O stretching vibration. Meanwhile, the peculiar absorbance at 1244 cm^{-1} related to an aromatic ring C=O stretch and 1512 cm^{-1} indicates the presence of lignin (Qi et al. 2009). Reduction in the intensity of the peaks in the pretreated sample demonstrates the partial removal of lignin.

3.6 *Fermentation*

The hydrolysate from AHP pretreated cassava pulp containing 12.4 g/L reducing sugar was fermented using *S. cerevisiae* and the ethanol produced after 40 h was 5.32 g/L and it is 83.5 % of theoretical yield. The ethanol produced was estimated to be 0.429 g/g sugar.

4 Conclusions

In the present study, the influence of H_2O_2 concentration, temperature and pretreatment time on reducing sugar yield was examined. The RSM was used to examine the interaction effect of these process parameters. The experimental results showed that H_2O_2 concentration and its interaction with pretreatment time had significant effect on the reducing sugar yield. The regression equation derived by RSM implies the optimal conditions for enhanced reducing sugar yield which was 4.9 % H_2O_2 with pretreatment time of 20 min at 42 °C. On fermentation, ethanol yield of 83.5 % was obtained. Henceforth, our study clearly demonstrates that RSM with BBD can be employed for studying the correlationship among significant factors influencing the reducing sugar yield from pretreated cassava pulp on enzymatic hydrolysis under microwave energy.

References

Agnemo, R., Gellerstedt, G.: The reactions of lignin with alkaline hydrogen peroxide. Part II. Factors influencing the decomposition of phenolic structures. Acta Chem. Scand. **33**, 337–342 (1979)

Amjed, M., Jung, H.G., Donker, J.D.: Effect of alkaline hydrogen peroxide treatment on cell wall composition and digestion kinetics of sugarcane residues and wheat straw. J. Anim. Sci. **70**, 2877–2884 (1992)

APHA: In: Eaton AD, Clesceri LS, Rice EW, Greenberg AE, Franson MAH (eds.) Standard methods for the examination of water and wastewater, 21st edn. American Public Health Association, American Water Works Association, Water Environment Federation, USA (centennial edition) (2005)

Azzam, A.M.: Pretreatment of cane bagasse with hydrogen peroxide for enzymatic hydrolysis of cellulose and ethanol fermentation. J. Environ. Sci. Health Part B **24**, 421–433 (1989)

Binod, P., Satyanagalakshmi, K., Sindhu, R., Janu, K.U., Sukumaran, R.K., Pandey, A.: Short duration microwave assisted pretreatment enhances the enzymatic saccharification and fermentable sugar yield from sugarcane bagasse. Renew. Energy **37**, 109–116 (2012)

Caputi, A.J., Ueda, M., Brown, T.: Spectrophotometric determination of ethanol in wine. Am. J. Enol. Viticulture **19**, 160–165 (1968)

Chintala, R., Wimberly, M.C., Djira, G.D., Tulbure, M.G.: Interannual variability of crop residue potential in the north central region of the United States. Biomass Bioenergy **49**, 231–238 (2013)

Euis, H., Djumali, M., Titi, C.S., Ono, S., Bambang, P.: Potential utilization of cassava pulp for ethanol production in Indonesia. Sci. Res. Essays **7**, 100–106 (2012)

Goering, H.K., Van Soest, P.J.: Forage fiber analyses (apparatus, reagents, procedures, and some applications). USDA agricultural handbook (1970)

Gong, G., Liu, D., Huang, Y.: Microwave-assisted organic acid pretreatment for enzymatic hydrolysis of rice straw. Biosyst. Eng. **107**, 67–73 (2010)

Gould, J.M., Freer, S.N.: High-efficiency ethanol production from lignocellulosic residues pretreated with alkaline H_2O_2. Biotechnol. Bioeng. **26**, 628–631 (1984)

Hari Krishan, S., Prabhakar, Y., Chowdary, G.V., Ayyanna, C.: Simultaneous saccharification and fermentation of pretreated sugar cane leaves to ethanol. Process Biochem. **33**, 825–830 (1998)

Hermiati, E., Azuma, J.I., Tsubaki, S., Mangunwidjaja, D., Sunarti, T.C., Suparno, O., Prasetya, B.: Improvement of microwave-assisted hydrolysis of cassava pulp and tapioca flour by addition of activated carbon. Carbohydr. Polym. **87**, 939–942 (2012)

Kim, I., Han, J.I.: Optimization of alkaline pretreatment conditions for enhancing glucose yield of rice straw by response surface methodology. Biomass Bioenergy **46**, 210–217 (2012)

Kosugi, A., Kondo, A., Ueda, M., Murata, Y., Vaithanomsat, P., Thanapase, W., Arai, T., Mori, Y.: Production of ethanol from cassava pulp via fermentation with a surface-engineered yeast strain displaying glucoamylase. Renew. Energy **34**, 1354–1358 (2009)

Miller, G.L.: Use of dinitrosalicylic acid reagent for determination of reducing sugar. Anal. Chem. **31**, 426–428 (1959)

Nair, M.P.D., Padmaja, G., Moorthy, S.N.: Biodegradation of cassava starch factory residue using a combination of cellulases, xylanases and hemicellulases. Biomass Bioenergy. **35**, 211–1218 (2011)

Pandey, A., Soccol, C.R., Nigam, P., Soccol, V.T., Vandenberghe, L.P., Mohan, R.: Biotechnological potential of agro-industrial residues. II: cassava bagasse. Bioresour. Technol. **74**, 81–87 (2000)

Qi, B., Chen, X., Shen, F., Su, Y., Wan, Y.: Optimization of enzymatic hydrolysis of wheat straw pretreated by alkaline peroxide using response surface methodology. Ind. Eng. Chem. Res. **48**, 7346–7353 (2009)

Rabelo, S.C., Fonseca, N.A., Andrade, R.R., Maciel Filho, R., Costa, A.C.: Ethanol production from enzymatic hydrolysis of sugarcane bagasse pretreated with lime and alkaline hydrogen peroxide. Biomass Bioenergy **35**, 2600–2607 (2011)

Ramadoss, G., Muthukumar, K.: Influence of dual salt on the pretreatment of sugarcane bagasse with hydrogen peroxide for bioethanol production. Chem. Eng. J. **260**, 178–187 (2015)

Silverstein, R.A., Chen, Y., Sharma-Shivappa, R.R., Boyette, M.D., Osborne, J.: A comparison of chemical pretreatment methods for improving saccharification of cotton stalks. Bioresour. Technol. **98**, 3000–3011 (2007)

Srinivas, T.: Industrial demand for cassava starch in India. Starch/Stärke **59**, 477–481 (2007). doi:10.1002/star.200700657

Subhedar, P.B., Gogate, P.R.: Alkaline and ultrasound assisted alkaline pretreatment for intensification of delignification process from sustainable raw-material. Ultrason. Sonochem. **21**, 216–225 (2014)

Sudha, A., Sivakumar, V., Sangeetha, V., Devi, K.P.: Enhancing fermentable sugar yield from cassava pulp for bioethanol production: microwave-coupled enzymatic hydrolysis approach. Bioprocess Biosyst. Eng. **38**, 1509–1515 (2015)

Production of Biodiesel from Neem Oil Feedstock Using Bifunctional Catalyst

N. Samsudeen, Sruti Dammalapati, Souvik Mondal and Lekshmi Unnithan

1 Introduction

With the advancement in technology, the demand for oil and gas is ever increasing and in the near future, the supply may not meet the requirements due to the complete exhaustion of oil. There have also been problems due to pollution caused by burning such oils. Environmental norms have been emphasising on air pollution control measures to solve these problems. This has made man look into other renewable sources that can produce 'green' fuel and promote sustainable development. Biodiesel has proved to be just the right solution (Jimenez et al. 2011; Shu et al. 2010). Biodiesels are monoalkyl esters of long chain fatty acids obtained from vegetable oils and animal fats.

A significant factor affecting the cost of biodiesel production is the selection of raw material. Soybean oil is widely used as a feedstock for biodiesel production. Other vegetable oils include corn oil, canola oil as well as palm oil and tallow, poultry fat and white grease are animal fat-based feed stocks used as a raw material for the biodiesel production. Feed stock like yellow grease and waste cooking oils are being researched with, to promote the reuse of such wastes which can also be economically benefitting by producing biodiesel at lower costs (Farooq et al. 2013).

N. Samsudeen (✉) · S. Dammalapati · S. Mondal · L. Unnithan
Department of Chemical Engineering, National Institute of Technology, Tiruchirappalli, Tiruchirappalli, India
e-mail: samsudeen@nitt.edu

S. Dammalapati
e-mail: sruti1996@gmail.com

S. Mondal
e-mail: souviknitt94@gmail.com

L. Unnithan
e-mail: leku1995@gmail.com

R. Mohan B. et al. (eds.), *Materials, Energy and Environment Engineering*,
DOI 10.1007/978-981-10-2675-1_22

However, the use of refined vegetable oil increases the production cost which make it difficult for biodiesel to compete with diesel and petrol prices in the global market (Shi et al. 2010; Wan et al. 2011). Waste oils and non-edible oils are available cheaply which can reduce the production costs by 60–70 % (Gurunathan and Ravi 2015). Generally, non-edible oils have high free fatty acid (FFA) content along with triglycerides and water. The presence of these may lead to higher purification costs and higher rate of catalytic degradation, thus, producing lesser transesterification reaction yields (Zhang et al. 2010). This is because the FFA molecules react with homogeneous catalysts in alkali catalysed transesterification reactions reducing the effectiveness of the catalyst (Ghadge and Raheman 2006; Juan et al. 2011; Kumar et al. 2007). Several pre-treatment methods such as steam distillation, solvent extraction and acid catalysed esterification procedures have been proposed to reduce the high content of FFA.

Many such alternatives have been suggested to overcome these problems but the most effective method till date is the branched two step production process. In this case, if the FFA content exceeds 1 wt%, the two step process involves an acid catalysed esterification reaction followed by a base catalysed transesterification reaction, thus eliminating extreme temperature and pressure pre-treatment requirements (Koh and Ghazi 2011; Leung et al. 2010). Essentially, water promotes saponification reactions during the transesterification step with the presence of a homogeneous catalyst forming foam which makes the separation process difficult (Kansedo et al. 2009; Yee et al. 2011). To avoid such difficulties, heterogeneous catalysts are used as they can also be recovered easily. Metal oxides, mixed or doped metal oxides of alkali metals are the commonly used heterogeneous catalysts for the transesterification reaction, where triglycerides react with methanol to form FAME and glycerol (Shi et al. 2010). With the need for production of biodiesel at lower costs, research into catalysts required has brought up a new interest in the use of bifunctional catalysts. As these catalysts possess dual functional groups and share properties of acidic and basic catalysts, the transesterification and esterification steps can be combined in a single step (Díaz and Borges 2012).

In this study, pumice, which is a porous volcanic rock mainly consisting of silica and alumina was used as a catalyst. Volcanic rocks are formed as a result of solidification of molten lava, when it comes in contact with the atmosphere. The trapped gases present inside the lava escape, leaving behind a porous rock structure. Being a geological raw material, pumice has many applications in dental, cement, cosmetic, textile and glass industries and it is also easily available inexpensively.

Neem tree is found in abundance in the Indian subcontinent and it is not soil specific. The oil contains three compounds that can be extracted from it. They are nimbin, nimbinin and nimbidin which are mostly responsible for its medicinal value. Neem oil has high FFA value accompanied by moisture content and some other impurities. These two effect the transesterification of glycerides with alcohol using catalyst. This study emphasises the behaviour of crushed pumice stone as a bifunctional catalyst for the production of biodiesel from neem oil. The effect of catalyst, temperature and time on the biodiesel yield was also studied.

2 Materials and Methods

2.1 Preparation of Catalyst

Pumice, a commercially available material was purchased from a local market in Tiruchirappalli, India. Sieved pumice particles of size range (<0.355 mm), obtained after milling using a ball mill, were used to catalyse the esterification and trans-esterification reactions. The pumice was treated with potassium hydroxide for activation and was used as a bifunctional heterogeneous catalyst. The catalyst showed good reusability, activity and also could be recovered easily. The pumice was used to investigate the optimum conditions required for maximum yield of biodiesel from neem oil. Neem oil used was extracted from neem seed kernels, the properties of which are given in Table 1. Ethanol and potassium hydroxide were bought from Merck Specialities Private Ltd., and directly used without further purification.

2.2 Method

The esterification reactions were carried out in a slurry reactor with 250 mL capacity. The reacting system was equipped with a water-cooled reflux condenser, a thermometer as well as a mechanical stirrer. A photograph of the experimental setup is shown in Fig. 1. 100 mL of Neem oil was added into the reactor and the system was heated to 65 °C to remove volatile matter present prior to start the reaction. After the desired temperature was reached, ethanol was added along with the potassium treated catalyst into the reactor. The temperature was maintained at 65 °C throughout the reaction. The reactor was operated for three reaction times (1.5, 2 and 2.5 h), ethanol to oil ratios [18:100, 20:100 and 22:100 (v/v)] and catalyst to oil ratios (10.65, 12.425 and 14.2 g for 100 mL of feedstock), to investigate their effect on the biodiesel yield. The resulting product mixture was centrifuged to separate the spent catalyst. The mixture of biodiesel and glycerol was then transferred to a separating funnel and was left undisturbed for few hours. The product mixture separated into two layers, the top layer being biodiesel and the bottom layer being glycerol. The flash-point, fire-point, calorific value and kinematic viscosity was measured as per standard method. The properties of the biodiesel obtained and standard are given in Table 2.

Table 1 Properties of neem oil used

Property	Standard	Obtained
Kinematic viscosity (c St)	35.83	33.38
Density (g/cc)	0.9	0.85
Flash point (°C)	152	152.4
Calorific value (MJ/kg)	44.65	40.625
Acid value (%FFA)	21 %	23.10 %

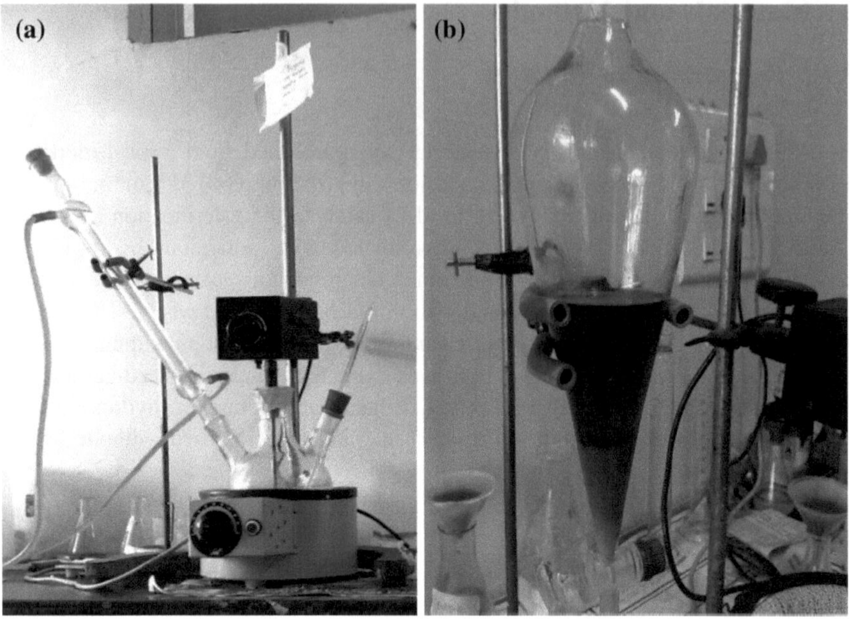

Fig. 1 Photograph of **a** experimental setup and **b** biodiesel and glycerol in the top and bottom layer of the separating funnel

Table 2 Specifications of biodiesel produced

Property	Standard	Obtained
Kinematic viscosity (cSt)	4.937	4.761
Density (g/cc)	0.92	0.932
Flash point (°C)	160	156.8
Calorific value (MJ/kg)	39	38.43
Carbon residue (%)	0.09	0.065

We have adopted the following method to calculate the free fatty acid content in neem oil feedstock: A weighed amount of the oil sample was added into a flask, and it was dissolved in a mixture of diethyl ether and ethanol. Phenolphthalein was used as an indicator. Then, titration with alkali (0.1 M KOH) as titrant gives the acid value of the oil. The acid value was calculated using the Eq. 1.

$$\text{Acid value} = \frac{\mathrm{V} \times \mathrm{N} \times 56.1}{\mathrm{W}} \tag{1}$$

where, V is the volume of the titrant solution (mL) consumed by the oil sample at the equivalence point, W is the weight of the oil sample taken (mg), 56.1 is the molecular weight of the solution employed for titration (mg mmol^{-1}), 0.1

(mmol mL^{-1}) is the KOH concentration used, N is the normality of titrant. The FFA conversion is defined as the ratio of the change of the acid value of the feedstock before and after reaction to the initial acid value as given in Eq. 2.

$$\text{FFA conversion} = \frac{(a_i - a_f)}{a_i} \times 100 \tag{2}$$

where, a_i is the acid value of feed stock (mg KOH/g), a_f is the acid value of the product mixture (mg KOH/g).

2.3 Catalyst Recovery

Heterogeneous catalysts are characterized by their recyclability, thus testing of their reusability is of significance in any process. Catalyst recovery was studied for optimization of the production process. Here, potassium treated pumice (less than 0.355 mm size) has been used as the bifunctional catalyst. Separation of treated pumice from the reaction mixture was done effectively to achieve a recovery of 99.3 %. The product obtained after transesterification reaction was centrifuged (REMI CM8 PLUS) to separate catalyst from biodiesel-glycerol mixture. The separated catalyst was thoroughly washed with detergent to remove residual oil from it. Consequently, it was dried inside a hot air oven to remove moisture present. Dried catalyst obtained could be reused.

3 Results and Discussion

3.1 Effect of Bifunctional Catalyst

Usually, processing of low quality feedstock containing high amounts of FFA and moisture requires high temperature and pressure conditions which are unfavourable. The presence of water in the feedstock results in foaming during the reaction which reduces the efficiency of catalyst, thus affecting the yield. Processing of low quality feedstock commonly requires an additional step of acid esterification before base catalysed transesterification. Bifunctional catalysts simultaneously catalyse both esterification and transesterification reactions which can effectively reduce the process time. In this study, treated pumice has been used as a bifunctional catalyst. Pumice, a volcanic material, being non-toxic, reusable and easily available, makes the process economical.

3.2 *Effect of Process Time*

The transesterification of neem oil using potassium treated pumice catalyst was carried out in the reactor at 65 °C and the process time was varied from 90 to 150 min. It was observed that with increase in process time the biodiesel yield decreases first and then increases as shown in Fig. 2. The maximum biodiesel yield of 90.5 % was achieved at 90 min with 15 % catalyst. From Fig. 2, where yield is plotted against time for different ethanol to oil ratios, it can be seen that the fall in yield is more pronounced between 90 and 120 min than between 120 and 150 min. This could be accounted to the fact that vaporization of ethanol makes its availability insufficient (Gurunathan and Ravi 2015). Ethanol has a boiling point of 78.4 °C and is polar in nature, having hydroxyl (OH-) groups. The high electronegativity of oxygen allows hydrogen bonding to take place with other molecules. At elevated temperatures, polarity of ethanol decreases and this effect is stronger with the increase in process duration.

3.3 *Effect of Catalyst*

Presence of moisture in the feedstock during reaction is known to cause froth formation which reduces the activity of the catalyst. However, the volcanic material is an efficient water resistant catalyst. The catalyst concentration was varied within the range of 15–20 % (weight of catalyst/weight of oil) and its influence on transesterification of neem oil was examined. The maximum yield of biodiesel was obtained when 15 % (w/w) of catalyst was used at 90 min. It has been demonstrated that with an increase in amount of catalyst, the biodiesel yield decreased as shown in Fig. 3.

It is noted that there is a fall in biodiesel yield except for the case corresponding to oil: ethanol ratio (100:22). This could be attributed to the diffusion limitation between oil and ethanol at increased solid interfaces in the reaction mixture at high

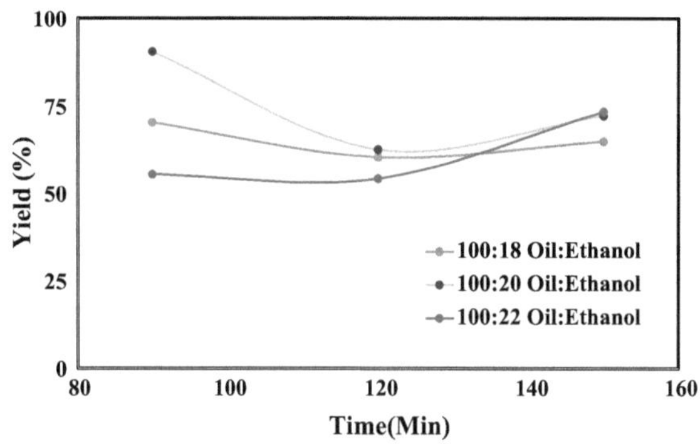

Fig. 2 Shows biodiesel yield as function of process time for oil–ethanol ratio

Fig. 3 Shows yield versus catalyst % (w/w) for different oil: ethanol ratios

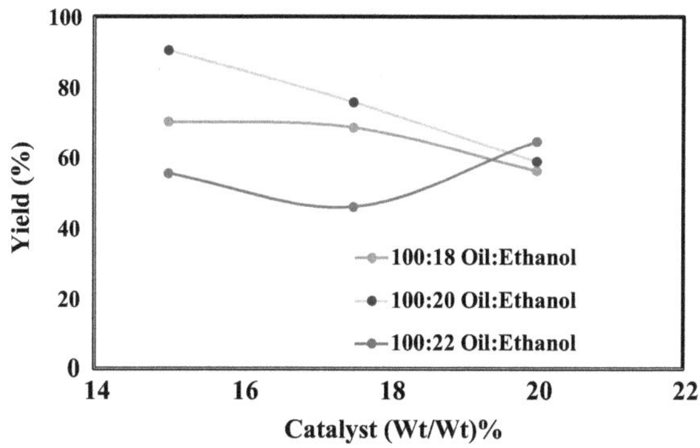

pumice concentration (Gurunathan and Ravi 2015). The anomaly arises due to the fact that higher quantity of ethanol more than compensates the diffusion limitation at oil-ethanol interface.

3.4 *Effect of Ethanol to Oil Ratio*

In this study, ethanol was used instead of conventionally used methanol because ethanol has better solvent properties as compared to methanol. Excess ethanol is required to shift the equilibrium in the forward direction for the maximum formation of ethyl esters (Gurunathan and Ravi 2015). The oil: ethanol ratio was varied from 100:18 to 100:22 (volume of oil/volume of ethanol) to investigate the effect of alcohol on biodiesel production. The maximum yield of biodiesel was obtained at 100:20 (oil: ethanol) ratio. From Fig. 4, it was observed that the biodiesel yield increased from 100:18 to 100:20 oil: ethanol ratio. This is because heterogeneous catalysed transesterification reaction requires higher amount of oil; methanol ratio. A maximum yield was, hence, found to occur at an optimum

Fig. 4 Shows yield versus ethanol: oil ratios for amounts of catalyst

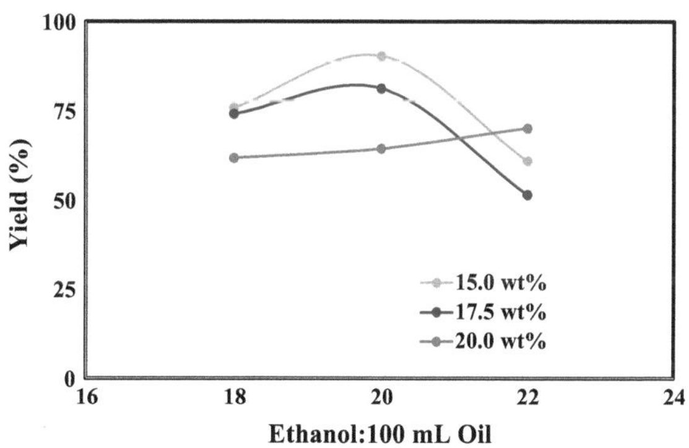

operation condition of 100:20 oil: ethanol ratio. Thereafter, there was a fall in yield. This might be due to the theoretical fact that ethanol accumulation occurs on the surface of the potassium treated pumice.

4 Conclusion

Pumice is an economically cheap material and has the bifunctional characteristics that are required for catalyzing esterification and transesterification reactions simultaneously. The effect of pumice on the production of biodiesel from neem oil at various process conditions (process time, oil–methanol ratio and oil–catalyst ratio) was investigated in this study. The maximum biodiesel yield of 90.5 % was achieved for the optimum conditions of 15 % (wt/wt) catalyst with 100:20 oil to ethanol ratio and a process time of 90 min respectively. However, further research to increase the biodiesel yield with the use of bifunctional catalyst is necessary. Finally, it is concluded that the use of bifunctional catalyst for the production of biodiesel can be made more economical by using low grade feedstock.

References

Díaz, L., Borges, M.E.: Low-quality vegetable oils as feedstock for biodiesel production using k-pumice as solid catalyst. tolerance of water and free fatty acids contents. J. Agricult. Food Chem. **60**(32), 7928–7933 (2012)

Farooq, M., Ramli, A., Subbarao, D.: Biodiesel production from waste cooking oil using bifunctional heterogeneous solid catalysts. J. Clean. Prod. **59**, 131–140 (2013)

Ghadge, S.V., Raheman, H.: Process optimization for biodiesel production from mahua (Madhuca indica) oil using response surface methodology. Bioresour. Technol. **97**(3), 379–384 (2006)

Gurunathan, B., Ravi, A.: Process optimization and kinetics of biodiesel production from neem oil using copper doped zinc oxide heterogeneous nanocatalyst. Bioresour. Technol. **190**, 424–428 (2015)

Jimenez, L.A., Jimenez-Morales, I., Santamaría-Gonzalez, J., Maireles-Torres, P.: Biodiesel production from sunflower oil by tungsten oxide supported on zirconium doped MCM-41 silica. J Mol. Catal. A Chem. **335**(1–2), 205–209 (2011)

Juan, J.C., Kartika, D.A., Wu, T.Y., Hin, T.-Y.Y.: Biodiesel production from jatropha oil by catalytic and non-catalytic approaches: an overview. Bioresour. Technol. **102**(2), 452–460 (2011)

Kansedo, J., Lee, K.T., Bhatia, S.: Biodiesel production from palm oil via heterogeneous transesterification. Biomass Bioenergy **33**(2), 271–276 (2009)

Koh, M.Y., Ghazi, T.I.M.: A review of biodiesel production from Jatropha curcas L. oil. Renew. Sustain Energy Rev. **15**(5), 2240–2251 (2011)

Kumar, T.A., Kumar, A., Raheman, H.: Biodiesel production from jatropha oil (Jatropha curcas) with high free fatty acids: an optimized process. Biomass Bioenergy **31**(8), 569–575 (2007)

Leung, D.Y.C., Wu, X., Leung, M.K.H.: A review on biodiesel production using catalyzed transesterification. Appl. Energy **87**(4), 1083–1095 (2010)

Shi, W., He, B., Ding, J., Li, J., Yan, F., Liang, X.: Preparation and characterization of the organic–inorganic hybrid membrane for biodiesel production. Bioresour. Technol. **101**(5), 1501–1505 (2010)

Shu, Q., Nawaz, Z., Gao, J., Liao, Y., Zhang, Q., Wang, D., Wang, J.: Synthesis of biodiesel from a model waste oil feedstock using a carbon-based solid acid catalyst: Reaction and separation. Bioresour. Technol. **101**(14), 5374–5384 (2010)

Wan, O., Nadyaini, W.N., Amin, Saidina, Aishah, N.: Optimization of heterogeneous biodiesel production from waste cooking palm oil via response surface methodology. Biomass Bioenergy **35**(3), 1329–1338 (2011)

Yee, K.F., Wu, J.C.S., Lee, K.T.: A green catalyst for biodiesel production from jatropha oil: Optimization study. Biomass Bioenergy **35**(5), 1739–1746 (2011)

Zhang, J., Chen, S., Yang, R., Yan, Y.: Biodiesel production from vegetable oil using heterogenous acid and alkali catalyst. Fuel **89**(10), 2939–2944 (2010)

Influence of Feed Vapour Fraction on the Performance of Direct Methanol Fuel Cell

Vineesh Ravi and Shiny Joseph

1 Introduction

Direct methanol fuel cells (DMFC) are promising power sources for portable power devices and transportation. It has received much attention because of its simplicity, higher energy density, quick refueling and easy transportation of fuel. The commercialization of DMFC is still an uphill task because of the problems associated with methanol crossover, slow reaction kinetics, cathode flooding and heat management.

In DMFC, methanol solution is fed to the anode where methanol and water molecules react to form electrons and protons by the electrochemical reaction (Xianglin et al. 2008)

$$CH_3OH + H_2O \rightarrow 6H^+ + 6e^- + CO_2$$

The generated protons reach to the cathode through the electrolyte and the electrons through the external circuit. In cathode, air combines with protons and electrons to form water.

$$\frac{3}{2}O_2 + 6H^+ + 6e^- \rightarrow 3H_2O$$

The overall reaction

$$CH_3OH + \frac{3}{2}O_2 \rightarrow 2H_2O + CO_2$$

V. Ravi · S. Joseph (✉)
National Institute of Technology Calicut, Kozhikode, India
e-mail: shiny@nitc.ac.in

R. Mohan B. et al. (eds.), *Materials, Energy and Environment Engineering*,
DOI 10.1007/978-981-10-2675-1_23

Generally liquid methanol with a lower concentration of methanol is fed on the anode side (active or passive) which reduces the methanol crossover, but it decreases the throughput power from the cell. The present study deals with the feasibility of methanol in lower concentration as well as in the pure form in the vapour phase as well as in a different fraction of vapour to the liquid ratio as fuel on the anode side. The study conducted by Shukla et al. (1995) used a quartz vaporizer for producing methanol vapour from 1 M solution. They reported that the performance of the DMFC using vapour feed is better than that of liquid feed. Scott et al. (1999) conducted flow visualisation studies on carbon dioxide evolution in liquid as well as vapour feed and also compared the performance. They observed a current density of 1.7 W/cm^2 for vapour feed compared to 0.12 W/cm^2 of liquid feed under the same operating conditions. Transient studies were conducted by Kallo et al. (2004a, b) in a vapour feed fuel cell. Sumio et al. (2011) conducted studies on methanol crossover in vapour feed and concluded that lower concentration of methanol solution gives a better result compared to higher concentration. They also compared the methanol crossover with liquid feed and CO_2 production on the cathode side is found to be half with vapour feed under same operating conditions. Nakagawa et al. (2012) used a PCP for generation of methanol vapour to feed DMFC and studied the effects of different variables like methanol concentration, temperature, methanol flow rate, air flow rate and humidity. Faghri and co-workers (2007, 2010, 2011, 2012) did extensive work on passive DMFC modelling and fuel delivery system. Their work focused on neat methanol in the vapor phase as feed using membranes. The developed steady state and transient model and found that the response time varies from 10 to 100 s if we change the current density from low to high.

From the literature cited it is seen that the work focused only on either liquid feed or vapour feed in active or passive feed operation. No attempt was made to study the effects of the proportion of vapour/liquid fraction as feed. The present study focuses on the effect of feed fraction, methanol concentration and use of neat methanol in the vapour phase as active feed on the performance of DMFC.

2 Experimental Set Up

The experimental setup consists of DMFC unit, cathode humidifier, electronic load bank, peristaltic pump and a temperature controller for keeping the cell temperature constant. The cell is a single fuel cell with an active area of 25 cm^2 equipped with serpentine channels (1 mm width, 1 mm depth), carbon cloth as diffusion layers with a platinum loading of 4 mg/cm^2 on cathode side and 4 mg/cm^2 of Pt/Rh on anode side, Nafion 117 membrane, gold plated current collecting plates and aluminum end plates. Methanol in the reservoir is supplied to the anode channels using a peristaltic pump (0.1–99.9 ml/min) via a 3 mm diameter silicon tubing and humidified air to the cathode side. In the liquid feed operation, the output from the peristaltic pump is directly fed to the anode side of the fuel cell. During vapor feed

(a) (b)

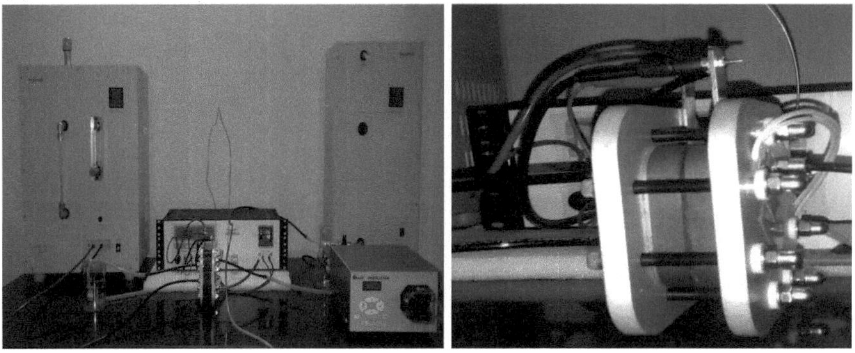

Fig. 1 **a** Experimental set up of DMFC, **b** single cell DMFC

operation, the methanol solution is pumped to the vaporizer which vaporizes the feed and sends to the anode inlet. Figure 1a shows the experimental setup of DMFC used in this study and Fig. 1b is the single cell DMFC.

3 Results and Discussion

3.1 Effect of Feed Vapor Fraction

Experiments were conducted for liquid feed, 25 % vapor feed, 50 % vapor feed, 75 % vapor feed and 100 % vapor feed under the conditions of 2 M methanol concentration, 1 ml/min methanol flow rate, 3 l/min air flow rate, cell temperature is kept constant at 60 °C and a relative humidity of 90 %. The Fig. 2a, b shows that the increase in the vapor fraction of the feed from liquid to total gas feed increases the performance of the cell. For liquid feed, the power density is 0.012 mW/cm^2 compared to 0.04 mW/cm^2 in the case of vapor feed. The increase in performance is mainly due to increase in reaction rates and power density outputs, high mass transfer rates, lesser methanol crossover and reduction in CO production which poisons the catalyst layer.

3.2 Effect of Methanol Concentration

In order to find out the optimum concentration, experiments were conducted for different methanol concentrations of 1, 2 and 3 M under the same operating conditions. Figure 3b shows that a maximum power density of 0.04 W/cm^2 is obtained with 2 M methanol solution as vapor feed. From the analysis, it is found that with increasing concentration, there is a decline in the performance due to increase in methanol crossover.

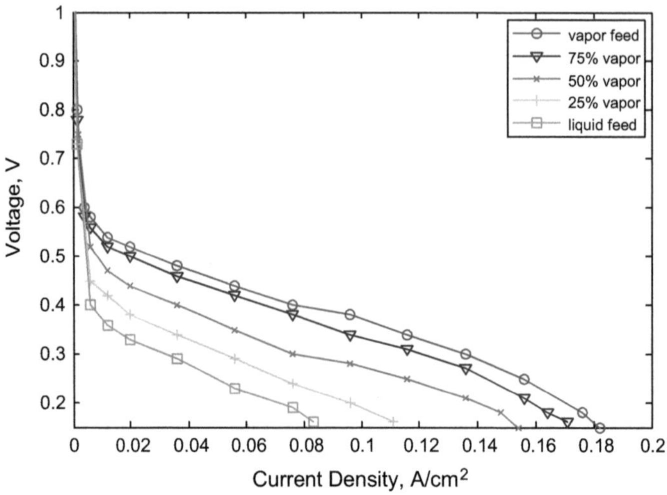

(a) Experimental polarization curve for varying feed vapor fraction

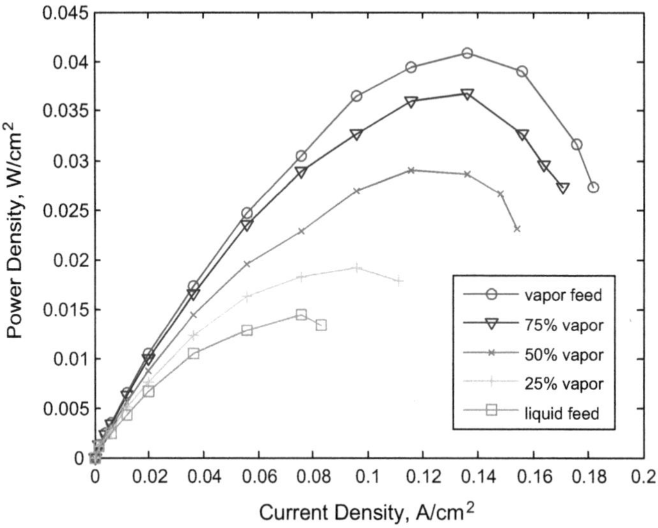

(b) Experimental power density curve for varying feed vapor fraction

Fig. 2 **a** Experimental polarization curve for varying feed vapor fraction. **b** Experimental power density curve for varying feed vapor fraction

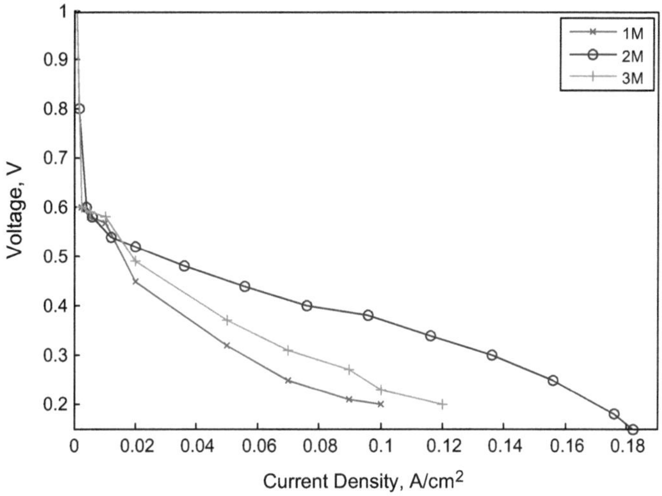

(a) Experimental polarization curves for varying methanol concentration

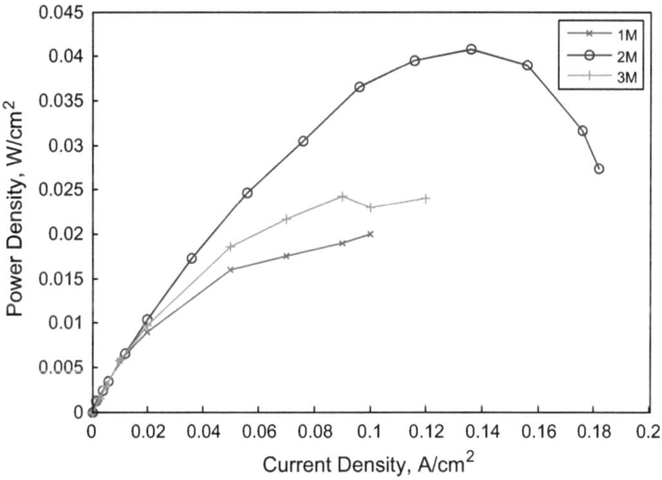

(b) Experimental power density curves for varying methanol concentration

Fig. 3 **a** Experimental polarization curves for varying methanol concentration. **b** Experimental power density curves for varying methanol concentration

3.3 *Comparison with Neat Methanol*

Figure 4a, b shows the comparison of the performance of DMFC with 2 M solution of methanol and neat methanol as vapour feed. Experiments were conducted under

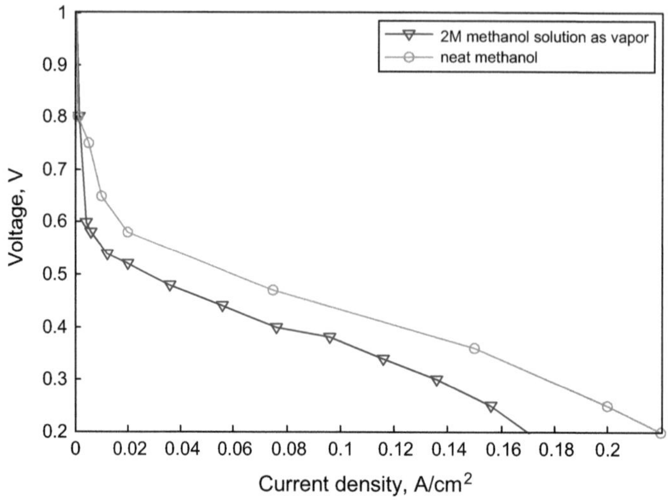

(a) Experimental polarization curve of neat methanol on comparison with 2M methanol solution

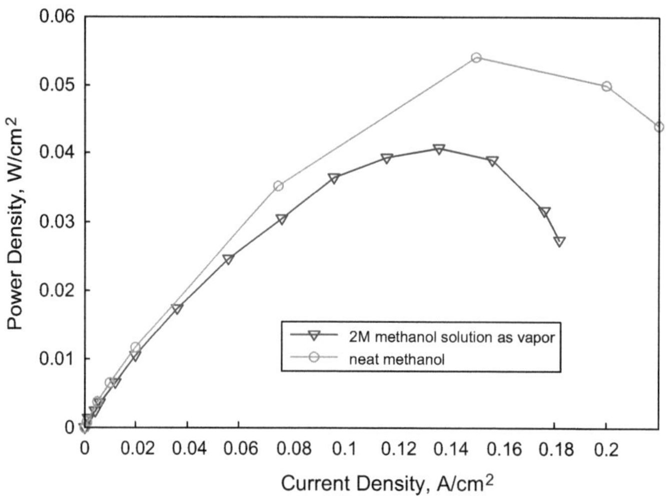

(b) Experimental power density curve of neat methanol on comparison with 2M methanol solution

Fig. 4 **a** Experimental polarization curve of neat methanol on comparison with 2 M methanol solution. **b** Experimental power density curve of neat methanol on comparison with 2 M methanol solution the same operating conditions. A maximum power density of 0.054 W/cm^2 is obtained with neat methanol as vapour feed compared to 0.04 W/cm^2 when we use 2 M methanol solution as vapor as a feed which happened to be the optimum out of all the previous experimental values. The use of neat methanol reduces the cathode flooding since the water required for the anode reaction is coming from cathode via diffusion through the membrane. On decreasing the relative humidity from 100 % to lower value results in lower cell performance due to lack of water for anode reaction. So it is recommended to use relative humidity between 80 and 100 % for good results.

4 Conclusion

A single fuel cell of active area 25 cm^2 is investigated for the effect of vapor fraction on the performance of fuel cell. The results show that the increase in vapor fraction of feed increases the performance of the cell due to high diffusivity of methanol vapor, high activity of methanol vapor which increases the reaction rate and low permeability of methanol which decreases the methanol crossover. Even though it gives higher performance with an increase in vapor fraction, there may be a chance that during longer operation the membrane may get dry out and lower the performance. The tests conducted with a change in concentration shows that 2 M concentration is the optimum concentration and further increase causes methanol crossover in the cell, which decreases the performance of the cell. On using neat methanol as feed, a maximum power density of 0.054 W/cm^2 is obtained.

References

Bahrami, H., Faghri, A.: Start-up and steady-state operation of a passive vapor-feed direct methanol fuel cell fed with pure methanol. Int. J. Hydrogen Energy **37**, 8641–8658 (2012)

Chao, X., Faghri, A., Li, X.: Improving the water management and cell performance for the passive vapor-feed DMFC fed with neat methanol. Int. J. Hydrogen Energy **36**, 8468–8477 (2011)

Guo, Z., Faghri, A.: Vapor feed direct methanol fuel cells with passive thermal-fluids management system. J. Power Sources **167**, 378–390 (2007)

Kallo, J., Kamara, J., Lehnert, W., Helmolt, R.: Cell voltage transients of a gas-fed direct methanol fuel cell. J. Power Sources **127**, 181–186 (2004a)

Kallo, J., Lehnert, W., Helmholtz, R.: Conductance and methanol crossover investigation of Nafion membranes in direct methanol fuel cell. J. Electrochem. Soc. **150**(6), 765–769 (2004b)

Nakagawa, N., Tsujiguchi, T., Sakurai, S., Aoki, A.: Performance of an active direct methanol fuel cell fed with neat methanol. J. Power Sources **219**, 325–332 (2012)

Scott, K., Taama, W.M., Argyropoulos, P.: Engineering aspects of the direct methanol fuel cell system. J. Power Sources **79**, 43–59 (1999)

Shukla, A.K., Christensen, P.A., Hamnett, A., Hogarth, M.P.: A vapour-feed direct-methanol fuel cell with proton-exchange membrane electrolyte. J. Power Sources **55**, 87–91 (1995)

Sumio, K., Noboru, K., Shaharom, N.H.: Methanol crossover reduction by vapor feed direct methanol fuel cell. Int. J. Chem. Environ. Eng. **2**, 1–6 (2011)

Xianglin, L., Yalin, H., Benhao, Y., Zheng, M., Xiaoyue, L.: Exergy flow and utilization of direct methanol fuel cells based on a mathematical model. J. Power Sources **178**(1), 344–352 (2008)

Xu, C., Faghri, A., Li, X.: Development of a high performance passive vapor-feed DMFC fed with neat methanol. J. Electrochem. Soc. **157**, B1109–B1117 (2010)

Electrocatalytic Borohydride Oxidation by Supported Tungsten Oxide Nanoclusters Towards Direct Borohydride Fuel Cells

Aarti Tiwari and Tharamani C. Nagaiah

1 Introduction

Sodium borohydride has emerged as an aqueous fuel in the past decade, revolutionizing the achievable power output from fuel cells. A recent sub-type of fuel cell namely, direct borohydride fuel cell (DBFC) has been receiving constant attention due to promising specific capacity (5.67 A h g^{-1}), high theoretical open circuit voltage (1.64 V) and lofty specific energy density (9.3 W h g^{-1}) with 10.6 wt% hydrogen content as quoted by Leon et al. (2006) and Tamasiunaite et al. (2014). Aqueous alkaline borohydride serves as the fuel in DBFCs, besides the final product, borate can be potentially recycled back to borohydride encouraging further development due to its self-sustainability as shown by Kojima and Haga (2003). Additionally it practically delivers superior specific energy density versus methanol fuel and also high power density when compared with the direct methanol fuel cell (DMFC) quoted by Ma et al. (2010), the preferred alternative amongst acid and alcohol fuel cells. Even though DBFCs were first introduced by Indig and Synder (1962) but research in this arena remained stagnant till Amendola et al. (1999) demonstrated the working of DBFCs and further more since 2000. DBFC development was neglected on account of exhaustive focus being laid upon polymer electrolyte fuel cells (PEFCs). But since PEMFCs are powered by hydrogen fuel, which is difficult to handle and store besides requiring a pure feed to avoid catalyst poisoning presently researchers are seeking for alternatives.

As represented in Fig. 1, DBFC is fed with an aqueous alkaline solution of borohydride fuel which undergoes oxidation (Eq. 1) at the anode catalyst. The resulting electrons follow an external pathway, driving applied load to reach cathode where it facilitates oxygen reduction (Eq. 2) and the anion exchange

A. Tiwari · T.C. Nagaiah (✉)
Department of Chemistry, Indian Institute of Technology Ropar,
Rupnagar, Punjab 140001, India
e-mail: tharamani@iitrpr.ac.in

R. Mohan B. et al. (eds.), *Materials, Energy and Environment Engineering*,
DOI 10.1007/978-981-10-2675-1_24

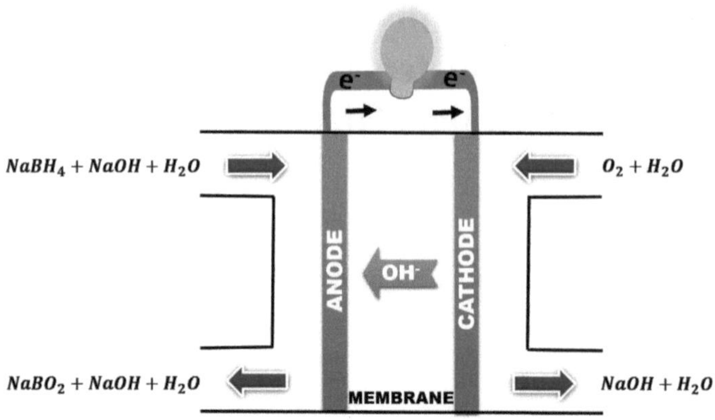

Fig. 1 Schematic representation of direct borohydride fuel cell (DBFC)

membrane shuttles the resulting hydroxide ions to anode forming the borate end product. As shown by Santos and Sequeira (2011) the overall reaction is given as (Eq. 3);

$$\textbf{Anode}: \mathbf{BH_4^- + 8\,OH^- \rightarrow BO_2^- + 6\,H_2O + 8e^-} \quad E^\circ = -1.24\ \text{V (vs. SHE)} \tag{1}$$

$$\textbf{Cathode}: \mathbf{2\,O_2 + 4\,H_2O + 8\,e^- \rightarrow 8\,OH^-} \quad E^\circ = +0.40\ \text{V (vs. SHE)} \tag{2}$$

$$\textbf{Overall}: \mathbf{BH_4^- + 2\,O_2 \rightarrow BO_2^- + 2\,H_2O} \quad E^\circ = +1.64\ \text{V (vs. SHE)} \tag{3}$$

Borohydride oxidation occurring at the DBFC anode is a complex reaction proceeding via multiple steps in an alkaline solution to prevent fuel hydrolysis. The complete borohydride oxidation without contribution from concomitant side reactions ideally yields 8 electrons. This complexity in the 8-electron oxidation of the borohydride fuel, poses difficulty in establishing the mechanism involved and hence rationalizing an appropriate active catalyst is a challenging task. The complete hydrolysis of borohydride is represented in Eq. 4, the extent of which is governed by the pH of the electrolyte as shown by Liu et al. (2009) but rather, is an integral part in the prevalent circumstances;

$$\mathbf{BH_4^- + 2\,H_2O \rightarrow BO_2^- + 4\,H_2} \tag{4}$$

Therefore, the need of the hour is an oxidatively competent, hydrolysis sturdy and concomitantly cost-effective anode catalyst for DBFCs to aid its competitive commercialization. In this regard Pt and Au have been the most intensively investigated systems as reviewed by Liu and Li (2009). Alternatively Jiménez et al. (2012) have also reviewed activity of other noble and transition metal catalysts like Ag, Ru, Pd, Ni, Zn, Os, Cu and hydrogen storage alloys. The noble metals were usually found to be inactive towards undesired hydrolysis but simultaneously suffer from incompetent kinetics urging further exploration of possible anodic electrocatalyst. To the best of our knowledge, this is the first instance of exploring nitrogen containing mesoporous carbon supported tungsten-oxide (WOx/MNCx) as a potential anode catalyst for DBFCs. We have synthesized WOx/MNCx catalyst via hard-template methodology followed by its physicochemical characterization and electrochemical activity towards borohydride oxidation.

2 Experimental

The WOx/MNCx composite catalyst was synthesized as per our previous work (Tiwari et al. 2015), briefly 4 g P123 triblock copolymer (P123; Aldrich, MW = 5800, EO20PO70EO20) structure-directing agent was dissolved in 30 g water under rapid stirring. Further 2.5 mL (4.7 wt%) of aqueous $Na_2WO_4.2H_2O$ was added and stirred for 1 h followed by addition of 2 M HCl (120 g) and 9.1 g of TEOS and continued stirring at 40 °C for 24 h. The obtained product was filtered, dried and calcined at 550 °C yielding tungsten oxide coated SBA-15 (WOx/SBA-15). The calcined product (1 g) was refluxed with ethylenediamine (4.5 g) and CCl_4 (11 g) mixture for 6 h. This refluxed product was further pyrolyzed under inert atmosphere at 600 °C and treated with NaOH solution (2.5 wt%) under refluxation and subsequently washed with a 1:1 mixture of ethanol/water till the filtrate became neutral. The dried product obtained was WOx/MNCx catalyst.

SEM analysis was performed using FEI Quanta 3D for morphology analysis and XRD was performed over D8 Advance Bruker AXS diffractometer with CuKα radiation at λ = 1.5418 A° in θ–2θ geometry with a position-sensitive detector for low-angle powder XRD patterns. Electrochemical analysis was performed over Autolab 302 N modular potentiostat/galvanostat operated by the Nova 1.11 software. Glassy carbon disk electrode (Ø 3 mm), Pt mesh and Hg/HgO were employed as the working, counter and reference electrodes respectively in a single-compartment three-electrode electrochemical glass. The catalyst slurry was prepared by dispersing 1.25 mg of WOx/MNCx in 1 mL deionised water (18 MΩ Millipore water) followed by ultrasonication for 30 min. Linear sweep voltammetry (LSV) was carried out in different concentration of $NaBH_4$ in 2.0 M NaOH purged with N_2.

3 Results and Discussion

The WOx/MNCx electrocatalyst was characterized by using small-angle XRD measurements as shown in Fig. 2a. The peaks obtained could be ascribed to (100), (110) and (200) planes belonging to p6mm space group (Zhao et al. 1998). These peaks correspond well with that of SBA-15 template reflecting the hexagonally ordered mesoporous structure of the prepared composite catalyst. The inset of Fig. 2a depicts the energy dispersive X-ray (EDAX) analysis spectra confirming the

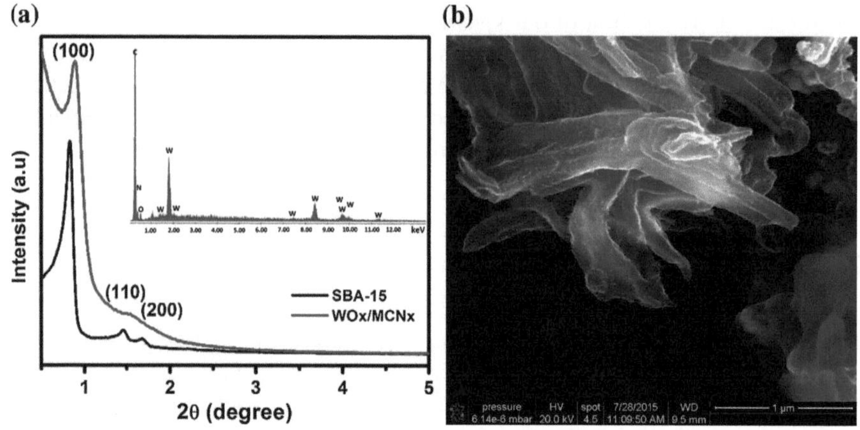

Fig. 2 **a** Small angle XRD and EDAX spectra (*inset*) of WOx/MNCx; and **b** SEM image of WOx/MNCx catalyst

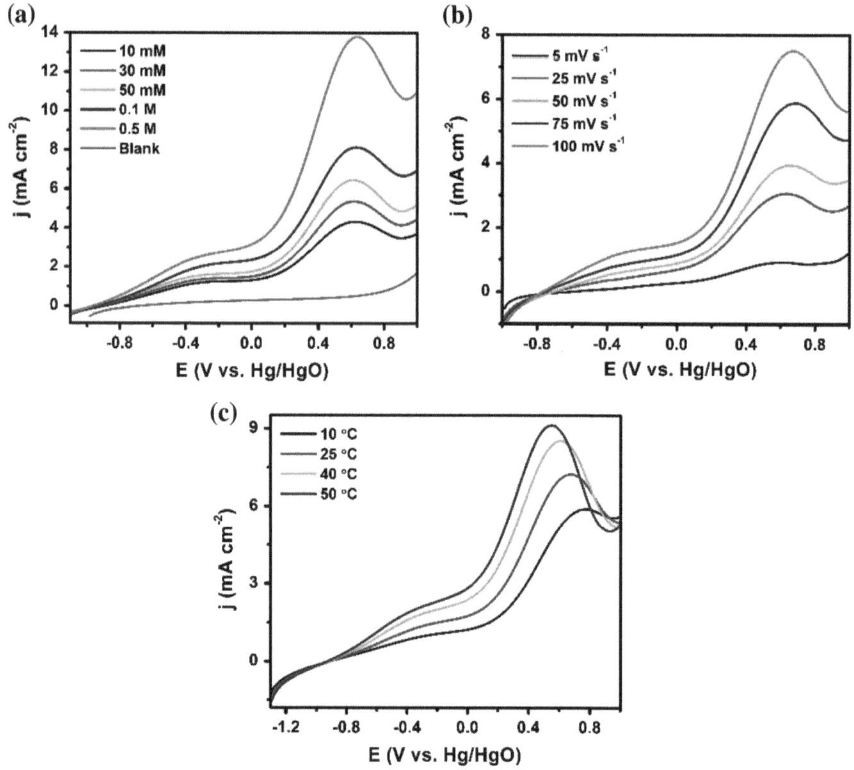

Fig. 3 Linear sweep voltammograms of WOx/MNCx at **a** different concentrations of $NaBH_4$ (10 mM to 0.5 M) in 2 M NaOH; **b** different scan rates (5–100 mV s^{-1}) and **c** different temperatures (10–50 °C) in alkaline (2 M NaOH) 0.1 M $NaBH_4$ electrolyte at 50 mV s^{-1} under N_2 atmosphere. *CE:* Pt mesh, *RE:* Hg/HgO

presence of tungsten oxide in the supported catalyst synthesized. The microstructural appearance of the as-synthesized WOx/MNCx catalyst is illustrated in Fig. 2b, showing mesoporous structure with WOx clusters over it. The WOx/MNCx electrocatalyst prepared was analysed towards its borohydride oxidation ability by performing LSV measurements in deaerated aqueous alkaline sodium borohydride solution. As shown in Fig. 3a, the linear sweep voltammograms depict oxidation of sodium borohydride upon varying its concentration in 2 M NaOH in the potential range of −1.1 to 0.9 V at a sweep rate of 50 mV s^{-1} under Ar purging. Increasing $NaBH_4$ concentration form 10 mM to 0.5 M leads to a gradual increase in the observed anodic peaks centred at −0.35 and 0.57 V (vs. Hg/HgO). The onset potential for the oxidation of borohydride was observed to be −0.88 V, quite negative compared to various other transition metal catalysts and even few noble metal catalysts (Çelikkan et al. 2007). The observed increase in peak current density against increasing concentration could be attributed to the improved accessibility of the catalytically active sites, favouring mass transfer and simultaneously borohydride oxidation kinetics.

The onset potential for borohydride oxidation was found to be −0.88 V which demonstrates the first oxidation peak and can be assigned to direct $NaBH_4$ oxidation, whereas the other steep oxidation peak appearing at 0.04 V results due to further oxidation of both hydrolysis and oxidative intermediates yielding borane which rapidly dimerises forming diborane.

The concentration studies were followed by varying the sweep rates at a $NaBH_4$ concentration of 0.1 M in 2 M NaOH solution. It can be observed from Fig. 3b that incremental sweep rates have a positive influence upon the oxidative current density at the same potential. The WOx/MNCx catalyst was cycled at different scan rates viz. 5, 25, 50, 75 and 100 mV s^{-1}, and the corresponding increase in the current density could be attributed to greater charge labialization.

Effect of varying temperature over the borohydride oxidation ability was subsequently taken up and is represented in Fig. 3c. The LSVs were recorded in a jacketed cell containing 0.1 M $NaBH_4$ (in 1.0 M NaOH electrolyte) in the temperature range of 10–50 °C at a sweep rate of 100 mV s^{-1}. As depicted, upon increasing the temperature both the oxidative peaks were found to shift cathodically with an increase in the overall current density. These observation are consistent with decreased viscosity of the solution, facilitating enhanced kinetics.

4 Conclusion

The electrochemical oxidation of sodium borohydride was studied over WOx/MNCx, a non-noble, cost-effective and catalytically active electrocatalyst. It was found to exhibit an appreciable current density of 2.0 mA cm^{-2} at quite cathodic −0.88 V potential. The onset potential observed was comparable to the presently prevalent noble-metal alternatives at 50 mV s^{-1} for 0.1 M $NaBH_4$ in 2 M aqueous NaOH electrolyte under ambient conditions. This has been the first instance where tungsten oxide has been analysed for its potential activity towards borohydride oxidation and the onset potential is sufficiently cathodic as desired by DBFC anode. This activity was further found to hold its ground even at higher temperatures reinforcing its stability and sustained catalytic activity. We further aspire to extend this analysis by compiling a direct borohydride fuel cell with this anode catalyst and also pave way for further studies by combining it with other metallic species to further enhance the activity towards borohydride oxidation.

References

Amendola, S.C., Onnerud, P., Kelly, M.T., Petillo, P.J., Sharp-Goldman, S.L., Binder, M.: A novel high power density borohydride-air cell. J. Power Sources **84**(1), 130–133 (1999)

Çelikkan, H., Sahin, M., Aksu, M.L., Veziro˘glud, T.N.: The investigation of the electrooxidation of sodium borohydride on various metal electrodes in aqueous basic solutions. Int. J. Hydrogen Energy **32**, 588–593 (2007)

Indig, M.E., Snyder, R.N.: Sodium borohydride, an interesting anodic fuel (1). J. Electrochem. Soc. **109**(11), 1104–1106 (1962)

Jiménez, I.M., León, C.P., Shahb, A.A., Walsha, F.C.: Developments in direct borohydride fuel cells and remaining challenges. J. Power Sources **219**, 339–357 (2012)

Kojima, Y., Haga, T.: Recycling process of sodium metaborate to sodium borohydride. Int. J. Hydrogen Energy **28**(9), 989–993 (2003)

Leon, C.P., Walsh, F.C., Pletcher, D., Browning, D.J., Lakeman, J.B.: Direct borohydride fuel cells. J. Power Sources **155**, 172–181 (2006)

Liu, B.H., Li, Z.P.: Current status and progress of direct borohydride fuel cell technology development. J. Power Sources **187**, 291–297 (2009)

Liu, B.H., Yanga, J.Q., Lib, Z.P.: Concentration ratio of [OH −]/[BH4 −]: A controlling factor for the fuel efficiency of borohydride electro-oxidation. Int. J. Hydrogen Energy **34**(23), 9436–9443 (2009)

Ma, J., Choudhury, N.A., Sahai, Y.: A comprehensive review of direct borohydride fuel cells. Renew. Sust. Energ. Rev. **14**, 183–199 (2010)

Santos, D.M.F., Sequeira, C.A.C.: Sodium borohydride as a fuel for the future. Renew. Sust. Energ. Rev. **15**, 3980–4001 (2011)

Tamasiunaite, L.T., Radomskis, A., Antanaviciute, K., Jablonskiene, J., Balciunaite, A., Zieliene, A., Naruskevicius, L., Kondrotas, R.: Graphene supported platinum–cobalt nanocomposites as electrocatalysts for borohydride oxidation. Int. J. Hydrogen Energy **39**, 4282–4290 (2014)

Tiwari, A., Nagaiah, T.C., Bordoloi, A.: Electrocatalytic activity of tungsten oxide nanoclusters grafted on mesoporous nitrogen-rich carbon material in the dioxygen reduction reaction. ChemPlusChem (2015). doi:10.1002/cplu.201500253

Zhao, D., Huo, Q., Feng, J., Chmelka, B.F., Stucky, G.D.: Nonionic triblock and star diblock copolymer and oligomericsurfactant syntheses of highly ordered, hydrothermally stable, mesoporous silica structures. J. Am. Chem. Soc. **120**, 6024–6036 (1998)

Optimal Off-Grid Hybrid Options for Power Generation in Remote Indian Villages: HOMER Application and Analysis

Naveen Kumar Vasudevan and D. Ruben Sudhakar

1 Introduction

Rural electrification remains a far cry in certain developing countries, like India. Hybrid power frameworks are mix of two or more conversion equipment (e.g., power generators or capacity devices), or two or more energizes for the same device. Hybrid Energy System (HES) commonly involves two or more Renewable Energy Systems (RES) such as Photo-Voltaic, wind, hydro-thermal, biomass together with conventional equipment, such as diesel generator, fuel cells and gas turbines (Alliance for Rural Electrification (ARE) 2008). HES are suitable candidates for remote village electrification where grid electricity is infeasible. Table 1 presents a brief literature summary on some important studies on hybrid energy systems for small villages.

The main aim of the work is to design an optimal off-grid hybrid system to meet the sustained electricity load at a cost equal or almost equal to the grid supply cost, based on the analysis of various electricity demands and consumption patterns of the villages due to their presence in very different geographic locations. Four Indian villages from four different directions, representative of North, East, West and South zones of the sub-continent is chosen to study the various possible scenarios of nature of energy resource available, electricity demand patterns, impact of location on system performance. HOMER tool is used for the study and analysis taking into consideration the cost of erection, operation and the cost of electricity generated. Its

N.K. Vasudevan
Department of Chemical Engineering, Alagappa College of Technology, Anna University, Chennai, India
e-mail: naveenovan@gmail.com

D.R. Sudhakar (✉)
Department of Chemical Engineering, National Institute of Technology Karnataka, Surathkal, India
e-mail: rubensudhakar@nitk.ac.in

R. Mohan B. et al. (eds.), *Materials, Energy and Environment Engineering*,
DOI 10.1007/978-981-10-2675-1_25

Table 1 Summary of important studies on hybrid energy systems for village electrification

Authors	Work summary	Energy resources	Region	Objective
Sen and Bhattacharya (2014)	Off grid power generation with RET in India	PV, small scale hydro, wind turbines, bio-diesel generators	Palari, India	24 h supply as an off-grid source
Kumar and Manoharan (2014)	Economic study of mixed solar and diesel systems in different climatic zones of Tamilnadu	Solar PV, diesel generators, biomass	Zones of Tamilnadu, India	Economic analysis of hybrid systems
Bhattacharjee and Dey (2014)	Technological and economic performance analysis of grid with PV and biomass mixed power generation for rice mills	Solar PV, biomass	Tripura, India	Techno-economic evaluation
Murphy et al. (2014)	New routine for analyzing grid integrated PV, diesel and mixed distributed electricity systems considering an unoptimised electric grid, in Uganda	PV, diesel and mixed distributed electricity systems	Uganda	Grid connected stable hybrid distributed system
Mohammed et al. (2013)	Optimal layout design of solar PV systems using HOMER for Sohar, Oman	Solar PV	Oman	Determination of PV size for lighting load
Abdulqadiri and Ahmed (2012)	Comparative technological and economic study of mixed solar/diesel and mixed wind/diesel energy generation	Solar PV, diesel generation, wind	Nigeria	Studying mixed solar and diesel and wind/diesel systems

streamlining and affectability examination calculations permit one to assess the monetary and specialized possibility of an extensive number of innovative alternatives and to represent varieties in vast expenses and vitality asset accessibilities.

HOMER is a micro-power optimization modelling tool, it makes the task of testing models of both off-grid and grid-connected energy systems for a wide set of arrangements simple. when one models a system, one might have to make many decisions regarding the scope and arrangement of the system like which component to use, where, how, size and count of the components etc. The wide range of technology options and the wide array in technology costs and presence of energy resources make it difficult. This is where HOMER comes in handy. Its Optimization and sensitivity analysis algorithms, is profound and it enables one to decide the perfect grid based on various factors and decide upon a system configuration (Getting started guide for HOMER Version 2.1 2005).

2 Methodology and Data Used

Four villages are chosen by studying their house-holds size, current population and taking into account the variety in accessible significant energy resources, its electrical load pattern, land area, and population. The list of villages and their details used in the study are provided in Table 2.

The various usages points typical to a village environment like, irrigation, medical facilities, agriculture and household uses were given priority and the consumption of whole village was projected as an even distribution over all households. The load data is considered in the following pattern: Primary load 1—household consumption, Primary load 2—other consumption including schools, hospitals etc., and Deferrable load—Agricultural usage. Figure 1 shows the primary load and D-map of Chhachhohar village. Energy resources considered for study for the hybrid system are wind, solar, and diesel. Table 3 gives the various electrical loads during summer and winter seasons that were considered for the HOMER simulation of Chhachhohar village to configure the optimal hybrid energy system. Energy resources considered for study for the hybrid system are wind, solar, and diesel. The wind speed and solar insolation data available in the villages under study are obtained from NREL via Synergyenviron (Web-4 2014) and NASA database (Web-3 2014) respectively.

Table 2 Village details and location data (Web 1 2014; Web 2 2014)

Particulars	Village and its details			
Village name	Chhachhohar (1)	Belanta (2)	Ankevaliya (3)	Sirupagal (4)
Sub-district	Sardulgarh	Tigira	Barwala	Kanchipuram
District	Mansa	Cuttack	Ahmadabad	Kanchipuram
State	Punjab	Orissa	Gujarat	Tamil Nadu
Latitude	29.837993 N	20.434917 N	22.620764 N	12.834756 N
Longitude	75.390767 E	85.485249 E	71.793385 E	79.826873 E
Total population	459	643	441	272
Males	237	337	226	130
Females	222	306	215	142
Households	80	113	62	57

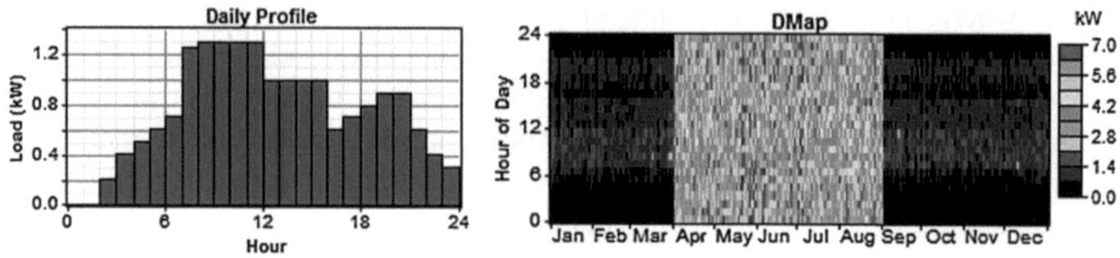

Fig. 1 Primary load 1 profile and D-map of Chhachokar village

Table 3 Electrical load details of Chhachhohar Village

S. no	Load	No. in use	Power (W)	Summer		Winter	
				h/day	W-h/day	h/day	W h/day
Domestic purpose							
1	CFL	1	20	6	120	7	140
2	CFL	1	20	6	120	7	140
3	CFL	1	11	5	55	6	66
4	Ceiling fan	1	10	3	30	4	40
5	Table fan	1	30	15	450	0	0
6	Radio	1	15	9	135	0	0
	Total	–	–	–	910	–	386
A	No. of houses	80	–	–	72,800	–	30,880
Industrial/commercial/community purpose							
1	Shops	5	500	8	20,000	7	17,500
2	Community center	1	1000	8	8000	6	6000
3	SMU	2	3000	12	72,000	10	60,000
4	Street lights	4	30	10	1200	12	1440
B	Total	–	–	–	101,200	–	84,940
Agricultural/irrigation purpose							
1	Water pumps	4	745	5	14,900	3	8940
2	Irrigation pumps	2	1500	6	18,000	4	12,000
3	Well	1	745	4	2980	2	1490
C	Total	–	–	–	35,880	–	22,430
Medical centre							
1	CFL	2	20	4	160	6	240
2	Ceiling fan	2	30	6	360	0	0
3	Refrigirator	0	600	20	0	16	0
D	Total	–	–	–	520	–	240
School							
1	CFL	2	20	2	80	4	160
2	Fan	2	30	6	180	0	0
3	Computer	1	300	2	600	2	600
4	Television	1	100	2	200	2	400
E	Total	–	–	–	1060	–	2600

3 HOMER Analysis

HOMER performs the reproduction for various imminent outline arrangements. In the wake of analysing each configuration, one that meets the heap with the framework requirements at least life cycle expense is chosen. HOMER performs its enhancement and affectability examination over all said segments and their assets, specialized and cost parameters, and framework limitations and affectability information over a scope of exogenous variables.

The grid layout for each village is first designed with the components which includes all the loads and available resources presented below. The system is designed with minimum possible sensitivities to ensure less deviation.

1. Primary load 1—household consumption	5. Solar PV
2. Primary load 2—other consumption including schools, hospitals etc.	6. Diesel generator
3. Deferrable load—agricultural usage	7. Convertor
4. Wind turbine	8. Battery

4 Results and Discussion

4.1 Optimal Hybrid Energy System Architecture

Figure 2 gives the grid layout of the optimal mixed energy system configuration for Chhachhokar village. This configuration is chosen as optimum, based on the

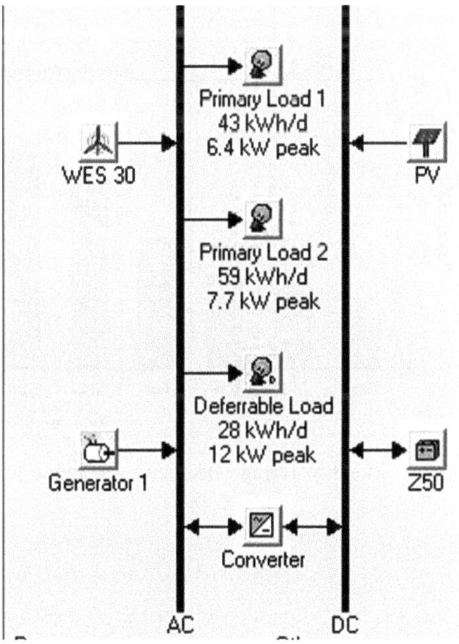

Fig. 2 Grid layout of the mixed energy system configuration for Chhachhokar village

Table 4 Optimal hybrid energy system architecture with actual capacities for all the villages

Resources	Villages			
	Chchhokar	Belanta	Ankevaliya	Sirupagal
Solar PV	20 kW–23 %	6.1 kW–2 %	6.1 kW–4.5 %	25 kW–25 %
Wind turbines	WES 30 (250 kW)–75 %	E33 (330 kW)–96 %	E33 (330 kW)–91.35 %	WES 30-1 (250 kW)–72 %
Diesel engines	10 kW–2 %	10 kW–2 %	10 kW–4.15 %	10 kW–3 %
Converters	70 kW (i&r)	100 kW (i&r)	100 kW (i&r)	50 kW (i&r)
Battery	ZESS 50-2 (50 kWh)	ZESS 50-2 (50 kWh)	ZESS 50-1 (50 kWh)	ZESS 50-2 (50 kWh)

accounts of lowest Net Present Cost (NPC) and the Cost of Electricity (CoE). The layout includes the system load, resources accounted in and their corresponding labels to scale them. The grid view designates both AC and DC power sources, and consumption points. The actual capacities of the different individual energy systems forming the optimal hybrid system for all villages The analysis is carried out considering the life span of the systems to be 25 years, at the end of which they are considered for salvage.

Table 4 gives the actual capacities and the loading levels of the various parts of the mixed system for all the villages under study. It can be observed that for Belanta and Ankevaliya the main source of power remains to be from the wind turbines for most of the time. In the case of Chchhokar and Sirupagal, approximately three fourth of the electricity comes from wind and the rest is from solar photovoltaic system. The contribution from diesel generator is highest for Ankevaliya village, increasing the operational cost of the total hybrid system.

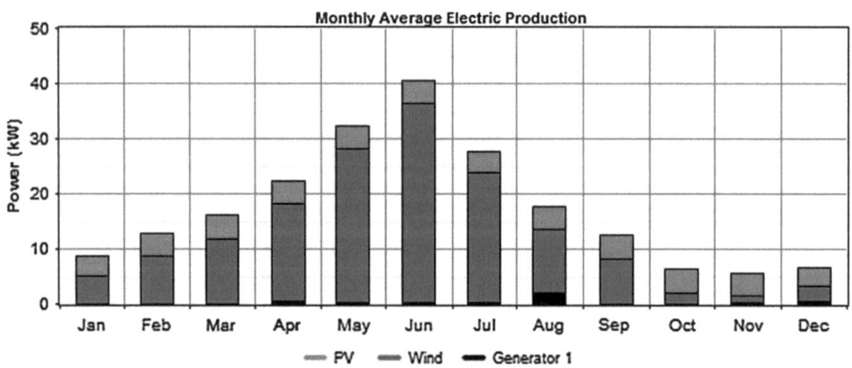

Fig. 3 Monthly average electricity generation for Chhachhokar village

Table 5 NPC, COE, OC of the optimal energy system for the villages under study

Factors	Villages			
	Chchhokar	Belanta	Ankevaliya	Sirupagal
NPC	$97,723	$102,719	$143,871	$101,503
Levelized COE	$0.162/kWh	$0.118/kWh	$0.207/kWh	$0.157/kWh
OC	$1903/yr	$3185/yr	$4605/yr	$2426/yr

Figure 3 gives the monthly average electricity generation from wind, PV and the diesel generator for Chhachhokar village. It can be observed that the contribution from wind energy system is the high both in summer and winter. This generation is relatively very low during winter months when compared to summer because of the relatively low availability of wind during winter.

Table 5 gives the total Net Present Value, Levelized COE and the operational cost for the optimum hybrid system configuration for all the villages under study. NPC is the amount HOMER uses to assess the life-cycle expense of the framework. This incorporates the beginning capital expense of the framework parts, the expense of any segment substitutions that happen inside of the task lifetime, the expense of upkeep and fuel.

The optimal hybrid system configuration for Chchhokar has the lowest NPC and is of comparable order with Belanta and Sirupagal villages. The NPC of the system stands to be the highest for Ankevaliya among the villages, partly because of (i) the use of diesel generator to make-up for the intermittency from PV and wind system is highest, (ii) the operational cost of the hybrid system is highest among the villages under study. This is also reflected in highest COE in Ankevaliya. The optimal hybrid energy system of Belanta with the highest contribution coming from the wind energy offers the lowest COE of 0.118 $/kWh.

Emissions from the optimal hybrid energy systems for all the villages under study are given in Table 6. It can be seen that the particulate level, CO SO_2, CO_2 emissions stand to be the highest in Ankevaliya village because of the relatively large quantity of diesel combustion in the diesel generators.

Table 6 Emissions from the optimal energy system configuration for all the village

Pollutants/villages	Emission kg/yr			
	Chachhokar	Belanta	Ankevaliya	Sirupagal
CO_2	2735	5541	8696	4771
CO	6.75	13.7	21.5	11.8
Unburned hydrocarbons	0.748	1.52	2.38	1.3
Particulate matter	0.509	1.03	1.62	0.888
SO_2	5.49	11.1	17.5	9.58
NO_X	60.2	122	192	105

5 Conclusions

Considering the current study, the following conclusions are obtained:

(1) The different geographic locations of the villages influence considerably the configuration of the optimal hybrid energy system, operational cost and the emission levels.
(2) Among the villages, Ankevaliya in the western region of the country experiences the highest NPC, COE, operational cost and emission levels.
(3) The seasonal changes and the system load stand to be the critical factors of influence in the modelling of an offsite-grid mixed energy system for inaccessible villages in India.

References

Alliance for Rural Electrification (ARE): Hybrid power systems based on renewable energies: A suitable and cost competitive solution for electrification. ARE Position Paper (2008)
Abdulqadiri, A.B, Ahmed, E.A.: Comparative technoeconomic analysis of hybrid pv/diesel and hybrid wind/diesel energy generation for commercial farm land in Nigeria, Int. J Eng. Adv. Technol. **2**, (1), 2249–8958 (2012)
Bhattacharjee, S., Dey, A.: Techno-economic performance evaluation of grid integrated PV-biomass hybrid power generation for rice mill. Sustain. Energy Technol. Assessments **7**, 6–16 (2014)
Getting started guide for HOMER Version 2.1, National Renewable Energy Laboratory (2005)
Kumar U.S., Manoharan, P.S.: Economic analysis of hybrid power systems (PV/diesel) in different climatic zones of Tamil Nadu. Energy Convers. Manage. **80**, 469-476 (2014)
Mohammed, M., Aziz, A., Kazem H.A, Alwaeli A.H.: Optimal sizing of photovoltaic systems using homer for Sohar, Oman. Int. J. Renew. Energy Res. **3**(2), 388–398 (2013)
Murphy, P.M., Twaha, S., Murphy, I.S: Analysis of the cost of reliable electricity: A new method for analyzing grid connected solar, diesel and hybrid distributed electricity systems considering an unreliable electric grid, with examples in Uganda. Energy **66**, 523–534 (2014)
Sen, R., Bhattacharyya, S.C.: Off-grid electricity generation with renewable energy technologies in India: An application of HOMER. Renew. Energy **62**, 388–398 (2014)

Websites

Web-1: http://www.censusindia.gov.in/pca/Searchdata.aspxconsulted on 14 June 2014
Web-2: http://www.mynasadata.larc.nasa.gov/latitudelongitude-finder/ consulted on 15 June 2014
Web-3: https://eosweb.larc.nasa.gov/sse/RETScreen/ consulted on 17 June 2014
Web-4: http://www.synergyenviron.com/tools/wind_data.asp consulted on 17 June 2014

Experimental Studies on Electricity Production and Removal of Hexavalent Chromium in Microbial Fuel Cell

N. Samsudeen, Arvind Pari and B. Soundarya

1 Introduction

In the recent decades, consumption of energy within the world has had a prosperous trend. Energy sources can be classified into three major categories—nuclear, renewable and fossil fuels. Currently the non-renewable sources account for about 87 % of the total energy economy. However, this trend cannot continue forever with the sources depleting drastically. Countries around the world have made remarkable efforts to find a solution for this energy crisis, by turning towards renewable energy sources such as energy produced from sun, wind and water. MFC is an alternative renewable and sustainable energy generation from the wastewater (Rabaey and Verstraete 2005). A MFC is a biological reactor that bio-electrochemically reduces the chemical energy present in the bonds of organic compounds to electrical energy, through the microbial metabolism under anaerobic conditions (Logan et al. 2006). The electrical current produced by bacteria was first observed by Potter in 1911, but results that were actually feasible were limited in this area for the next 50 years (Pant et al. 2010).

When it comes to environmental applications, only one half cell of the MFC has been put to use so far, i.e. the anode chamber, through bio-catalytic reactions is used to remove the pollutants. The cathode chamber only serves the purpose of closing the MFC circuit. The MFC performance is mainly depends on the types of electron acceptors prevailing in the cathode chamber. Due to its high redox potential and abundance, Oxygen is a commonly preferred electron acceptor potential (Li et al. 2009; Min and Logan 2004). Other examples of efficient electron acceptors used in the MFC include potassium ferricyanide, permanganate, ferric iron and nitrate MFC (Pandit et al. 2011). Recently, certain pollutants (e.g. Nitrate and trichloroethene, Cr^{6+} etc.) have also been used as electron acceptors owing to their high electro-

N. Samsudeen (✉) · A. Pari · B. Soundarya
Department of Chemical Engineering, National Institute of Technology, Tiruchirappalli, Tiruchirappalli, India

R. Mohan B. et al. (eds.), *Materials, Energy and Environment Engineering*,
DOI 10.1007/978-981-10-2675-1_26

chemical redox potentials. The usage of these contaminants as electron acceptors in the cathode chamber, would make MFCs more versatile and serve multiple functions simultaneously. Several industrial such as leather, tanning, and electroplating are known to discharge chromium containing wastewater (Wang et al. 2008). Chromium exists in aqueous phase either as hexavalent chromium [Cr(VI)] or trivalent chromium [Cr(III)]. Cr(VI) owing to its mutagenic and carcinogenic properties is highly harmful to public health. There are several chemical and electrochemical methods to reduce Cr(VI) to Cr(III), which is non-toxic. Earlier work done on the reduction of Cr(VI) had focussed on a broad range of pH values, but the current research focusses on a more narrow range (acidic) in which high efficiency has been predicted (Gangadharan and Nambi 2015). In this study, we investigated the performance of a dual chamber MFC in terms of power production and Cr(VI) treatment using synthetic wastewater as electrons acceptors (catholyte). The effect of pH and the concentration of chromium content on the performance of MFC was investigated.

2 Materials and Methods

2.1 *MFC Construction*

A MFC contains anode and the cathode chamber which are separated by a Proton Exchange Membrane (PEM) that ensures the transfer of only hydrogen ions and not the other ions. The anode and the cathode chamber has the working volume of 210 mL each in 250 mL of empty bed volume. The membrane was pre-treated using 5 % H_2O_2 followed by 0.5 M of H_2SO_4 and stored in the deionized water prior to use. The plain graphite plate with the dimension having an area of 4 cm × 6 cm × 10 cm (projected surface area of 50 cm^2) on the either side of the chamber was first immersed in the deionized water prior to use. The anode chamber is completely sealed with epoxy sealant to keep anaerobic environment while the cathode chamber is left open to the atmosphere through a small circular opening. The photograph of the experimental setup as shown in Fig. 1. In the anode compartment, the biocatalyst oxidizes the carbon source or the substrate to generate protons and electrons. Since the oxygen in the anode chamber will inhibit the production of electricity, a practical system must be designed to ensure the isolation of bacteria from oxygen. This is achieved by hermetically sealing the chamber.

2.2 *MFC Operation*

The anode compartment was filled with 4 % (by Vol.) of distillery effluent and synthetic solution (50 mM of potassium dichromate solution) was used as the catholyte. 13.6 g/L solution of potassium di-hydrogen phosphate (KH_2PO_4) was

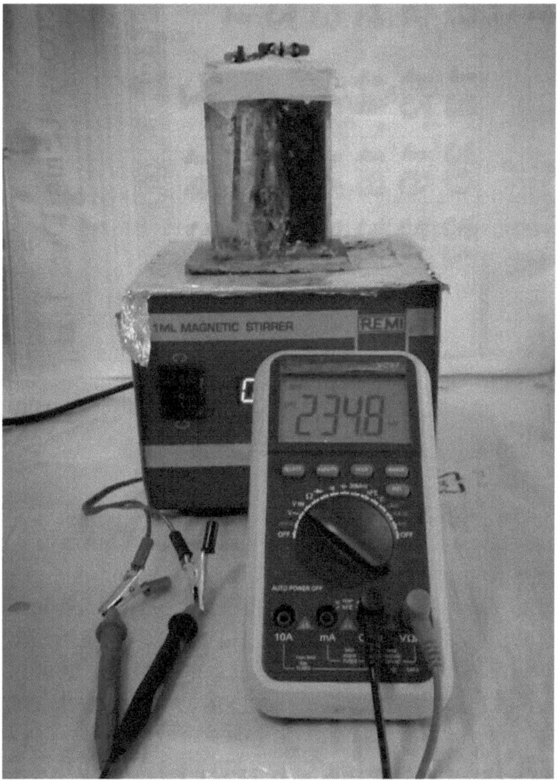

Fig. 1 Photograph of microbial fuel cell setup

added as the buffer and the pH value was adjusted to 7 using 1 N of NaOH. The anode side was hermetically sealed while the cathode side was kept open. A magnetic stirrer was used in the anode chamber for continuous distribution of the solution and formation of a uniform bio-film on the anode side PEM. Initially, the experiment was carried out with 50 mM of $K_3Fe(CN)_6$ solution as the catholyte, till the cell voltage stabilised and a proper uniform bio-film was formed. Once this was achieved, the catholyte was replaced with the prepared synthetic effluent solution, without disturbing the bio-film on the anode side.

2.3 *Measurement and Analysis*

The voltage generation in the MFC was measured using digital multimeter. The current and power density was determined by using the Ohm's law expression. Polarization data was taken during every cycle using resistors of the range between 56 and 4000 Ω and a multimeter. The amount of reduction of Cr^{6+} was calculated from the absorbance data (ranges between 190 and 1100 nm) of the initial and final solution (taken after 48 h) using a double beam UV-vis spectrophotometer.

3 Results

3.1 *Effect of PH*

Since pH can greatly affect the transfer of ions, experiments were conducted to validate the same. With concentration fixed at 100 mg/L, the MFC performance was investigated at three different pH values of 2, 4 and 6.7.

$$Cr_2O_7^{2-} + 14H^+ + 6e^- \longrightarrow 2Cr^{3+} + 7H_2O \tag{1}$$

From the above equation, it is evident that for reduction of every mole of Cr(VI), 7 mol of H^+ ions were consumed. This quite clearly highlights the strong dependency of the performance of the MFC on pH, favouring the pH in the smaller range. The maximum power density of 84, 135.5 and 57.1 mW/m^2 with corresponding current densities of 460, 520 and 338 mA/m^2 was observed at the pH values of 6.7, 4 and 2 respectively as shown in Figs. 2 and 3. This unusual trend of an increase and then a decrease is due to the interplay of several phenomena, each imposing its own restriction on the dynamics of electron and proton transfer. Though the electron transfer increases with pH, every enzyme has an optimum pH at which its activity is the maximum. On either side of this peak value, the performance of the enzyme will decrease. Also, it has to be noted that when the catholyte becomes too acidic, it prevents the transfer of electrons and harms the PEM membrane, inadvertently reducing its performance. Thus, as a result of these two competing phenomenon, a compromise between electron transfer rate and the enzyme activity will be achieved at moderately acidic pH conditions. The results obtained are consistent with this ideology as a similar trend of high current density was obtained at an intermediate pH of 4.

From Fig. 3, a positive slope can be observed for the curves initially, which can be attributed to two conditions that prevail at the start of the reduction process; the first one being the large reserves of H^+ ions and Cr(VI) in the catholyte at the starting stage. The second one is the abundant organic matter initially present in the

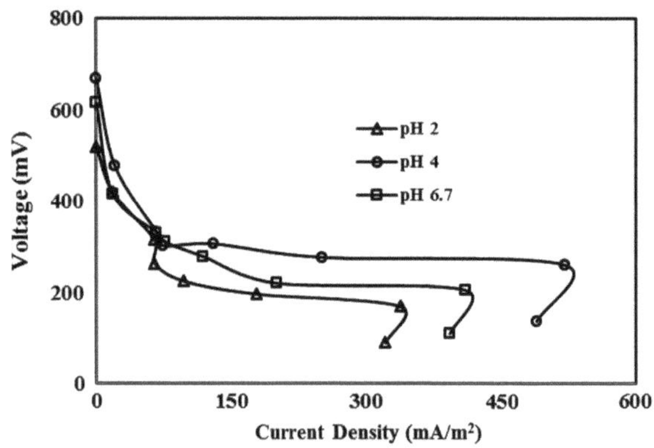

Fig. 2 Polarization behaviour at different pH of the dichromate solution

Fig. 3 Power density behaviour at different pH of the dichromate solution

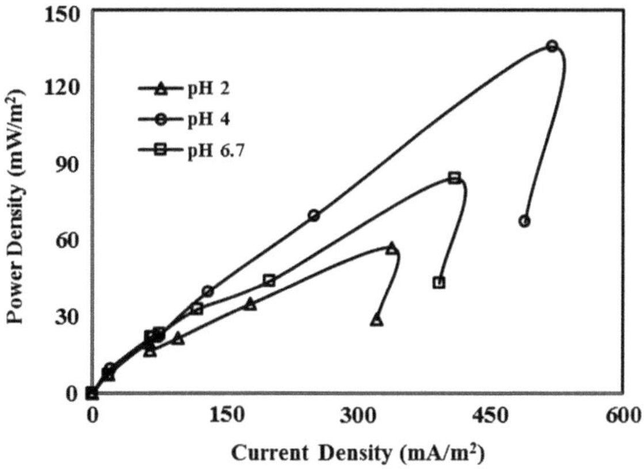

distillery effluent in the other half-cell, that promotes the growth of a healthy and uniform layer of biofilm on to the anode. This ensures a high rate of organic waste decomposition and resulting electron production. The interplay of these two factors culminated into the high rate of Cr(VI) reduction, that was observed initially, which is depicted by the linear curves with a positive slope. But as the reduction proceeds, due to the depletion of H^+ ions and Cr(VI) in the catholyte and organic wastes in the anolyte, there is a significant decrease in the driving potential for the reduction to occur. Subsequently, the rate of Cr(VI) reduction decreases during this stage, which is represented by linear curves with a negative slope in Fig. 3. The point at which the shift from the curves with positive slope to the curves with negative slope occurs marks the peak power production point, for each of the three pH values. This change in the slope of the curve depending upon the availability of H^+ ions in the catholyte is a proof of the pH dependency of the performance of the MFC.

3.2 *Effect of Concentration*

It is essential to know the concentration at which maximum power can be produced, as in the practical scenario, the concentration of metal ions in the effluent will have a wide range of values. Therefore, knowledge of the range in which optimum power production with a high reduction rate occurs is imperative. Three concentrations were chosen to investigate their influence on the power density at the previously determined optimum pH of 4. The peak power density of 135.5 mW/m^2 corresponding to a current density of 520 mA/m^2 was obtained for the concentration of 100 mg/L as shown in Fig. 4. While the power produced increases with increase in concentration, presence of too much Chromium obstructs the movement of protons and thus increases the resistance to its flow (reduction in mobility), thereby decreasing the power production at higher concentration levels of Cr^{6+}. This trend is similar to that observed for the effect of pH, wherein an intermediate value

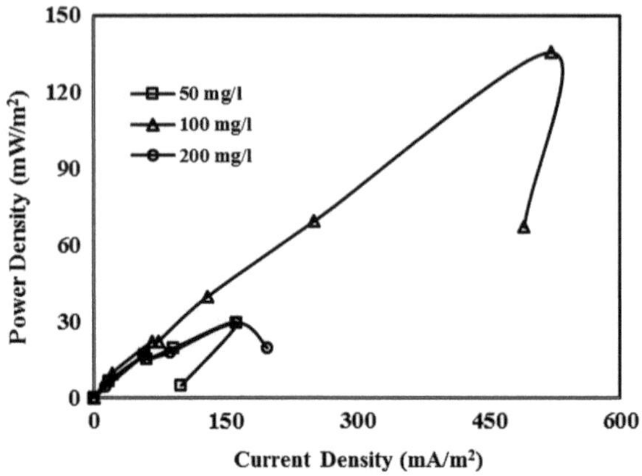

Fig. 4 Power density behaviour for different concentration of $K_2Cr_2O_7$

produced the maximum output energy. It was also observed that the phenomenon of fouling of the membrane was accelerated by the presence of high levels of Cr^{6+}.

3.3 Chromium Reduction

The reduction efficiency of Cr(VI) to Cr(III) at the optimum concentration of 100 mg/l and the optimum pH of 4 was observed to be 13.8 %, determined using the absorbance data obtained from a double beam UV-vis spectrophotometer. The electrons released in the anodic chamber due to the metabolism of organic matter by the microorganisms is transferred to the cathodic chamber through the external circuit. This manifests as the electricity generated by the MFC. $Cr_2O_7^{2-}$ in the cathodic chamber being an electron acceptor, accepts the incoming electrons, thus reducing Cr(VI) to Cr(III). Oxygen present in the anode chamber, due to the prevailing aerobic conditions, also makes use of the incoming electrons and reduces to form molecules of water.

$$O_2 + 4H^+ + 4e^- \longrightarrow 2H_2O \quad (2)$$

4 Discussion

Cr(VI) was reduced to a lesser toxic form Cr(III) in the cathode chamber of the MFC, while simultaneously generating power up to 135.5 mW/m^2 (at a current density of 520 mA/m^2). It has been observed that much higher power density of 453 mW/m^2 can be achieved using dichromate (10 mM, pH 2, at 0.079 mA/cm^2) as an electron acceptor (Wang et al. 2008).

In reality, theoretical calculations alone are not reliable for the prediction of the actual performance of the MFC. Once the MFC is scaled up to the industrial level, many more factors come into play. This includes the rate of electron transfer and its mechanism, fluctuation in the temperature, the design parameters and dimensions of the chambers, mode of operation (batch or continuous), the activity and efficiency of the microbes to name a few. The influence of some of these factors can be determined by intuition, to begin with, but has to be later supported by relevant experiments.

It is possible to verify the validity of the results obtained in this study by the use of a theoretical equation, i.e. the Nernst Equation. It has been shown in this study that the power production (power density) depends upon the pH and concentration of the catholyte. Accordingly, a critical analysis of the Nernst equation given below, would be required to appreciate the trends obtained in this study.

$$E_{cat} = E^{0}_{cat} + \frac{RT}{nF}\ln\frac{[Cr^{3+}]^{2}}{[Cr_2O_7{}^{2-}][H^{+}]^{14}} \tag{3}$$

where

E^0_{cat}	represents the standard electrode potential of the cathode;
R	is the ideal gas constant whose value in SI units is 8.314 J/mol K;
T	is the temperature in the Kelvin scale;
n	represents the total number of electrons exchanged in the redox reaction (n = 6 in this case);
F	denotes the Faraday's constant whose value is 96,485 C/mol;
$[Cr^{3+}]$, $[H^{+}]$ and $[Cr_2O_7{}^{2-}]$	denote the concentrations of the respective ions

The cell potential (E_{cell}) for the fuel cell is given by the difference between the cathode potential and the anode potential;

$$E_{cell} = E_{cathode} - E_{anode} \tag{4}$$

From Eqs. 3 and 4, it can be deduced that the cell potential (E_{cell}) is higher for a catholyte with high concentration of Cr(VI) and H^+ ions. Higher concentration of H^+ ions implies lower pH. This is the same trend that has been observed in our studies, till the peak power generation point was reached, beyond with enzyme activity and overcrowding of ions limited and reduced the efficiency of the MFC.

The experimental studies described here are only but a basic and fundamental approach to understand the behaviour of the MFCs (in terms of Cr(VI) reduction and power generation) at a superficial level. Further experiments and research is highly vital to understand their finer nuances for the improvement of design and efficiency, making them more economically feasible. Once the feasibility of using this system on a large scale is determined, proper survey and analysis of the effluent from the tanneries and the concentration range of Cr(VI) ions in it is necessary to predict the amount of electricity that can be generated by the application of this technology on an industrial level.

5 Conclusion

As the first stepping stone, the work discussed in this paper achieves the successful removal of Cr^{6+} (for a pH of 4 and an optimum concentration of 100 mg/L without the use of any buffer solution) with the simultaneous production of power for a cycle of 48 h duration. The maximum power density achieved in this case is 135 mW/m^2. The effects of buffer solution, pH of catholyte and concentration of Cr^{6+} on the removal efficiency as well as the power production has also been critically analyzed to arrive at the optimum values. Based on the results, efficiency seems to be high at an intermediate pH value of 4 and concentration of 100 mg/L. Further research with an exhaustive number of experiments have to be undertaken to arrive at an explanation for the trend that has been observed. At these conditions, the Chromium removal was observed to be 13.8 %. A possible scope for the effective utilization of this technology lies in the tannery industry, whose effluent is rich in pollutants like Cr^{6+}.

References

Gangadharan, P., Nambi, I.M.: Hexavalent chromium reduction and energy recovery by using dual-chambered microbial fuel cell. Water Sci. Technol. **71**(3), 353–358 (2015)

Li, Y., Lu, A., Ding, H., Jin, S., Yan, Y., Wang, C., Zen, C., Wang, X.: Cr(VI) reduction at rutile-catalyzed cathode in microbial fuel cells. Electrochem. Commun. **11**(7), 1496–1499 (2009)

Logan, B.E., Hamelers, B., Rozendal, R., Schröder, U., Keller, J., Freguia, S., Aelterman, P., Verstraete, W., Rabaey, K.: Microbial fuel cells: methodology and technology. Environ. Sci. Technol. **40**(17), 5181–5192 (2006)

Min, B., Logan, B.E.: Continuous electricity generation from domestic wastewater and organic substrates in a flat plate microbial fuel cell. Environ. Sci. Technol. **38**(21), 5809–5814 (2004)

Pandit, S., Sengupta, A., Kale, S., Das, D.: Performance of electron acceptors in catholyte of a two-chambered microbial fuel cell using anion exchange membrane. Bioresour. Technol. **102** (3), 2736–2744 (2011)

Pant, D., Van Bogaert, G., Diels, L., Vanbroekhoven, K.: A review of the substrates used in microbial fuel cells (MFCs) for sustainable energy production. Bioresour. Technol. **101**(6), 1533–1543 (2010)

Rabaey, K., Verstraete, W.: Microbial fuel cells: novel biotechnology for energy generation. Trends Biotechnol. **23**(6), 291–298 (2005)

Wang, G., Huang, L., Zhang, Y.: Cathodic reduction of hexavalent chromium [Cr(VI)] coupled with electricity generation in microbial fuel cells. Biotechnol. Lett. **30**(11), 1959–1966 (2008)

Experimental Studies on Performance of Single Cell PEM Fuel Cell with Various Operating Parameters

Shaik Shadulla, K. Satish Raj and S.V. Naidu

1 Introduction

With the increasing concern about sustainable energy and environmental issues, hydrogen and fuel cell technology is attracting more and more attention from academic research, industry and governments as shown by Wang and Chen (2011). Fuel cell is an energy conversion device in which hydrogen and oxygen are used as input gases to generate electricity with the aid of electro catalysts. A significant amount of research has been devoted to the development of the PEM fuel cell for both mobile and stationary applications owing to its better energy conversion efficiency and near zero emission as shown by Zhong et al. (2009).

Sreenivasulu et al. (2013) experimentally verified the performance of a PEM fuel cell using different back pressures and flow field designs. Experiments were conducted with three different flow field plates, viz., 4-serpentine, inter-digitated and dual inlet single outlet, at three different back pressures created by immersing the outlet tube in a cylinder at three different depths with each flow field plate. An increase in output power was obtained with an increase in back pressure; which is the highest in the case of serpentine flow field design. Leelamanohar et al. (2010) studied the influences of gas pressure and flow rate parameters on the performance of a PEM fuel cell. Experimental data gathered was analyzed by statistical sensitivity analysis by checking the effect of one variable parameter on the other. The mixed interaction between the factors was also considered along with the main interaction to explain the model developed using design of experiments.

S. Shadulla (✉) · K. Satish Raj · S.V. Naidu
Fuel Cell Technology Laboratory, Department of Chemical Engineering,
AU College of Engineering (A), Andhra University, Visakhapatnam, India
e-mail: shadullashaik@ymail.com

R. Mohan B. et al. (eds.), *Materials, Energy and Environment Engineering*,
DOI 10.1007/978-981-10-2675-1_27

Misran et al. (2013) investigated the influence of operating temperatures (40, 50, 60 °C), pressures (1 and 1.5 bar), flow rates of hydrogen (3 lpm) and air (6 lpm) on water transport properties. As the membrane hydration decreased along the channel, the electro-osmotic drag coefficient declined and the ionic conductivity of the membrane dropped. A large water concentration gradient between the anode and cathode resulted in a negative value for the water back-diffusion flux. Using this strategy, flooding conditions in the PEM fuel cell were successfully averted. Khazaee and Ghazikhani (2012) experimentally studied the impact of operating parameters such as inlet hydrogen and oxygen temperatures, cell temperature, inlet gas pressures, flow rates of oxygen and hydrogen on the output performance of a PEM fuel cell. They observed that an increase in inlet gas pressure enhances the output which results from a decrease of ohmic and concentration losses and an increase of more efficient fuel transport from the gas diffusion layers and the chemical reaction at the catalyst surfaces and exchange current density. The impact of inlet gas temperature on the performance of the cell is so significant that by increasing the temperature of hydrogen and oxygen, the cell performance is enhanced due to the a decrease of activation over potential and an increase in the electrochemical reaction.

Mughal and Li (2006) experimentally studied the effect of operating parameters including cell temperature, hydrogen and air humidification temperature, flow rates and inlet gas pressure on the performance of a PEM fuel cell with an active area of 50 cm^2, 20 % Pt/C loading. The experimental results demonstrated that the output performance was primarily affected by the cell temperature, humidification and flow rates of cathode. Pressure mostly does not affect the cell performance. Shan and Hongtan (2014) also studied the variation of local current density in the lateral direction across the land and channel areas in a PEM field cell using a serpentine flow field geometry. When oxygen is used in the cathode, local current densities were always higher than when using air at any location and at all cell voltages. Considering all the studies, the local current density was the highest under the centre of the land area and declined toward the centre of the channel area. Zhang et al. (2015) studied the influence of varying cathode catalyst loading on the local current density distribution in a PEM fuel cell developing optimal catalyst loading variation so as to minimize the current density variation along the cell, while maximizing the power density. The optimum catalyst loading was found to increase with increase in cell voltage and the desired uniformity.

In the present work, we fabricated an MEA with an active area of 25 cm^2 and investigated the impact of different operating parameters such as cell temperature, inlet gas pressure, humidification temperatures of anode and cathode gases and flow rates of anode and cathode gases on the performance of our PEM fuel cell.

2 Experimental

2.1 Preparation of Catalyst Ink and Fabrication of MEA

MEA preparation consisted of two steps: (i) electrode preparation (ii) thermal pressing of the Nafion® 212 membrane embedded between the electrodes. To prepare the electrode, initially catalyst ink was prepared by mixing together sufficient quantities (0.5 mg/cm^2 loading × 25 cm^2) of the 10 % Pt/C catalyst (Johnson-Mathey), DI water, Isopropanol (Merck) and 5 wt% Nafion (Aldrich). This mixture was sonicated for 20 min in a bath sonicator at a temperature below 25 °C. Two batches of ink were prepared, one for each electrode. The catalyst ink was coated onto the wet-proofed GDL (micro-porous layer teflonized carbon paper) and dried using a hot air dryer. The catalyst ink coating procedure was repeated till the required loading of catalyst was achieved on the surface of the GDL, computed using the difference in initial weight and final weight of the GDL. After preparing both electrodes, a 1:1 mixture of Nafion: Isopropanol was applied onto the surface of the electrodes and air-dried for 24-h prior to thermal pressing. The MEA was then fabricated by thermally pressing the Nafion® 212 membrane between the two electrodes.

The fuel cell comprises of two graphite plates with grooved flow channels (4-serpentine geometry) with the membrane electrode assembly (MEA) embedded between the plates. These flow channels ensure maximum contact time between the reactant gases and the membrane electrode assembly for efficient utilization of the reactants. Hydrogen gas flows through the anode flow channel and oxygen gas flows through the cathode channel. Current collector plates and end plates fixed on either side using long bolts makes the fuel cell assembly complete and ready for the testing phase.

2.2 Fuel Cell Tests

The experimental apparatus consists of a 1 kW fuel cell test station (K-Pas Instronic Engineers-Chennai) containing instruments to quantify voltage, current, power and resistance. Mass flow controllers regulate the flow of hydrogen and oxygen and the in-built humidification chambers ensure that the hydrogen and oxygen are humidified before they enter the fuel cell. Humidification of the inlet gases ensures that the membrane used in the MEA does not dehydrate and crack under stress. A proprietary PEM fuel cell testing software is used to control and monitor the experimental setup through a desktop computer. Operating parameters such as cell operating temperature, pressures of inlet gases, humidification temperatures of inlet gases, flow rates of inlet gases were controlled using the software. After testing the flow of inlet gases and checking for cross flows and leakages, desired values of cell operating temperature, inlet gas pressures, humidification temperatures, flow rates are set to desired values. The software is operated under "constant" voltage mode by setting a fixed value of cell voltage (V) and noting down the corresponding readings of current (A) and current density (A/cm^2). This process is repeated by varying one of the operating parameters while keeping the values of other operating parameters constant.

3 Results and Discussion

3.1 *Effect of Operating Temperature*

Temperature has a strong effect on the transport phenomena inside the fuel cell and open circuit voltage. Experimental results are obtained by varying the cell operating temperatures from 303 to 318 K. During the experimental runs, the flow rates of both hydrogen and oxygen are maintained at 0.5 lpm for MEA with 10 % Pt/C with a loading of 0.5 mg/cm^2.

Figure 1 shows that the output performance of the fuel cell is enhanced as the cell temperature is increased from 303 to 318 K. It can be observed from Fig. 1 that the exchange current density increases as the fuel cell operating temperature is increased, which minimizes activation losses. At low operating temperatures, the membrane in the catalyst layer is partially hydrated causing a decrease in the active surface area of the catalyst. With an increase of operating temperature, water production increases causing better hydration of the membrane in the catalyst layer, increasing the active surface area of the catalyst layer and leading to improved fuel cell performance. The polarization curves for various fuel cell temperatures are shown in Fig. 1.

It can be observed from Fig. 1 that the exchange current density increases as cell operating temperature is increased from 303 to 318 K, indicating improved mass transport due to increase in diffusivity with an increase in cell operating

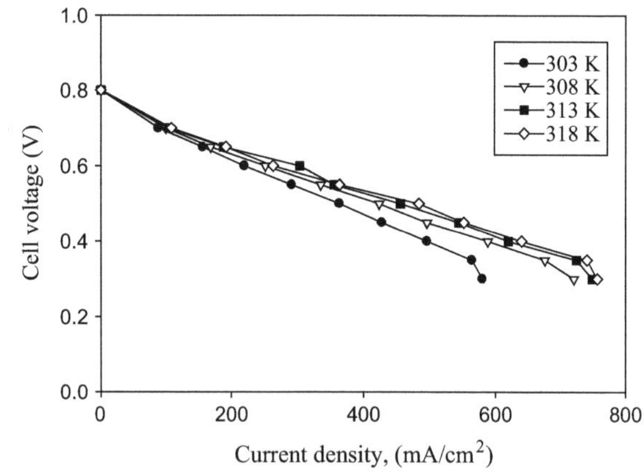

Fig. 1 Polarization curves for different fuel cell temperatures. Pressure is 1 atm, anode and cathode humidification temperatures are 303 K, and flow rates of anode and cathode gases are 0.5 lpm each

temperature. Wang et al. (2003) reported similar conclusion on the effect of fuel cell temperature on cell output performance.

3.2 Effect of Operating Pressure

Polarization curves with different operating pressures ranging from 1 to 2.5 atm are shown in Fig. 2. The pressure of both cathode and anode gases were kept the same. From Fig. 2, it can be inferred that the output performance of fuel cell is enhanced as the operating pressure is increased from 1 to 2.5 atm. It is observed from Fig. 2 that by increasing pressures of both anode and cathode gases, the limiting current and power are enhanced.

The improved performance is observed due to an increase in partial pressure of the reactant gases in the catalyst layer with increasing operating pressure. A decrease of ohmic and concentration losses and efficient transport of fuel from the GDL to the Catalyst Layer (CL) may account for this phenomenon. Although higher operating pressures improve cell output performance, the downside is increased complexity in designing gaskets to prevent leakages (Liu et al. 2015).

3.3 Effect of Anode Humidification Temperature

To understand the effect of anode humidification temperature on fuel cell performance, experimental values are obtained for the same system by varying the anode humidification temperature of the inlet gases from 303 to 318 K while other parameters are kept constant.

Polarization curves shown in Fig. 3 for varying anode humidification temperatures demonstrate that the fuel cell performance increases as the anode

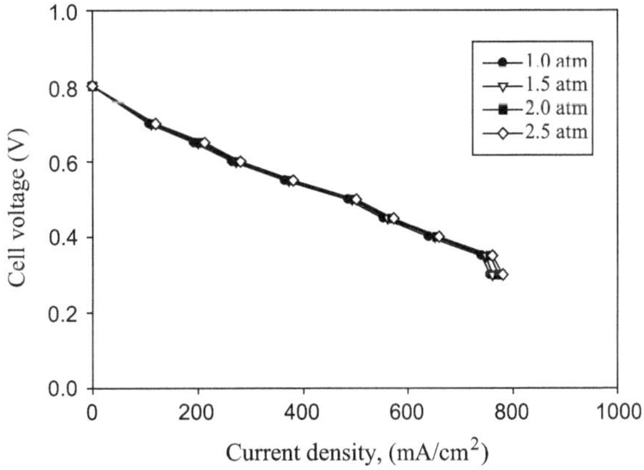

Fig. 2 Polarization curves for different pressures. The cell operating temperature is 318 K, anode and cathode humidification temperatures are 303 K, and flow rates of anode and cathode gases are 0.5 lpm each

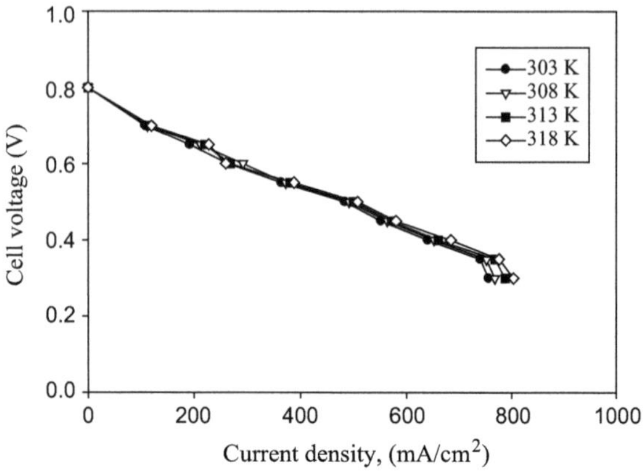

Fig. 3 Polarization curves for different anode humidification temperatures. Fuel cell temperature is 318 K, pressure is 2.5 atm, cathode humidification temperature is 303 K, and flow rates of anode and cathode gases are 0.5 lpm each

humidification temperature is enhanced from 303 to 318 K. Due to increased anode humidification temperature, the membrane becomes fully hydrated as a result of which proton conductivity increases and ohmic over potential decreases, while the limiting current density increases. Hence, we can observe an improvement in cell performance with an increase in anode humidification temperature.

3.4 *Effect of Cathode Humidification Temperature*

To understand the effect of cathode humidification temperature, experimental values are obtained for the same system by varying the cathode humidification temperature from 303 to 318 K while other parameters are kept constant.

Polarization curves shown in Fig. 4 for varying cathode humidification temperature demonstrate that there is no visible change in fuel cell performance with a change in the cathode humidification temperature. A decrease in limiting current

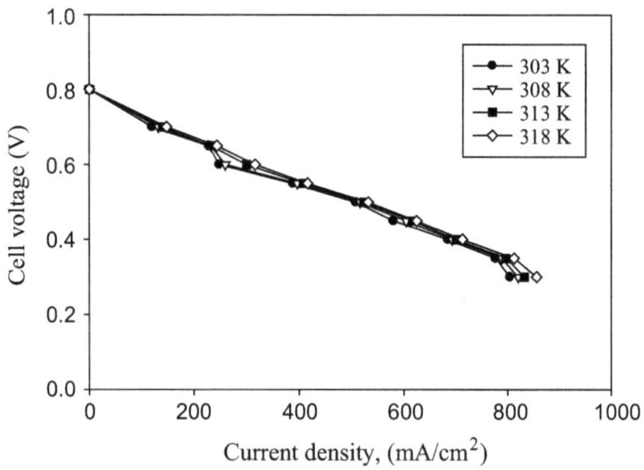

Fig. 4 Polarization curves for different cathode humidification temperatures. Fuel cell temperature is 318 K, pressure is 2.5 atm, anode humidification temperature is 318 K, and flow rates of anode and cathode are 0.5 lpm each

density with an increase of cathode humidification temperature can be observed from the graph. A decline in the effective porosity of the gas diffusion layers and a reduction in reactant concentration may have contributed to this phenomenon (Wang et al. 2003).

3.5 *Effect of Anode Gas Flow Rate (H_2)*

To understand the effect of anode gas flow rate, experimental results are obtained by varying the anode gas flow rates (H_2) between 0.5 and 1.2 lpm respectively while the other operating parameters are kept constant.

The impact of anode gas flow rate on cell performance is shown in Fig. 5. An increase in anode gas flow rate from 0.5 to 1.2 lpm results in an increase in fuel cell performance as shown in the graph. By increasing the flow rate of hydrogen, more fuel is transported from the GDL to the Catalyst Layer (CL) and the electrochemical reaction is enhanced. A sudden drop in voltage indicates lack of hydrogen (zero concentration) on the catalyst surface. The current obtained at the point of zero concentration of hydrogen is called the limiting current. A fuel cell output cannot exceed the limiting current since no reactants exist at the catalyst surface beyond the point of zero concentration. One can increase the limiting current and power by increasing the flow rate of hydrogen. The polarization curve showing the impact of anode flow rate on fuel cell performance is shown in Fig. 5.

3.6 *Effect of Cathode Gas Flow Rate (O_2)*

To understand the effect of cathode gas flow rate, experimental results are obtained by changing the cathode gas flow rates (O_2) between 0.5 and 1.2 lpm respectively while the other operating parameters are kept constant.

Fig. 5 Polarization curves for different anode flow rates. Fuel cell temperature is 318 K, pressure is 2.5 atm, anode and cathode humidification temperatures are 318 K, and cathode flow rate is 0.5 lpm

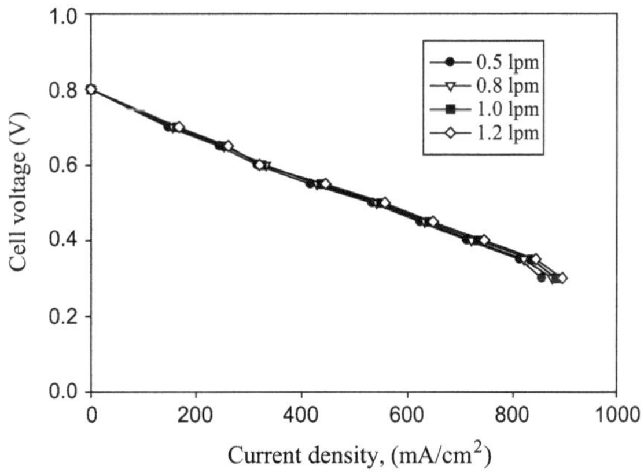

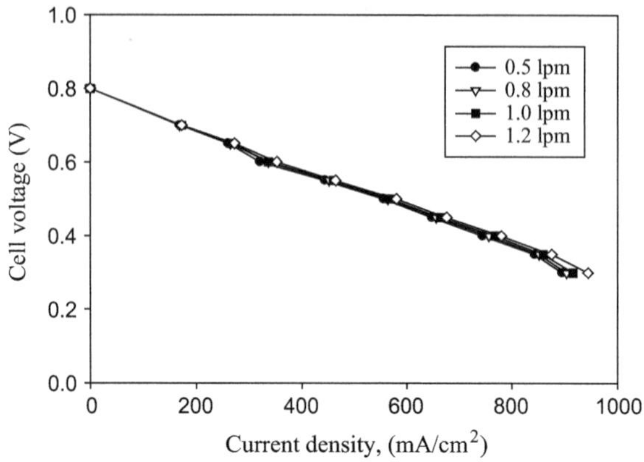

Fig. 6 Polarization curves for different cathode flow rates. Fuel cell temperature is 318 K, pressure is 2.5 atm, anode and cathode humidification temperatures are 318 K, and anode flow rate is 1.2 lpm

The impact of cathode gas flow rate on cell performance is shown in Fig. 6. An increase in cathode gas flow rate from 0.5 to 1.2 lpm results in an increase in fuel cell performance as shown in the graph. From Fig. 6, it could be inferred that an increase of oxygen flow rate results in an increase in the output cell voltage. This phenomenon is due to the high carryover of water by oxygen in the cathode catalyst layer as its flow rate is increased. The polarization curves for different fuel cell temperatures are shown in Fig. 6.

4 Conclusions

Experimental studies were carried out for a 25 cm^2 single PEM fuel cell fitted with an MEA fabricated using 10 % Pt/C with a Pt-loading of 0.5 mg/cm^2. The primary objective was to observe the impact of Pt-loading on fuel cell performance under the influence of variable operating parameters. Based on the study, the following inferences were made:

- The cell performance is enhanced as the operating temperature goes from 303 to 313 K. Increase of membrane conductivity, gas diffusivity and exchange current density at higher operating temperatures may have contributed to this phenomenon.
- Fuel cell performance increases with an increase in operating gas pressure due to an increase in exchange current density and improved partial pressure of reactant gases.
- Changes in anode gas humidification temperature has considerable impact on the cell performance. At anode humification temperature values higher than cell operating temperature, the cell performance increases due to increased membrane conductivity, limiting current density and decreased ohmic over potential.

- Changes in cathode gas humidification temperature has minimal impact on the cell performance. A decrease in the effective porosity of the GDL and a reduction in reactant concentration may have contributed to this phenomenon.
- The cell performance is increased when the flow rates of H_2 and O_2 gases are increased from 0.5 to 1.2 lpm. It can be inferred that an increase of hydrogen and oxygen gas flow rates results in an increase in the limiting current and power.

Acknowledgments The authors express their gratitude to the financial support extended by TEQIP-II Centre of Excellence (CoE) at AU College of Engineering (A), Andhra University for conducting research in the Fuel Cell Technology Laboratory, Department of Chemical Engineering.

References

Khazaee, I., Ghazikhani, M.: Experimental investigation of operating parameters and control the performance of a PEM fuel cell. Int. J. Energy Environ. **3**(3), 461–470 (2012)

Leela Manohar, A., Naidu, S.V., Dharma Rao, V., Sreenivasulu, B.: Parameter analysis of a 500 W PEM fuel cell stack using design of experiments. IUP J. Chem. Eng. 2 (2010)

Liu, H., Li, P., Hartz, A., Wang, K.: Effects of geometry/dimensions of gas flow channels and operating conditions on high-temperature PEM fuel cells. Int. J. Energy Environ. Eng. **6**, 75–89 (2015)

Misran, E., Hasan, N.S.M., Daud, W.R.W., Majlan, E.H., Rasli, M.I.: Water transport characteristics of a PEM fuel cell at various operating pressures and temperatures. Int. J. Hydrogen Energy **38**(22), 9401–9408 (2013)

Mughal, A., Li, X.: Experimental diagnostics of PEM fuel cells. Int. J. Environ. Stud. **63**(4), 377–389 (2006)

Shan, J., Hongtan, L.: Direct measurement of lateral current density distribution in a PEM fuel cell with a serpentine flow field. Int. J. Hydrogen Energy **39**(3), 1449–1456 (2014)

Sreenivasulu, B., Vasu, G., Dharma Rao, V., Naidu, S.V.: Effect of back pressure and flow geometry on PEM fuel cell performance—an experimental study. Int. J. Appl. Sci. Eng. **11**(1), 1–11 (2013)

Wang, L., Husar, A., Zhou, T., Liu, H.: A parametric study of PEM fuel cell performances. Int. J. Hydrogen Energy **28**, 1263–1272 (2003)

Wang, Y., Chen, K.S.: A review of polymer electrolyte membrane fuel cells: technology, applications, and needs on fundamental research. J. Appl. Energy **88**(4), 981–1007 (2011)

Zhang, Y., Smirnova, A., Verma, A., Pitchumani, R.: Design of a proton exchange membrane (PEM) fuel cell with variable catalyst loading. J. Power Sour. **291**, 46–57 (2015)

Zhong, Z., Chen, J., Zhuang, P.: Enhancement of proton exchange membrane fuel cell performance using a novel tapered gas channel. Chin. J. Chem. Eng. **17**(2), 286–297 (2009)

A Study on Utilization of Latex Processing Effluent for Treatment and Energy Recovery in Microbial Fuel Cell

Debabrata Das, Shweta Singh and Srimanta Ray

1 Introduction

Rubber latex is a stable dispersion of colloidal polymeric rubber particles in aqueous phase. The aqueous phase of rubber latex contains 2–2.5 % protein and 1–1.5 % sugars (Roberts 1998; Mathew 2009). Conventionally, rubber is separated from the rubber latex by reducing the pH through addition of acids, mostly organic acid and in some cases sulfuric acid (Mathew 2009). Thus, characteristic process effluent from latex processing has huge environmental footprint and requires thorough effluent treatment prior to disposal. But the treatment of LPE is a challenge due to acidic nature, high nitrogen content and high concentration of organic carbon (Jamatia et al. 2014). The requirement of pretreatment stages for neutralization and extended aeration for nitrogen stabilization in aerobic treatment system makes the treatment of LPE energy intensive and uneconomical (Mohammadil et al. 2010; Nguyen and Luong 2012). On the other hand, rubber is an essential commodity worldwide and almost 50 % of the rubber products are produced from rubber latex. India is the fourth largest rubber producer and Tripura (in northeast India) is the second most rubber producing state of India (Mathew 2009). Almost 72 % of the rubber latex is processed to sheet rubber which produces around 25 l of effluent per kilogram of latex processed (Seneviratne 1997). Accordingly defining a sustainable treatment route for rubber is a research priority in rubber processing regions worldwide. The present study is aimed at treatment of the latex processing

D. Das · S. Singh
Department of Civil Engineering, National Institute of Technology Agartala, Barjala, Jirania, Tripura (West) 799055, India

S. Ray (✉)
Department of Chemical Engineering, National Institute of Technology, Agartala, Barjala, Jirania, Tripura (West) 799055, India
e-mail: rays.nita@gmail.com

R. Mohan B. et al. (eds.), *Materials, Energy and Environment Engineering*,
DOI 10.1007/978-981-10-2675-1_28

effluent (LPE) in a sustainable manner in an effluent treatment process with minimum processing stages.

Microbial degradation of waste is a conventional treatment route (Eckenfelder 1989). In wastewater treatment, the microbial treatment process is responsible for degrading more than 80 % of the organic waste. Typically, in a biological waste treatment process, the organic wastes are oxidized to liberate electrons with microbes as biocatalyst. In absence of an electron harvesting mechanism a pool of energy remain unutilized in biological waste treatment process. Conventionally, the unutilized electrons from organic waste act as a source of energy for microbial processes and are diverted towards cell synthesis, otherwise known as sludge generation. The sludge generated has associated disposal hazard. Hence, directly harvesting electrons liberated from organic waste in biological oxidation process can be a promising technique of energy recovery (Rabaey and Verstraete 2005). The device used for harvesting electricity from reduced organic carbon source (fuel) using microbe mediated oxidation-reduction reaction has gained popularity in recent years as microbial fuel cell (MFC) (Rabaey and Verstraete 2005; Lovely 2006). In an MFC, microbial oxidation of reduced organic compounds at the anode is coupled to a reduction reaction at the cathode and the chemical energy stored in the organic waste (typically reduced carbon compounds) is harvested as electrical energy (Lovely 2006; Reddy et al. 2010). The microbial electricity production provides an alternative route of waste treatment and energy recovery (Rabaey and Verstraete 2005; Mohan et al. 2009). In addition, waste degradation and energy recovery in MFC has several advantages, which include, direct conversion of chemicals to electricity and potential for application in locations lacking electrical infrastructures. A vast array of carbon sources such as waste-water, agricultural by-products and industrial pollutants has been exploited in MFC for energy recovery, but utilization of LPE is yet not reported (Pant et al. 2010). The present study is focused in that direction and is aimed to treat LPE without pre-treatment in MFC with concurrent energy recovery.

2 Materials and Methods

The MFC used in the study (Fig. 1) containing an anode chamber and cathode chamber constructed was fabricated in-house from electrically nonconductive polymeric material. The anode was constructed of polymeric non-woven mesh with stainless steel support, with effective area of 0.003 m^2. The MFC was inoculated with an uncharacterized anaerobic microbial consortium that represents the diversity of microbes found in anaerobic wastewater treatment systems or in the anoxic sediment. Microorganisms are restricted in the anode compartment by the separator. The anode chamber was hermetically sealed having provision for purging (liquid or head space) and liberation of off-gas. The anode, maintained under anaerobic

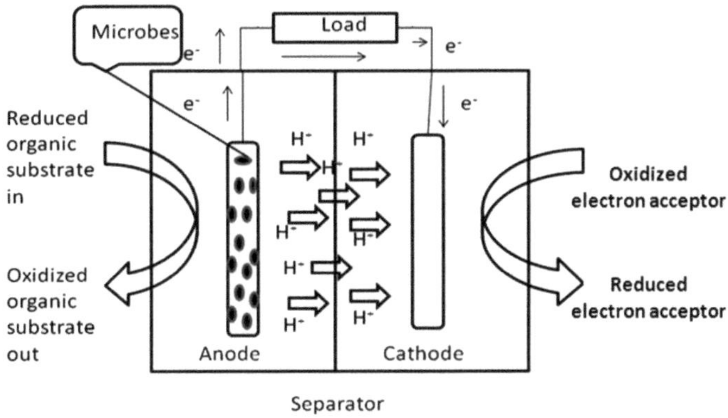

Fig. 1 The schematic of a MFC

conditions, was agitated mechanically. The cathode was constructed of conductive carbon without costly noble metal (platinum) catalyst with effective area of 0.003 m^2. The electrodes were connected to the external circuit by conductive wires. The cathode was maintained aerobic by a continuous purge of air. The anode and the cathode were separated by a membraneless separator constructed from carbon and porous silica-clay (with impurities of metal ions, including but not limited to, sodium, potassium, magnesium, aluminum, and iron). The external circuit was composed of a variable resistance, the resistance value of which can be altered to attain the internal resistance of the MFC. A potentiostat (SP-150, Biologic, France) was connected across the resistance to measure the potential drop (voltage) across the resistance. Current and power outputs were then computed using Ohm's law. The latex processing effluent (LPE) used in the study as substrate for the MFC. The characteristics of the LPE used in the study is shown in Table 1. MFC was operated in batch mode. The carbohydrate and metabolites were analyzed using high performance liquid chromatograph (Flexar, Perkin-Elmer, Singapore) fitted with AMINEX 87H column using 5 mM H_2SO_4 as eluent. The total organic carbon and total kjeldahl nitrogen was analyzed using total organic carbon (TOC) analyzer (TOC-LPH, Shimadzu Asia Pacific, Singapore) and kjeldahl nitrogen analyzer (KEL, Pelican Equipments, India) respectively.

Table 1 The chemical characteristics of the LPE

Sl. No.	Parameter	Value
1	pH	4.6 ± 0.3
2	TKN (Total Kjeldahl nitrogen)	399 (mg/l)
3	TOC (Total organic carbon)	4423 (mg/l)

3 Results and Discussion

3.1 Contaminant Removal

The MFC was operated with LPE in batch mode. The microorganisms in the anode compartment oxidize the organic matter in the LPE and transfer the electrons to the anode. The protons (hydrogen ions) produced by the oxidation half reaction (Eq. 1) pass through the separator to the cathode compartment where oxygen accepts electrons from cathode is reduced with addition of protons to a reduced product (the reduction half reaction (Eq. 2)). As a result of these two half reactions, a potential difference develops between the anode and the cathode and current/electrons flows in the external circuit. The degradation of LPE (substrate) was recorded against time (Fig. 2a) and around 80 % of the LPE was observed to degrade within 18 days time.

$$\text{Anode}: (CH_2O)_n + nH_2O \rightarrow nCO_2 + 4nH^+ + 4ne^- \quad (1)$$

$$\text{Cathode}: 4nH^+ + 4ne^- + nO_2 \rightarrow 2n(H_2O) \quad (2)$$

However, around 22 % of the total organic carbon was observed to mineralize within the said period (Fig. 2b). The earlier studies reported that anaerobic degradation of organic matter consists of several interdependent, sequential and parallel biological reactions resulting in transformation of organic matter mainly into a mixture of volatile fatty acids (VFA) in acidogenic route and which are subsequently degraded into carbon dioxide (Ray et al. 2010). VFA levels of the MFC was monitored and presented in Fig. 3b. Due to transformation of the organic matter into VFA, accordingly VFA level initially increased with time. However, with the further progress of time, VFA levels continued to decrease through the experimentation period. The high initial VFA level in LPE fed MFC was due to the fact that conventionally VFAs, namely formic acid and acetic acid, are used for separation of rubber particle from the natural latex by coagulation method. The increased VFA level in the LPE created an acidic environment in the anaerobic anode. The lower pH

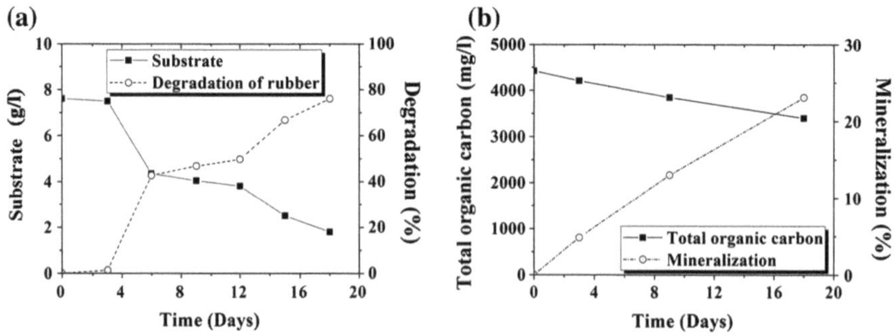

Fig. 2 The performance evaluation of MFC. **a** Degradation of substrate (LPE) and **b** mineralization (removal) of total organic carbon from LPE

was attributed for suppression of methanogenic activity in the anode compartment (confirmed by gas chromatography). Slower mineralization and substrate degradation can also be accounted to acidogenic stresses on the microbial metabolism (Ray et al. 2010; Jadhav and Ghangrekar 2009). The degradation of LPE (80 %) reported in this study is comparable with the degradation percent reported for wastewater in earlier studies (Jain et al. 2015; Ghangrekar and Shinde 2006).

LPE is known for high nitrogen content due to the presence of rubber protein (Roberts 1998). Hence, total Kjeldahl nitrogen was monitored in the LPE fed MFC (Fig. 3a). The removal of nitrogen was slower that the degradation of the substrate. Around 40 % nitrogen removal was noted in 18 days time and 66 % nitrogen was removed through batch operation of MFC with LPE feed in 50 days.

3.2 Energy Recovery

The MFC was observed to produce bio-electricity on feeding with LPE. Accordingly, the electrical parameters of the MFC were monitored to compute the energy recovery from the MFC fed with LPE. The cell potential was measured

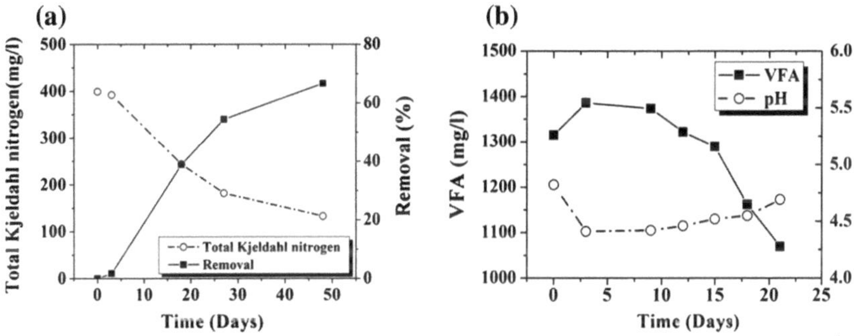

Fig. 3 The performance evaluation of MFC. **a** Removal of nitrogen in LPE and **b** VFA formation from LPE and pH level in MFC

Fig. 4 The performance evaluation of MFC in energy recovery from LPE

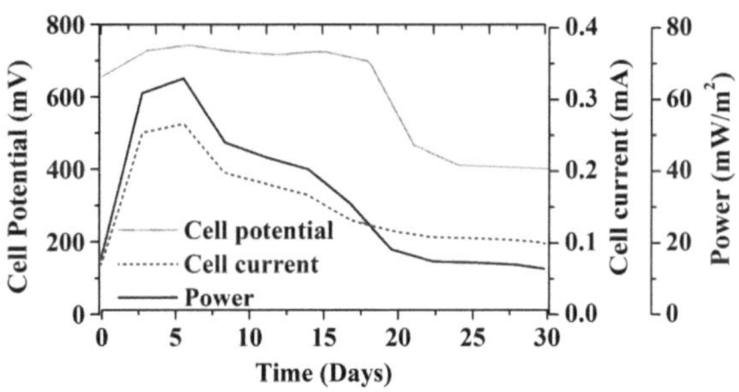

Table 2 The comparative study of the energy output with the literature studies

MFC configuration	Anode	Cathode	Substrate	Power density (mW/m^2)	Reference
Two chamber with clayware	Stainless steel (area = 0.026 m^2)	Carbon felt (area = 0.023 m^2)	Acetate (3.843 g/l)	51.65	Ghadge et al. (2014)
Two chamber with PEM	Carbon paper (2.0 cm × 5.0 cm)	Carbon paper, Pt loaded (0.4 mg/cm^2) (2.0 cm × 5.0 cm)	Glucose (1 g/l)	10.2	You et al. (2006)
Two chamber with nafion 117	Teflon treated carbon paper (area = 13.8 cm^2)	Carbon cloth with Pt catalyst (0.5 mg/cm^2) (area = 13.8 cm^2)	Cheese whey	18.4	Antonopoulou et al. (2010)
Two chamber with nafion 117	Plain carbon paper (area = 22.5 cm^2)	Carbon paper with Pt loading (0.35 mg/cm^2)	Ethanol (70 mg/l)	40	Kim et al. (2007)
Two chamber with nafion 117	Carbon paper (area = 22.5 cm^2)	Carbon paper with Pt loading (0.5 mg/cm^2) (areas = 22.5 cm^2)	Food processing wastewater	81	Oh and Logan (2005)
Two chamber with silica clay	Stainless steel (area 0.003 m^2)	Conductive carbon (area = 0.003 m^2)	Rubber effluent	64	This study

under open circuit condition through the operation of the MFC. The cell current was computed by monitoring the voltage drop against known resistance (Logan et al. 2006). The profile of cell potential and cell current is presented in Fig. 4. Max cell potential of around 0.7 V was observed for 20 days after feeding the MFC with LPE, thereafter the potential drops rapidly. The cell current was observed to vary through the operation of the MFC. The maximum current output of 0.3 mA (0.1 A/m^2) was recorded 5 days after feeding. The recorded power output was compared with the earlier reports where MFC was operated with wastewater (Table 2). The maximum power output reported in the study (64 mW/m^2) was found to be comparable with earlier reports (Kim et al. 2007; You et al. 2006; Antonopoulou et al. 2010). The total energy recovered through 30 days operation of the MFC was computed as 953 kJ/m^2 of electrode area or 7152 kJ per cubic meter of LPE treated.

4 Conclusions

The study reported fabrication of a low cost membrane less MFC without noble metal catalyst. The MFC was operated in batch mode with latex processing effluent (LPE) as feed substrate. Around 80 % degradation of substrate (LPE) was recorded in 18 days with 40 % total nitrogen removal. The degradation was comparable with the values reported in the literature for wastewater. The net power output recorded for the MFC in the present study with LPE is 64 mW per square meter of electrode area. The total energy recovered by the MFC in a batch operation from LPE is computed as 7152 kJ per cubic meter of effluent treated. The study successfully demonstrated the treatment of the LPE in a low cost membrane-less MFC with concurrent energy recovery and thereby present study demonstrates a sustainable utilization route of a problematic effluent, like LPE, using MFC. In addition the MFC treatment technique presented in this study is economical since cost due to aeration in conventional biological treatment can be evaded.

Acknowledgments The authors also like to express their gratitude to the, National Institute of Technology, Agartala (NITA), India for the infrastructure and facilities and TEQIP-II, NITA program for the financial support.

References

Antonopoulou, G., Stamatelatou, K., Bebelis, S., Lyberatos, G.: Electricity generation from synthetic substrates and cheese whey using a two chamber microbial fuel cell. Biochem. Eng. J. **50**, 10–15 (2010)

Eckenfelder, W.W.: Industrial water pollution control. McGraw Hill, New York (1989)

Ghadge, A.N., Sreemannarayana, M., Duteanu, N., Ghangrekar, M.M.: Influence of ceramic separator's characteristics on microbial fuel cell performance. J. Electrochem. Sci. Eng. **4**, 315–326 (2014)

Ghangrekar, M.M., Shinde, V.B.: Wastewater treatment in microbial fuel cell and electricity generation: a sustainable approach. In: Proceedings of 12th Annual International Sustainable Development Research Conference, Hong Kong, Apr 6–8, pp 1–9 (2006)

Jadhav, G.S., Ghangrekar, M.M.: Performance of microbial fuel cell subjected to variation in pH, temperature, external load and substrate concentration. Bioresour. Technol. **100**, 717–723 (2009)

Jain, R., Tiwari, D.C., Mishra, P.: Efficiency and stability of carbon cloth electrodes for electricity production from different types of waste water using duel chamber microbial fuel cell. J. Sci. Ind. Res. **74**, 308–314 (2015)

Jamatia, A., Chakraborty, S., Das, D., Jamatia, S.K., Ray, S., Das, M.K.: Evaluation of Physiochemical characteristics of disposed Rubber industry effluent: a case study of Bodhjungnagar industrial growth centre. IOSR J. of Eng. **4**, 44–50 (2014)

Kim, J.R., Jung, S.H., Regan, J.M., Logan, B.E.: Electricity generation and microbial community analysis of alcohol powered microbial fuel cells. Bioresour. Technol. **98**, 2568–2577 (2007)

Logan, B.E., Hamelers, B., Rozendal, R., Schroder, U., Keller, J., Freguia, S., Aelterman, P., Verstraete, W., Rabaey, K.: Microbial fuel cells: methodology and technology. Environ. Sci. Technol. **40**, 5181–5192 (2006)

Lovely, D.R.: Microbial fuel cells: novel microbial physiologies and engineering approaches. Curr. Opin. Biotech. **17**, 327–332 (2006)

Mathew N.M.: Natural rubber. In: De, S.K., White, J.R., Naskar, K. (eds.) Rubber Technologist's Handbook, 2nd edn, pp. 11–25. Smithers Rapra Press (2009)

Mohammadil, M., Man, H.C., Hassanl, M.A., Yee, P.L.: Treatment of wastewater from rubber industry in Malaysia. African J. Biotechnol. **9**, 6233–6243 (2010)

Mohan, S.V., Raghavulu, S.V., Peri, D., Sarma, P.N.: Integrated function of microbial fuel cell (MFC) as bio-electrochemical treatment system associated with bioelectricity generation under higher substrate load. Biosens. Bioelectron. **24**, 2021–2027 (2009)

Nguyen, N.H., Luong, T.T.: Situation of wastewater treatment of natural rubber latex processing in the Southeastern region. Vietnam. J. Viet. Env. **2**, 58–64 (2012)

Oh, S., Logan, B.E.: Hydrogen and electricity production from a food processing wastewater using fermentation and microbial fuel cell technologies. Water Res. **39**, 4673–4682 (2005)

Pant, D., Van Bogaert, G., Diels, L., Vanbroekhoven, K.: A review of the substrates used in microbial fuel cells for sustainable energy production. Bioresour. Technol. **101**, 1533–1543 (2010)

Rabaey, K., Verstraete, W.: Microbial fuel cells: novel biotechnology for energy generation. Trends Biotechnol. **23**, 291–298 (2005)

Ray, S., Saady, N.M.C., Lalman, J.A.: Diverting electron fluxes to hydrogen in mixed anaerobic communities fed with glucose and unsaturated C18 long chain fatty acids. J. Env. Eng. (ASCE) **136**, 568–575 (2010)

Reddy, L.V., Kumar, S.P., Wee, Y.-J.: Microbial fuel cells—a novel source of energy for new millennium. Curr. Res. Technol. Edu. Topics Appl. Microbiol. Microbial Biotechnol. **2**, 956–964 (2010)

Roberts, A.D.: Natural Rubber Chemistry and Technology. Oxford University Press (1998)

Seneviratne, W.M.G.: Waste water from raw rubber processing industry in Sri Lanka and related environmental aspects. Bull. Rubber Res. Inst. Sri Lanka **35**, 42–48 (1997)

You, S., Zhao, Q., Zhang, J., Jiang, J., Zhao, S.: A microbial fuel cell using permanganate as the cathodic electron acceptor. J. Power Sour. **162**, 1409–1415 (2006)

Effect of Traditionally Synthesized Carbon Nano Particles as Bio-Fuel Blend on the Engine Performance

Shyama Prasad Sajankila, Vinayaka B. Shet, Keshava Joshi and N. Lokeshwari

1 Introduction

Energy crisis especially concerned with fossil fuel availability is one of the major concerns of emerging technology. In this regard, an effort is on to look for alternative sources of fuels such as biodiesel to reduce the use of fossil fuels by blending with diesel. Burning of fossil fuels by combustion engines are the major contributors for environmental pollution. Engine modifications and/or fuel alteration such as generation of various types of blends could a practical approach being tried out to improve the engine efficiency and to reduce pollution. Blending of biodiesel with conventional diesel is in practice (Raja et al. 2015). Use of metal and metal oxide nanoparticles seem to be a promising approach in enhancing the engine efficacy (Saraee et al. 2015). The innate characters of nanoparticles such as surface area, reactivity, thermal conductivity are thought to have role in enhancing the combustion.

In the current investigation, a one step, non-hazardous methods (Mohanty et al. 2007) was employed to generate uniform shape and sized Carbon nanoparticles (Gujar et al. 2013). Blending of carbon nanoparticles with biodiesel was carried out to study the engine characteristics.

S.P. Sajankila (✉) · V.B. Shet
Department of Biotechnology Engineering, NMAMIT (VTU Belagavi),
Nitte 574110, Karnataka, India
e-mail: shyama.sajankila@nitte.edu.in

K. Joshi · N. Lokeshwari
Department of Chemical Engineering, SDMCET (VTU Belagavi),
Dharwad 580002, Karnataka, India

R. Mohan B. et al. (eds.), *Materials, Energy and Environment Engineering*,
DOI 10.1007/978-981-10-2675-1_29

2 Materials and Methodology

2.1 Materials

Cotton wick and earthen lamp was bought from local vender. Freshly prepared ghee from cow's milk (Ai-Khalifah et al. 1993) was used for the investigation.

2.2 Traditional Method of Synthesizing Carbon Nanoparticles

Earthen lamp was filled with 20 ml ghee. Cotton wick was dipped in ghee to achieve uniform cohesion and then placed in the earthen lamp. Lamp was lit to achieve the blue flame. Ghee reaching the tip of wick by capillary action was undergoing combustion in presence of air and size reduction lead to produce smoke containing carbon (Gujar et al. 2013; Mohanty et al. 2007). Smoke was made to deposit on the surface of bronze plate, placed above the flame. In order to prevent excessive heating of the plate, 50 ml of water was poured into the plate. Due to the heat transfer, water was evaporated and water level was maintained by addition of water. Smoke containing the carbon deposit form the heap. Set up was not disturbed till the combustion of entire ghee.

2.3 Materials Characterization of Carbon Nanoparticle

Structural characterization of the carbon nanoparticle was carried out by SEM.

2.4 Blending of Carbon Nanoparticles with Diesel

The carbon nanoparticles blending with biodiesel fuel is prepared by mixing the biodiesel and carbon nano-particles using an ultrasonicator. The ultrasonicator technique is the best suited method to disperse the nano-particles in the base fuel, as it facilitates possible agglomerate nanoparticles back to nanometre range. The 25 mg of carbon nanoparticles are dispersed in the biodiesel with the aid of ultrasonicator at a frequency of 40 kHz for 30 min. The nanoparticles blended diesel is named as BD + 25NP. The same procedure is adopted for the mass fraction of 50 mg of carbon nano particles to prepare the carbon nanoparticles blended diesel fuel (BD + 50NP). Table 1 presents the properties of biodiesel blended with carbon nano particles.

Table 1 Properties of diesel blended with carbon nano particles

Properties	Diesel	BD + 25NP	BD + 50NP
Density, kg/m^3	840	898	910
Calorific value, kJ/kg k	42,390	34,056	33,010
Kinematic Viscosity at 40 °C (cSt)	3.5	4.7	4.8
Flash point in °C	56	160	154

Table 2 Specifications of engine

SI. No	Engine specification	Parameters engine
1	Type of engine	Kirloskar make single cylinder four stroke direct injection diesel engine
2	Nozzle opening pressure	200–205 bar
3	Rated power	5.2 kW (7 HP) @ 770 rpm
4	Cylinder diameter (bore)	87.5 mm
5	Stroke length	110 mm
6	Compression ratio	17.5:1

2.5 Experimental Setup

The performance and emission level for the engine of single cylinder four stroke water cooled direct injection was tested using diesel, biodiesel and diesel blended with carbon nanoparticles fuels.The engine has developed 3.7 kW at 770 rev/min, with a displacement volume of 662 cc, compression ratio of 17.5:1. The specification of the engine is presented in Table 2. The engine injector is each having an orifice diameter of 0.3 mm provided with three holes. The injector opening pressure and the static injection timing are 205 bar and 230 bTDC respectively as per manufacturer specifications. The engine was provided with a hemispherical combustion chamber and cooling was accomplished by circulating water through the jackets on the engine block and cylinder head. The cylinder pressure was measured by a piezoelectric pressure transducer mounted with the cylinder head surface. The smoke was determined by hartridge smoke meter during engine steady state operation and the un-burnt hydrocarbons, carbon monoxide, nitric oxide emissions from engine were measured by five gas analyzer (Saraee et al. 2015).

3 Results and Discussion

3.1 Characterization of Carbon Nanoparticle

The average size of the nanoparticle was found to be in the range of 80.3–88.6 nm with 100 % pure carbon as in Fig. 1.

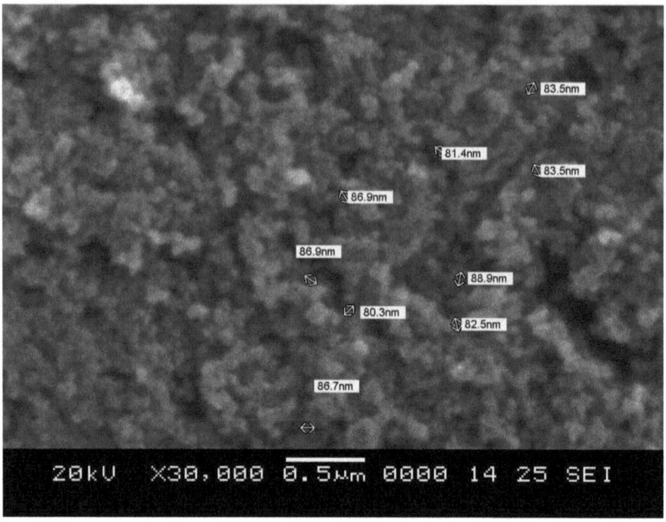

Fig. 1 SEM analysis of the CNP

Fig. 2 Variation of brake thermal efficiency for diesel (*D*), biodiesel (*BD*), BD + 25NP, BD + 50NP

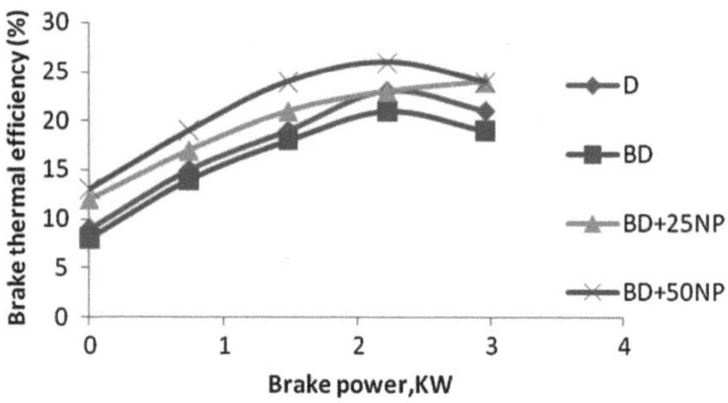

3.2 *Characterization Variation of Brake Thermal Efficiency (BTE)*

The brake thermal efficiency variations for diesel (D), biodiesel (BD) and BD-carbon nanoparticle blended fuels are as shown in Fig. 2. The higher Brake Thermal Efficiency of the engine of 25 % is observed for BD + 50NP fuel. The efficiency of engine is improved by using nanoparticle blended BD when compared to diesel and biodiesel. Biodiesel showed less efficiency as it is having lower volatility and calorific value. The high surface area and reactive surfaces of the nanoparticles might have resulted in higher efficiency (Banapurmath et al. 2014; Bhagwat et al. 2015). The nano-particles provide higher surface area and hence in the case of BD + 50NP fuel, the catalytic activity is enhanced due to the high dosage of carbon nano-particles when compared to BD + 25NP fuel. With increase

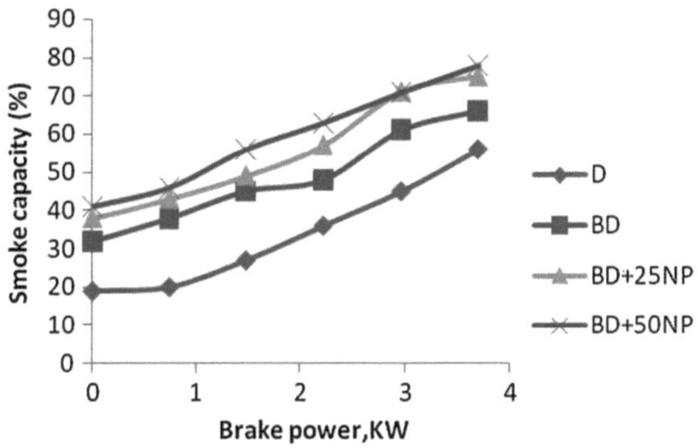

Fig. 3 Effect of smoke capacity for diesel (*D*), biodiesel (*BD*), BD + 25NP, BD + 50NP

in surface area in liquid fuel droplets formed on the nano-particles could also be responsible for the higher efficiency (Banapurmath et al. 2014). These properties provide increased reactivity and faster burning rates of fuel.

3.3 *Effect of Smoke Capacity*

The variation of smoke capacity for biodiesel blended with nano-particles, biodiesel and a diesel fuel is shown in Fig. 3. It is observed that the smoke capacity emissions have increased in biodiesel blend in comparison with diesel. The solid carbon particles were generated from the exhaust of diesel engine in the fuel-rich zones within the cylinder during combustion. The smoke emission capacity has increased for carbon nano particle blending fuel as it contains solid carbon particles.

3.4 *Effect of Smoke Capacity*

The hydrocarbon emission variations for BD and BD-carbon nanoparticle blended fuels are shown in Fig. 4. Due to the lower brake thermal efficiency resulting from incomplete combustion, the HC emission for diesel operation was higher of 100 ppm compared to BD (90 ppm). However HC emissions were marginally lower for the BD-carbon blended fuels compared to BD alone operation due to increased catalytic activity and the improved combustion characteristics of carbon NPs (Banapurmath et al. 2014). BD + 50NP Carbon showed better performance with comparatively lower HC as compared to BD + 25NP Carbon due to the increased dosing level of carbon nano-particles that provided higher surface area resulting in improved combustion characteristics.

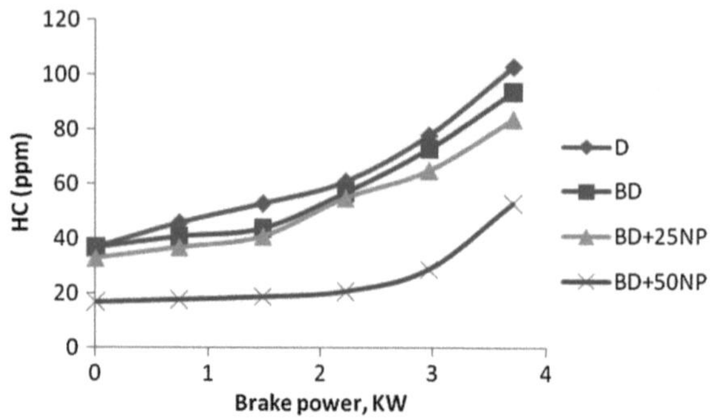

Fig. 4 Variation of HC Emission for diesel (*D*), biodiesel (*BD*), BD + 25NP, BD + 50NP

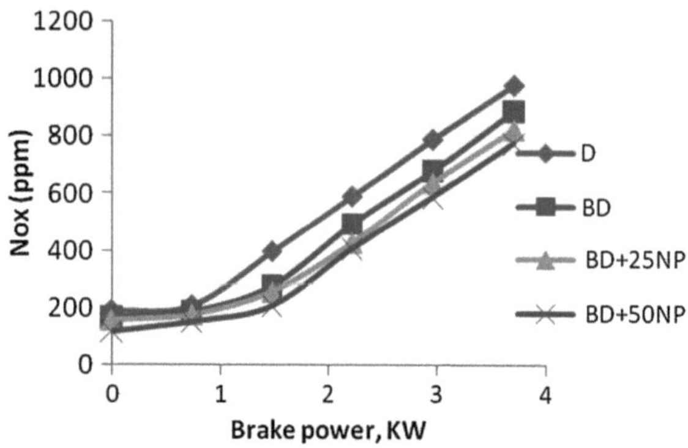

Fig. 5 Variation of NOx Emission for diesel (*D*), biodiesel (*BD*), BD + 25NP, BD + 50NP

3.5 Variation of NOx Emission

NOx emission variations for diesel (D), biodiesel (BD), BD-carbon blended fuels are as shown in Fig. 5. Heat release rates of BD were lower during premixed combustion phase. The lower peak temperatures prevailing inside the combustion chamber leads to lower NOx emissions for BD compared to the diesel operation (Bhagwat et al. 2015). The BD-carbon nanoparticle blended fuels produced lower NOx emission compared to that of BD, due to higher premixed combustion observed with BD- carbon nanoparticles blends. The BD + 50NP Carbon showed lower NOx compared to the BD + 25NP Carbon.

4 Conclusion

The current investigation has revealed that the traditional method of nanoparticle synthesis is economical, and uniform size range (80.3–88.6 nm) and shape could be achieved. The performance of the engine and the emission characteristics of BD, BD-carbon nano-particles (BD + 25NP; BD + 50NP) blended fuels were tested on a single cylinder, constant speed, direct-injection diesel engine. Based on the experimentation, BD resulted in poor performance 20 % at maximum break power tested, in terms of reduced brake thermal efficiency. However BD performance was enhanced with carbon nano-particle additives to 25 % with BD + 50NP. Performance was further improved with higher dosing level of carbon nano-particles in biodiesel. BD alone had about 95 ppm of HC content, while the Nanoparticle addition (BD + 50NP) caused significant reduction in emission except CO was observed.

Acknowledgments The authors wish to thank, Department of Biotechnology Engineering, NMAM Institute of Technology, Nitte, Karkala and Department of Chemical Engineering, SDMCET, Dharwad, Karnataka and for providing research facilities, support and encouragement for carrying out the research work.

References

AI-Khalifa, H.A., AI-Kahtani, H.: Composition of ghee (SamnBa1Ti's) from cow's and sheep's milk. Food Chem. **46**, 373–375 (1993)

Banapurmath, N., Narasimhalu, T., Anand H., Radhakrislman, S., Mohammad Hussain, R., Narasimhan, A., Rohan, K.: Effect of silver nano-particle blended biodiesel and swirl on the performance of diesel engine combustion. Int. J. Sustain. Green Energy. **3**(6), 150–157 (2014)

Gujar, K.N., Gambhire, M.S., Gambhire, V.M., Shelkel, G.T.: Preparation of carbon nanoparticle by lamp soot method and loading of ciprofloxacin hydrochloride. J. Pharm. Assoc. India. **1**(1), 26–30 (2013)

Mohanty, B., Verma, A.K., Claesson, P., Bohidar, H.B.: Physical and antimicrobial characteristics of carbon nanoparticles prepared from lamp soot. Nanotechnology **18**, 445102 (2007)

Raja, V., Kumar, A., Sakthinathan, G.: A review of nano additive in diesel. J. Chem. Pharma. Sci. JCHPS Special Issue **7**, 400–403 (2015)

Saraee, S.H., Jafarmadar, S., Taghavifar, H., Ashrafi, S.J.: Reduction of emissions and fuel consumption in a compression ignition engine using nanoparticles. Int. J. Environ. Sci. Technol. **12**, 2245–2252 (2015)

Vishwajit, A., Bhagwat, C.P., Banapunnath, N.: Graphene nanoparticle—biodiesel blended diesel engine. Int. J. Eng. Res. Technol. (IJERT) **4**(2), 75–78 (2015)

Optimization of Chitosan Nanoparticles Synthesis and Its Applications in Fatty Acid Absorption

Ritu Raval, Raj H. Rangnekar and Keyur Raval

1 Introduction

Chitosan is a linear polysaccharide composed of β-(1-4)-linked D-glucosamine and N-acetyl-D-glucosamine. It is obtained commercially by chemically treating the chitin obtained from shrimp and other crustacean shells with the alkali; sodium hydroxide (Austin et al. 1981). Owing to the cationic polyamines present on the chitosan, it has many commercial and biomedical applications. The ability of the biopolymer to limit fat absorption has been exploited in the present research. The studies have involved the hypercholesterolemic mice model treated with chitosan varying in degree of polymerization (DP) and acetylation (DA) (Baker et al. 2009; Patti et al. 2015; Stefan et al. 2014; Sugano et al. 1988; Zhang et al. 2008). The usage of chitosan in fat scavenging has also been extended to human patients. The bottleneck in the fat absorption has been the linear morphology of chitosan. This manuscript addresses and improvises on this limitation by making nanoparticles of chitosan. The spherical morphology of the nanoparticle would improve the surface area and thus the ability to absorb fats. In making these nanoparticles, the Ionotrophic gelation method is used. This method makes use of the electrostatic interaction between the amine group of chitosan (dissolved in the organic acid) and

R. Raval · R.H. Rangnekar
Manipal Institute of Technology, Manipal 576104, Karnataka, India
e-mail: ritu.raval@manipal.edu

R.H. Rangnekar
e-mail: rajrd12@gmail.com

K. Raval (✉)
National Institute of Technology Karnataka, Surathkal 575025, Karnataka, India
e-mail: keyurnraval@gmail.com

R. Mohan B. et al. (eds.), *Materials, Energy and Environment Engineering*,
DOI 10.1007/978-981-10-2675-1_30

negatively charge group of polyanion such as Tri-polyphosphate (Krishna et al. 2011). The interaction can be controlled by the charge density of TPP and Chitosan, which is dependent on the pH of the solution The above method holds an advantage of being not only fast but is also economical.

2 Materials and Methods

2.1 Preparation of Chitosan Nanoparticles

Chitosan at a concentration of 1 mg/ml was dissolved in 3 % organic acid. The nanoparticles were made by dropwise addition of Na-TPP solution (0.4 mg/ml) to the beaker containing chitosan solution. The resulted opalescent solution was then ultrasonicated for 10 min (pulse of 30 s on and 30 s off).

2.2 Testing of Size of Chitosan Nanoparticles Using Zeta Analyzer

The samples were placed in Malvern Zetasizer Nano ZS and the particle size distribution (PSD) and zeta potential (ZP) of the Nanoparticles were analyzed.

2.3 Testing for Fat Absorption of Chitosan Nanoparticles

The chitosan nanoparticles (500 μl) were mixed with 500 μl of ω-3 fatty acid. Two set-up of the experiment were performed. In the first, the chitosan was kept for 0 min and in another it was incubated for a period of 24 h. The mixture was then vortexed for 5 min, followed by centrifugation at 1000 rpm for 2 min. The upper layer (300 μl) of chitosan nanoparticle was separated using a micropipette and later washed with 300 μl of hexane to absorb any fatty acid which might have been absorbed. This hexane extract was further esterified using methanol (94.8 %), concentrated HCl (4.8 %) and BHT (Butylated Hydroxy Toluene—0.4 %) mixture. 1 μl of methyl esters are then loaded in a Gas Chromatographic (GLC) Column to test for fat absorption. The GLC system equipped with a flame ionization detector (FID) and a capillary column were used. The oven temperature was raised from 50 to 400 °C. Nitrogen was used as the carrier gas and the sample run was 30 min (Vorbeck et al. 1961).

3 Results and Discussion

The main aim of the study was to prepare chitosan nanoparticles for fat absorption. The choice of the organic acid used for the preparation of nanoparticle was based on the future application as a fat absorber in food. Hence three acids viz acetic, citric and lactic acid were tested for dissolving chitosan. All three acids were used at a concentration of 3 %. In citric acid, the particles were in the range of 15–48 nm, while in acetic acid the range was 11–37 nm. In the case of lactic acid it was between 24 and 78 nm. It is found that size groups present in the acid had affected the both the size and the zeta potential of the nanoparticle as shown in Table 1.

It is observed that citric acid possessing three carboxylic groups gives the best results in terms of zeta potential and size. This is followed by acetic acid which has a methyl group attached to the carboxylic acid followed by lactic acid with the methyl and hydroxyl group present with the carboxylic acid.

All the formulations were then tested for the fat absorption capabilities. The experimental setup was tried with 24 h incubation of the nanoparticles with the fatty acid with a control kept of 0 min. 1 mg/ml of the chitosan nanoparticle were able to absorb 28.73 mg of fatty Acid after 24 h incubation time (Fig. 1). The peak at 9.67 min denotes the w-3 fatty acid absorbed by the chitosan nanoparticle. An increase in the peak height after 24 h denotes an improved absorption of the fatty acid. We find an improved fat absorption of the nanoparticles as compared to the already reported literature where Walke et al. has used chitosan and observed 706 % fat absorption (Walke et al. 2014).

Table 1 Zeta potential of the chitosan nanoparticles in the different organic acids

Chitosan nanoparticle solvent	Zeta potential (mV)
Citric acid solution	44.6
Acetic acid solution	41
Lactic acid solution	38.2

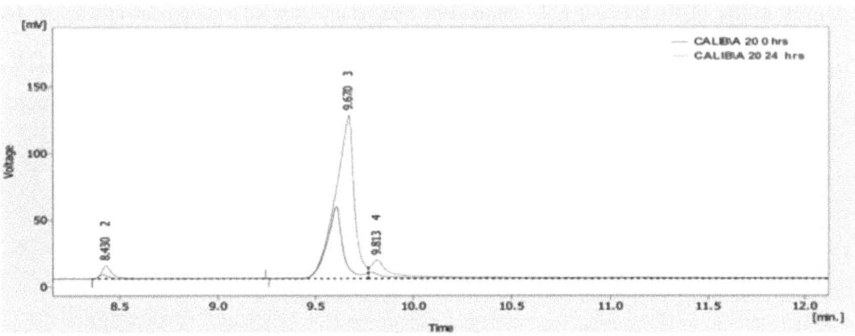

Fig. 1 The chromatogram showing the fat absorption by the chitosan nanoparticle obtained graph obtained after 0 and 24 h of incubation with the w-3 fatty acid sample

4 Conclusion

The ability of Chitosan to bind to the fat has been exploited in this study. Chitosan have been proved to be able to bind to the fats and render them unavailable to the body thus reducing the tendency of the body to absorb them. This study addressed the limitation of the chitosan morphology in its innateness to absorb fat by making nanoparticles. The size and the surface charge of the nanoparticles have been found to be dependent on the organic solvent in which the nanoparticles are prepared. In addition it was also observed that the capacity to absorb fats had also improved by 1.4 fold as compared to chitosan. Further studies have to be done to understand the mechanism of this absorption.

References

Austin, P.R., Brine, C.J., Castle, J.E., Zikakis, J.P.: Chitin: new facets of research. Science **212** (4496), 749–753 (1981)

Baker, W.L., Tercius, A., Anglade, M., White, C.M., Coleman, C.I.: A meta-analysis evaluating the impact of chitosan on serum lipids in hypercholesterolemic patients. Ann. Nutr. Metab. **55**, 368–374 (2009)

Krishna, A., Sailaja Amareshwar, P., Chakravarty, P.: Different techniques used for the preparation of nanoparticles using natural polymers and their application. Int. J. Pharm. Pharm. Sci. **3** (Suppl 2) (2011)

Patti, A.M., Katsiki, N., Nikolic, D., Al-Rasadi, K., Rizzo, M.: Nutraceuticals in lipid-lowering treatment a narrative review on the role of chitosan. Angiology **66**, 416–421 (2015)

Stefan, J., Lorkowska-Zawicka, B., Kaminski, K., Szczubialka, K., Nowakowska, M., Korbut, R.: The current view on biological potency of cationically modified chitosan. JPP **6**, 2 (2014)

Sugano, M., Watanabe, S., Kishi, A., Izume, M., Ohtakara, A.: Hypocholesterolemic action of chitosans with different viscosity in rats. Lipids **23**, 187–191 (1988)

Vorbeck, M.L., Mattick, L.R., Lee, F.A., Pederson, C.S.: Preparation of methyl esters of fatty acids for gas-liquid chromatography. Quantitative comparison of methylation techniques. Anal. Chem. **33**, 1512–1514 (1961)

Walke, S., Srivastava, G., Nikalje, M., Doshi, J., Kumar, R., Ravetkar, S., Doshi, P.: Physicochemical and functional characterization of chitosan prepared from shrimp shells and investigation of its antibacterial, antioxidant and tetanus toxoid entrapment efficiency. Int. J. Pharm. Sci. Rev. Res. **26**, 215–225 (2014)

Zhang, J., Liu, J., Li, L., Xia, W.: Dietary chitosan improves hypercholesterolemia in rats fed high-fat diets. Nutr. Res. **28**, 383–390 (2008)

Biosynthesis of Silver Nanoparticles Using Turmeric Extract and Evaluation of Its Anti-Bacterial Activity and Catalytic Reduction of Methylene Blue

Sneha Nayak, Louella C. Goveas and C. Vaman Rao

1 Introduction

Green synthesis of nanomaterials is a need in today's world to develop ecofriendly clean approaches for nanoparticles preparation using natural chemicals which are non-toxic and reducing agents which are renewable (Moritz and Moritz 2013). Synthesis of nanoparticles using biological methods is called green synthesis of nanoparticles. This has motivated researchers to develop better routes for synthesis that allows better control of size and shape for variety of nanotechnological applications (Saifuddin et al. 2009). Silver nanoparticles (Agnps) have been reported for Methylene blue dye reduction in the presence of arsine (Kundu et al. 2002) while Mallick et al. (2006), studied phenosafranin dye reduction. The present work aims at reducing a highly toxic (Ramsay et al. 2007) dye Methylene blue by TUAgnps. Thawatchai et al. (2008), produced bacterial cellulose by *Acetobacter xylinum* (strain TISTR 975), impregnated the same with silver nanoparticles and efficiently used it as antimicrobial wound dressings. In the present study, cotton bandage gauze was impregnated with TUAgnps and tested for its antimicrobial activity against *Bacillus subtilis* and *Pseudomonas aeruginosa* to see whether TUAgnps can be efficiently used in antimicrobial wound dressings to achieve better wound healing.

S. Nayak (✉) · L.C. Goveas · C. Vaman Rao
Department of Biotechnology Engineering, NMAMIT,
Udupi district, Nitte 574110, India
e-mail: snehanayak88@yahoo.com

R. Mohan B. et al. (eds.), *Materials, Energy and Environment Engineering*,
DOI 10.1007/978-981-10-2675-1_31

2 Methodology

2.1 *Preparation of Extract*

Turmeric tubers were washed in deionized water, 10 % (w/v) suspension was prepared by grinding the preweighed tuber in deionized water. The suspensions were kept in water bath maintained at 80 °C for half an hour, which was later filtered and stored in refrigerator (10 °C) till further use.

2.2 *Synthesis of TUAgnps*

25 ml of cell free extract was added to 225 ml of 1 mM aqueous $AgNO_3$ prepared in deionized water and mixed thoroughly, the resultant mixture is kept in the shaker at 100 rpm at room temperature.

2.3 *Characterization of TUAgnps*

2.3.1 UV-Visible Spectroscopic Characterization of TUAgnps

The colour change in the solution was measured at regular intervals by measuring absorbance at 360–700 nm in UV-visible spectrophotometer (UV1-Thermo electronic corporation, Merck).

2.3.2 FT-IR Spectroscopic Studies

The functional groups existing in the phytoconstituents found in turmeric tuber extract and their participation in the biosynthesis of TUAgnps was determined by recording the IR on Bruker Alpha.

2.3.3 Particle Size Distribution and Zeta Potential

Zetasizer nano instrument (Malvern) was used for data acquisition and for estimating the particle size distribution as well as zeta potential distribution of a colloidal solution of nanoparticles.

2.3.4 SEM and EDX Analysis

Size and morphology was examined by (SEM) JOEL JSM 6380 LA by sputtering, coupled with energy dispersive X-ray (EDX) for elementary composition. The sputtered samples were analyzed for its microstructure at 20 kV. SEM specimens were prepared by taking small amount of TUAgnp powder dispersed on carbon tape.

2.4 Effect of Biosynthesized TUAgnps on the Methylene Blue Reduction and Its Evaluation

For the study, Methylene blue was purchased from Loba chemie. Catalytic activity of synthesized TUAgnps was assessed by carrying out two reactions in a quartz cuvette of 3.5 ml capacity and absorbance was measured using UV-Vis spectrophotometer. A mixture of 1 ml of Methylene blue (1×10^{-4} M) along with 0.2 ml of aqueous tuber extract and 1.8 ml of water was prepared in the first reaction, which was then monitored after 15 min. In the second reaction, a mixture of 1 ml of Methylene blue with 0.2 ml of aqueous tuber extract and 1.8 ml of synthesized TUAgnps was prepared, and this reaction was monitored at 3 different time intervals, i.e., 15, 30 and 45 min. The total reaction volume of the mixture was made up to 3 ml and the absorption maxima (λ_{max}) values were checked in comparison with that of pure Methylene blue.

2.5 Immobilization of TUAgnps on Cloth and Disk Diffusion Studies

The antimicrobial activities of cotton bandage gauze impregnated with TUAgnps were investigated against *Bacillus subtilis* and *Pseudomonas aeruginosa* as model bacteria. Cotton bandage gauze was cut into 1.5 cm discs and impregnated with TUAgnps by dipping it in colloidal solution for 30 min followed by drying and autoclaving for 15 min at 120 °C which was then placed on 1 % nutrient agar plates which was spread with 100 µl of fresh bacterial culture. The culture plates were then incubated at around 37 °C for a day and zone of inhibition was then monitored.

3 Results

3.1 *Characterization of TUAgnps*

3.1.1 UV-Visible Spectrophotometer

Change in solution colour from light yellowish to brown indicated the formation of TUAgnps. UV-Visible spectrophotometer confirmed the production of TUAgnps at 430–460 nm (Fig. 1). The surface plasmon resonance peaks for Agnps produced by plant extracts are comparable to the literature reports (Mallikarjun et al. 2011).

3.1.2 FT-IR Spectroscopic Studies

Figures 2 and 3 show the FT-IR spectrum of turmeric tuber extract and TUAgnps respectively. The IR absorption spectra seen in the aqueous extract were also seen in phytocapped TUAgnps. This is a clear indication that the phytoconstituents like protein and amino acids present in extract lead to stabilization of nanoparticles and protect the TUAgnps from aggregation.

3.1.3 Particle Size Distribution and Zeta Potential

Figure 4 shows the particle size distribution of nanoparticles with the Z-average value of 141.4 nm, polydispersity index (PDI) of 0.5 which is low and Zeta potential of −0.88 for TUAgnps. This indicates that, the capping agents existing on the surface of TUAgnps are mainly comprising of negatively charged groups which may be responsible for the moderate stability of nanoparticles.

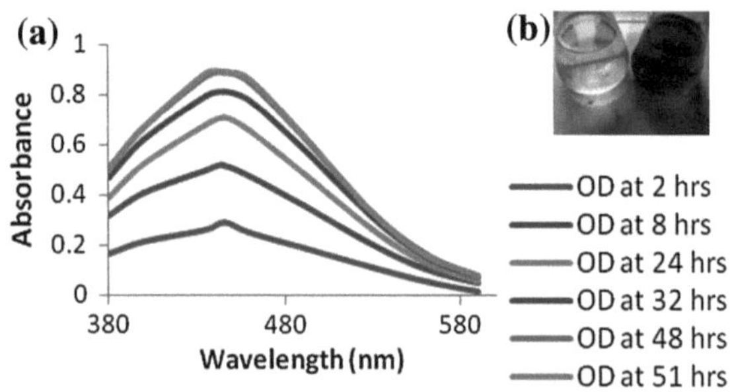

Fig. 1 UV-Visible absorption spectra of TUAgnps (**a**) and conical flasks confirming TUAgnps synthesis (**b**)

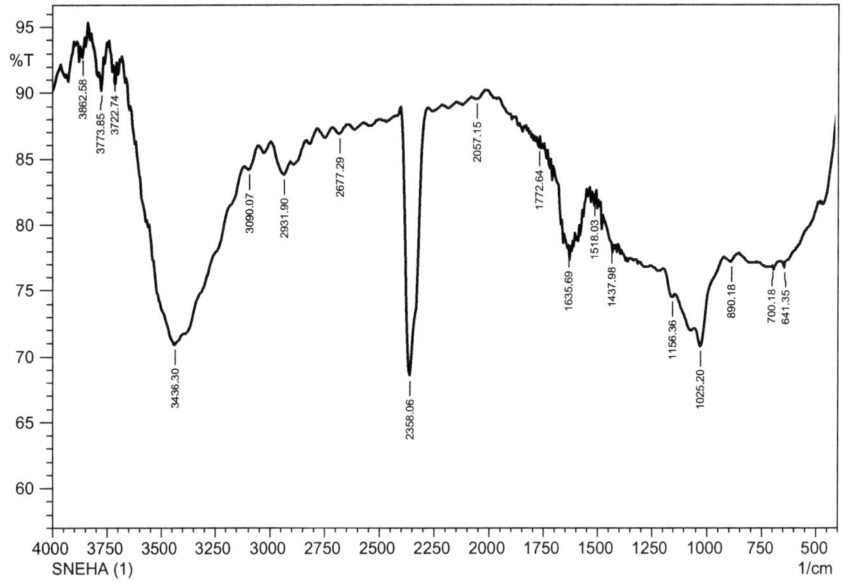

Fig. 2 FT-IR spectra of turmeric tuber extract

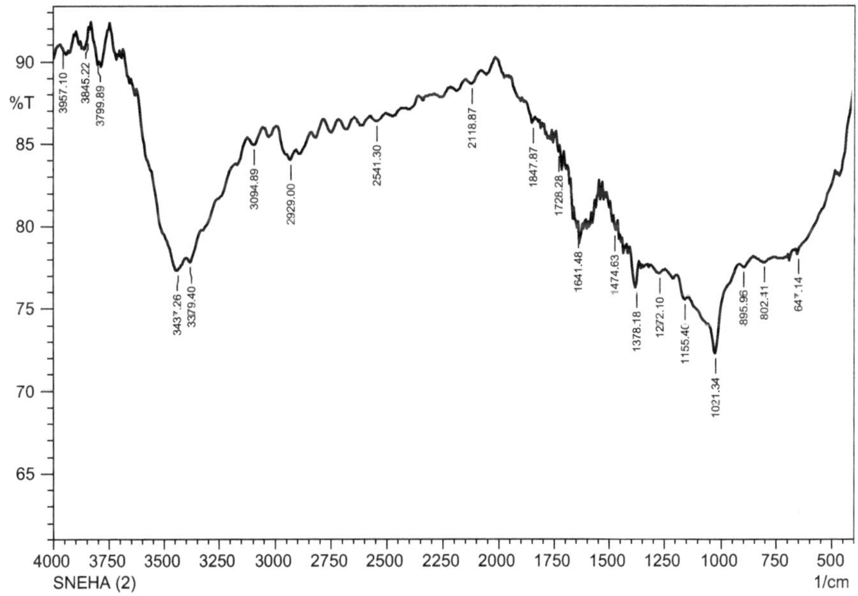

Fig. 3 FT-IR spectra of TUAgnps

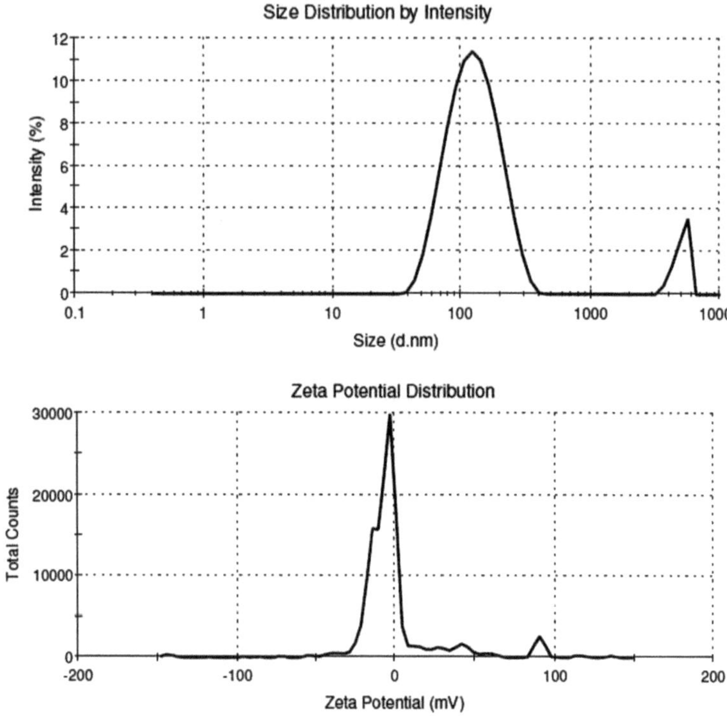

Fig. 4 Zetasizer and Zeta potential distribution of biosynthesized TUAgnps

3.1.4 SEM and EDX Analysis

Figure 5 reveals that the silver nanoparticles formed were spherical in shape, with size ranging from 120 to 160 nm and EDX confirmed the purity of the sample by giving its elemental composition (Fig. 5). In a study, AgNPs synthesized using *Candida albicans* had average size of 30 nm (Ahmad et al. 2013). *Ocimum sanctum* leaf extract induced synthesis of AgNPs had average size of 10–20 nm (Philip and Unni 2011). AgNPs synthesized using *Mentha piperita* leaf extract had average size of 90 nm (Ali et al. 2011). Compared to the earlier reports of biosynthesis of AgNPs, TUAgnps biosynthesized in the present study from turmeric extract were found to be quite large.

3.2 Methylene Blue Dye Reduction by TUAgnps and Its Catalytic Activity

The present study was carried out to see the Methylene blue reduction by the aqueous tuber extract of turmeric along with TUAgnps. Pure Methylene blue has

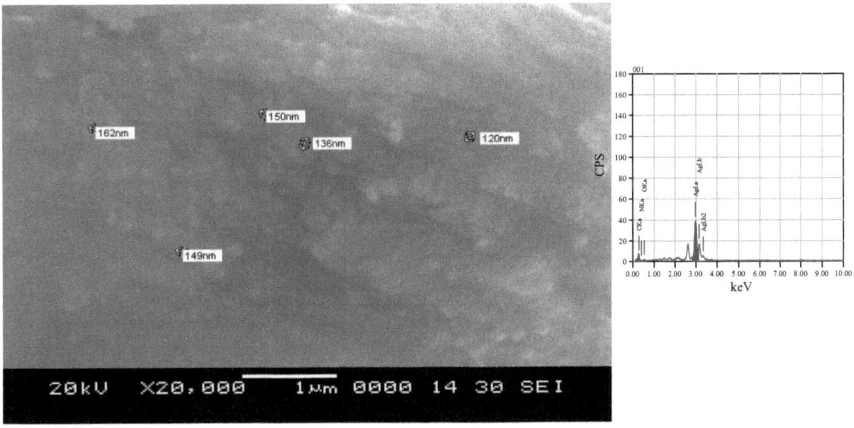

Fig. 5 Scanning electron micrograph and EDX image of TuAgnps impregnated cotton fiber

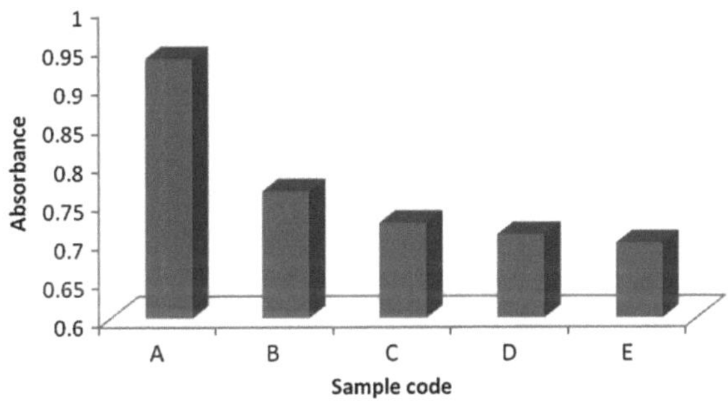

Fig. 6 Methylene blue degradation by synthesized TUAgnps. *A* = Methylene blue, *B* = Methylene blue + Extract after 15 min, *C* = Methylene blue + Extract + TUAgnps after 15 min, *D* = Methylene blue + Extract + TUAgnps after 30 min, *E* = Methylene blue + Extract + TUAgnps after 45 min

λ_{max} at 664 nm and 15 min after the addition of the aqueous tuber extract to the dye, a gradual decrease in absorbance is seen and λ gets shifted to higher wavelength. The decrease in absorbance of methylene blue by the phytoextract is indicative of degradation of dye (Fig. 6). The mixture of dye, TUAgnps and the extract showed a remarkable decrease in the absorbance peak of Methylene blue at the end of 45 min time interval and increase in the absorbance peak of TUAgnps (Table 1). This indicates that, TUAgnps acts as a redox catalyst for transfer of electrons between the tuber extract and the dye Methylene blue, which can be termed as electron relay effect (Gupta et al. 2011).

Table 1 UV-Vis readings of methylene blue dye reduction by Turmeric extract in the presence of TUAgnps

Wavelength (nm)	Methylene blue at 0 min	Methylene blue + Extract (15 min)	Methylene blue + Extract + TUAgnps		
			15 min	30 min	45 min
435	0.01	0.06	1.034	1.053	1.069
664	0.934	0.763	0.722	0.707	0.696

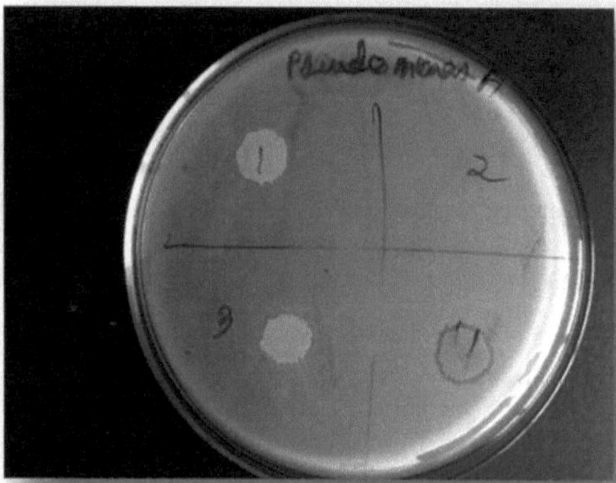

Fig. 7 Antibacterial inhibition zones by TUAgnps impregnated cotton bandage gauze against *Pseudomonas aeruginosa*

3.3 *Antimicrobial Activity Studies*

The antibacterial activity of TUAgnps impregnated cotton bandage gauze against *Bacillus subtilis* and *Pseudomonas aeruginosa* was measured by disc diffusion method where, inhibition zones were seen, i.e., 12 mm for *Bacillus subtilis and* 9 mm for *Pseudomonas aeruginosa*. No inhibition zone was observed with plain cotton bandage gauze and gauze with extract (Control) (Fig. 7).

4 Conclusions

By employing ecofriendly, economical and biological approach using aqueous extract of turmeric tuber Agnps capped with plant phytochemicals were synthesized. The synthesized TUAgnps could be effectively used for Methylene blue dye reduction and TUAgnps impregnated cotton bandage gauze could be effectively used as an antimicrobial wound dressing which was evident from the results.

References

Ahmad, T., Wani, I.A., Manzoor, N., Ahmed, J., Asiri, A.M.: Biosynthesis, structural characterization and antimicrobial activity of gold and silver nanoparticles. Colloids Surf B: Biointerfaces, **107**, 227–234 (2013)

Ali, D.M., Thajuddin, N., Jeganathan, K., Gunasekaran, K.: Plant extract mediated synthesis of silver and gold nanoparticles and its antibacterial activity against clinically isolated pathogens. Colloids Surf B: Biointerfaces, **85**, 360–365 (2011)

Gupta, N., Singh, H.P., Sharma, R.K.: Metal nanoparticles with high catalytic activity in degradation of methyl orange: an electron relay effect. J. Mol. Catal. A: Chem. **335**, 248–252 (2011)

Kundu, S., Ghosh, S.K., Mandal, M., Pal, T.: Silver and gold nanocluster catalyzed reduction of methylene blue by arsine in micellar medium. Bull. Mater. Sci. **25**, 577–579 (2002)

Mallick, K., Witcomb, M., Scurrell, M.: Silver nanoparticle catalysed redox reaction: an electron relay effect. Mater. Chem. Phys. **97**, 283–287 (2006)

Mallikarjun, K., Narsimha, G., Dillip, G., Praveen, B., Shreedhar, B., Lakshmi, S.: Green synthesis of silver nanoparticles using Ocimum leaf extract and their characterization. Dig. J. Nanomater. Biostruct. **6**, 181–186 (2011)

Moritz, M., Moritz, M.G.: The newest achievements in synthesis, immobilization and practical applications of antibacterial nanoparticles. Chem. Eng. J. **228**, 596–613 (2013)

Philip, D., Unni, G.: Extracellular biosynthesis of gold and silver nanoparticles using Krishna tulsi (*Ocimum sanctum*) leaf. Physica E **43**, 1318–1322 (2011)

Ramsay, R.R., Dunford, C., Gillman, P.K.: Methylene blue and serotonin toxicity: inhibition of monoamine oxidase A (MAO A) confirms a theoretical prediction. Br. J. Pharmacol. **152**, 946–951 (2007)

Saifuddin, N., Wong, C.W., Yasumira, A.A.N.: Rapid biosynthesis of silver nanoparticles using culture supernatant of bacteria with microwave irradiation. E-J. Chem. **6**, 61–70 (2009)

Thawatchai, M., Seiichi, T., Ratana, R.: Impregnation of silver nanoparticles into bacterial cellulose for antimicrobial wound dressing. Carbohydr. Polym. **72**, 43–51 (2008)

Comparison of Metal Oxide Nanomaterials: Humidity Sensor Applications

CH. Ashok, K. Venkateswara Rao and CH. Shilpa Chakra

1 Introduction

Nanostructured materials were showing variety of applications because of its properties such as dimensionality, high surface to volume ratio and surface plasmon resonance. Semiconductor metal oxide nanostructured materials used in many potential applications as optoelectronics, solar cells, catalytic, coatings and sensors (Qi et al. 2008; Anne et al. 2010). In this present work concentrated ZnO, CuO and TiO_2 metal oxide nanostructured materials synthesis and humidity sensor application. Zinc oxide is a metal oxide semiconductor material which belongs to II–VI semiconductor group. At room temperature ZnO has a wide band gap of 3.37 eV and 60 meV large exciton binding energy (Suhendro and Rosari 2011; Askarinejad et al. 2011). It also has high stability, piezoelectric nature, good electrical, optical properties, non-toxicity and low-cost (Kankanit et al. 2013). Copper oxide is a material which belongs to p-type semiconductor material; it has good electrical, magnetic, optical, physical properties and band gap ranging from 1.2 to 2.0 eV, depends on the all above properties it is used in optoelectronic device applications (Meilong et al. 2012; Khorsand et al. 2011). Titanium dioxide is also the semiconducting material which has ranging from 2.85 to 3.15 eV band gap. Depending on the physical properties of TiO_2, its showing three different crystalline phases such as anatase, rutile and brookite. This is used in various applications due to its non-toxicity nature, controlling oxidation strength, high chemical stability and wide

CH. Ashok (✉) · K. Venkateswara Rao · CH. Shilpa Chakra
Centre for Nano Science and Technology, Institute of Science and Technology, Jawaharlal Nehru Technological University Hyderabad, Kukatpally, Hyderabad 85, Telangana, India
e-mail: chinthakuntlaashok@gmail.com

R. Mohan B. et al. (eds.), *Materials, Energy and Environment Engineering*,
DOI 10.1007/978-981-10-2675-1_32

band gap range (Suresh et al. 2010). The nanostructured metal oxide materials prepared by microwave-assisted technique using room temperature ionic liquids. Microwave-assisted process is very easy, simple and rapid heating technique. Room temperature ionic liquids having high ionic conductivity, heat capacity and low vapour pressure. ZnO, CuO and TiO_2 nanostructured metal oxides prepared and studied the humidity sensor application using controlled humidity chamber, hygrometer and digital multi meter.

2 Experimental Details

Zinc acetate, Copper acetate and Titanium tetra isopropoxide (TTIP) have been used as initial precursor materials. Along with this Sodium hydroxide, Ethanol, Ionic liquids (1-butyl-3-methyl-imidazolium-tetrafluoro borate and 1-ethyl-3-methyl-imidazolium-ethyl sulphate) was utilised. Metal precursor is dissolved in water and made solution (In case of TTIP, ethanol is taken as solvent). 0.1 M of NaOH solution and Ionic liquid (0.8 mL) was added to the above solution drop by drop with simultaneous and vigorous stirring for 30 min at 45 °C. This solution beaker was covered with Teflon cap and kept in microwave oven for 5 min. The obtained suspension was cooled to room temperature, washed with water and ethanol several times. This solution is centrifuged at 1100 rpm for 10 min and dried at 60 °C for 90 min. The obtained nanomaterials has been converted into pellet form applying high pressure.

3 Results and Discussions

3.1 X-Ray Diffractometer

XRD-pattern of ZnO, CuO and TiO_2 nanostructured materials is shown in Fig. 1. The crystal structure, average crystallite size and lattice parameters are measured by XRD. The strong peaks are observed at 31, 34, 36, 47, 56, 62, 67, 69 and 77 along with (h k l) values (1 0 0), (0 0 2), (1 0 1), (1 0 2), (1 1 0), (1 0 3), (1 1 2), (2 0 1) and (2 0 2) respectively. The results match with JCPDS card number: 36-1451. ZnO is hexagonal with lattice parameters are a = b = 0.3249 nm and c = 0.5206 nm.

The CuO peaks are observed at 32, 35, 38, 48, 53, 58, 61, 66 and 68 with (h k l) values ($\bar{1}$ 1 1), (2 0 0), ($\bar{2}$ 0 2), (0 2 0), (2 0 2), ($\bar{1}$ 1 3), ($\bar{3}$ 1 1) and ($\bar{2}$ 2 0) respectively. The top bar of the (h k l) values indicate the negative direction at the plane. The results match with JCPDS card number: 45-0937. CuO is monoclinic with lattice parameters a = 0.4685 nm, b = 0.3425 nm and c = 0.5130 nm. The TiO_2 peaks are observed at 25, 38, 48, 53, 55 and 63 along with (h k l) values (1 0 1), (0 0 4), (2 0 0), (1 0 5), (2 1 1) and (2 0 4) respectively. These results match with

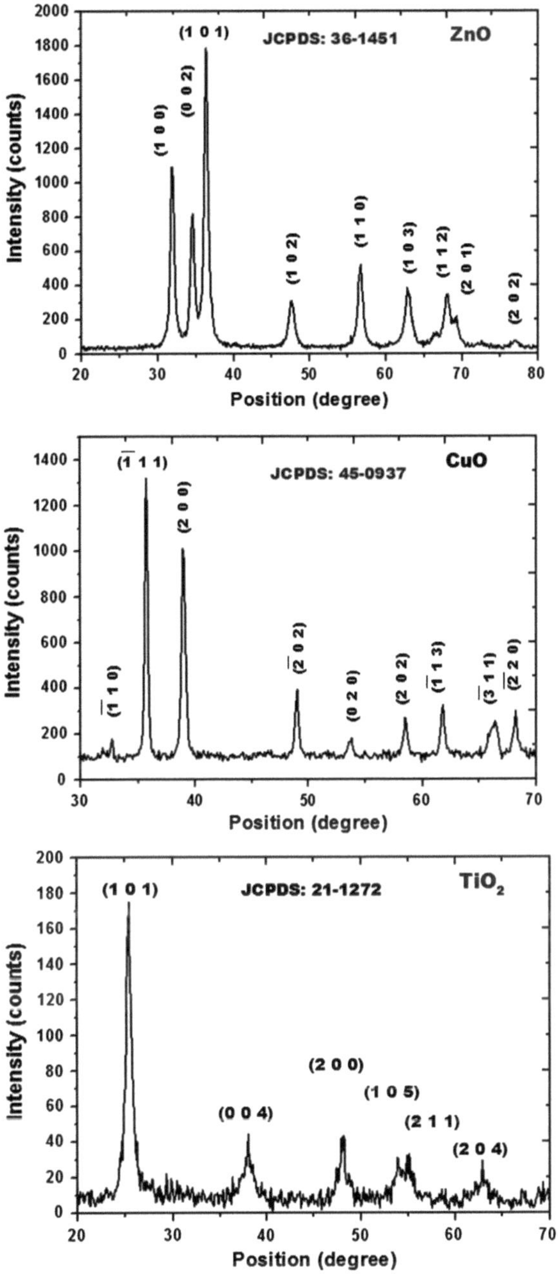

Fig. 1 XRD pattern of ZnO, CuO and TiO_2 nanostructured material

JCPDS card number: 21-1272. The structure of the TiO_2 is hexagonal and a = b = 0.3785 nm and c = 0.9513 nm are the lattice parameters. The sharpness of the nanostructured materials point to the crystallinity of the material.

The average crystallite size as measured by Debye-Scherrer's equation,

$$D = K.\lambda/\beta.\cos\theta \quad (1)$$

where, K-Scherer's constant (i.e. 0.9), λ-Wavelength of the radiation (for $CuK\alpha_1$-radiation i.e. 0.154 nm), β-Full width half maxima, θ-Bragg's angle. The average crystallite size of the ZnO, CuO and TiO_2 nanostructured material are obtained as 18, 27 and 23 nm respectively. Among all the above results the obtained materials are in the nano range.

3.2 *Particle Size Analyser*

Particles distribution of ZnO, CuO and TiO_2 nanostructured materials are shown in Fig. 2. The average particle size of the material has been attained from the particle size analyser. The nanostructured materials are dispersed in ethanol and ultrasonicated for 30 min.

This particle size analyser works on the principle of dynamic light scattering. In this dynamic light scattering, the Brownian motion of the solution is formed as histograms. The average particle size was obtained by the mean value of the histogram. The average particle sizes of the ZnO, CuO and TiO_2 are 22, 35 and 29 nm respectively. The average crystallite sizes of the nanostructured materials are less than the average crystallite sizes (Ashok and Venkateswara Rao 2014).

4 Humidity Sensor Application

The Sensitivity (S) of a humidity sensor is defined as change in resistance (ΔR) of a sensing element per unit change in relative humidity (ΔRH). Nanoparticles are made into pellets and these pellets are placed in between two Cu-electrodes, the digital multimeter is connected to the edges of the electrodes. This complete set-up has been kept in controlled humidity chamber. The resistance and sensitivity studies are measured from RH 10 % to RH 95 % and RH 15 % to RH 95 % respectively as shown in Fig. 3. The decrease in resistance obtained due to the increase of humidity (Presence of moisture). Table 1 inferred that resistance and sensitivity with respect to relative humidity.

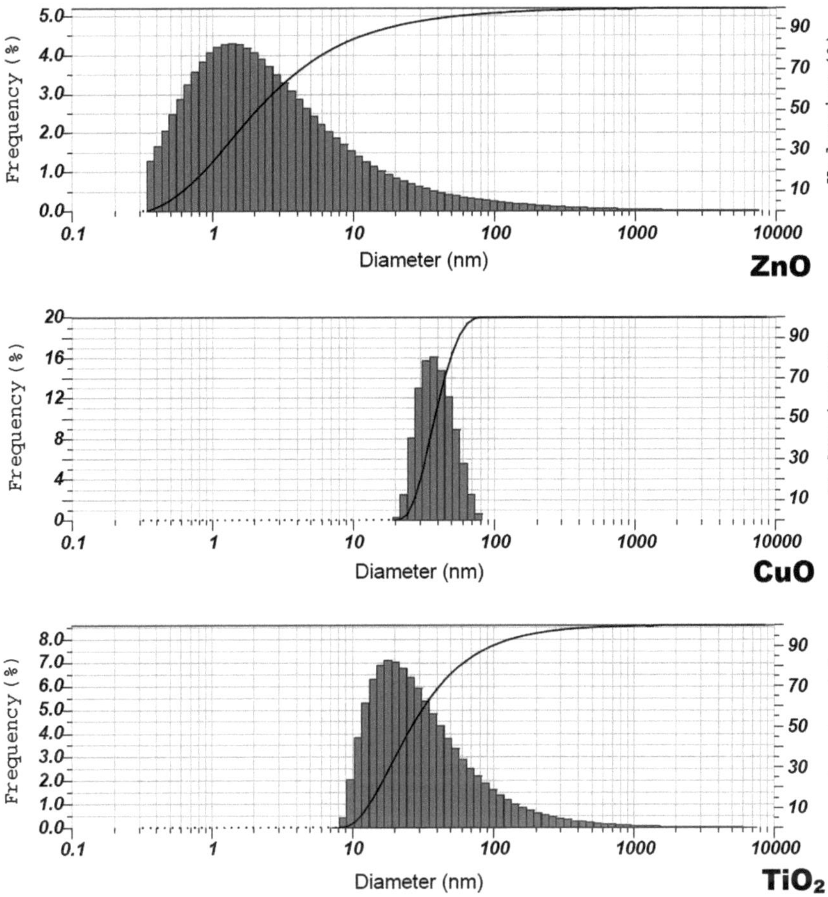

Fig. 2 Particles distribution of ZnO, CuO and TiO_2 nanostructured materials

The resistance value decreased at RH 10–95 % from 960 to 777, 840 to 702, and 920 to 746 MΩ for ZnO, CuO and TiO_2 respectively. The sensitivity has been increased from 1.4 to 2.9, 0.9 to 2.3 and 1.2 to 2.8 for ZnO, CuO and TiO_2 respectively. The average sensitivity of the ZnO, CuO and TiO_2 was 2.0, 1.5 and 1.9 respectively. Among three nanoparticles, ZnO is showing better sensing property. The porosity of the nanoparticles plays a significant role in the increasing of sensitivity. In this sensing material the cations are actively participating in the adsorption due to the high charge density. These ions are readily associated with the hydroxyl group of adsorbed water, due to this reason the electrical conduction has increased, and hence the resistance has decreased (Yadav et al. 2007).

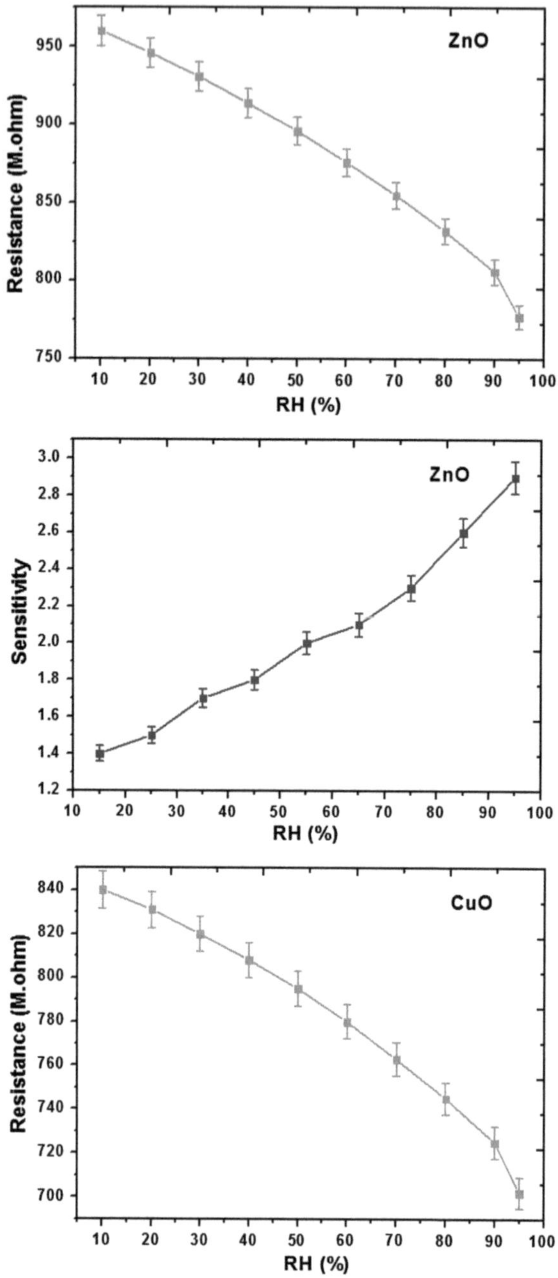

Fig. 3 Resistance and sensitivity curves

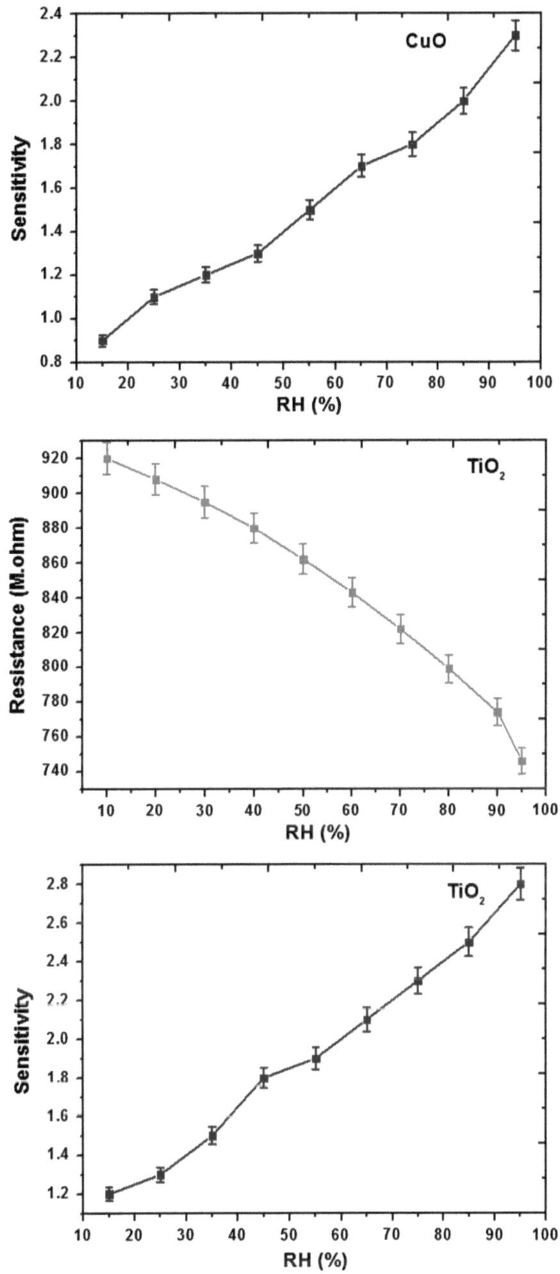

Fig. 3 (continued)

Table 1 Relative humidity with respect to resistance and sensitivity

RH (%)	R (ZnO) (MΩ)	R (CuO) (MΩ)	R (TiO_2) (MΩ)	RH (%)	S-ZnO	S-CuO	S-TiO_2
10	960	840	920	15	1.4	0.9	1.2
20	946	831	908	25	1.5	1.1	1.3
30	931	820	895	35	1.7	1.2	1.5
40	914	808	880	45	1.8	1.3	1.8
50	896	795	862	55	2.0	1.5	1.9
60	876	780	843	65	2.1	1.7	2.1
70	855	763	822	75	2.3	1.8	2.3
80	832	745	799	85	2.6	2.0	2.5
90	806	725	774	95	2.9	2.3	2.8
95	777	702	746	Avg.	2.0	1.5	1.9

5 Conclusion

The nanostructured ZnO, CuO and TiO_2 nanomaterials are successfully synthesized by microwave assisted method using room temperature ionic liquids. The XRD results inferred that the obtained materials are in the nano range. Both ZnO and TiO_2 structures has shown hexagonal, whereas CuO is monoclinic. The average crystallite sizes 18, 27 and 23 nm for ZnO, CuO and TiO_2 respectively. All the results match with corresponding JCPDS card numbers. The average particle sizes are 22, 27 and 29 nm for ZnO, CuO and TiO_2 respectively. The humidity sensor application has been investigated with the help of controlled humidity chamber, hygrometer and digital multi meter. The high sensitivity is shown in ZnO nanostructured materials compared with CuO and TiO_2.

Acknowledgments The authors expressed special thanks to '**University Grants Commission**' for providing financial support.

References

Anne, A., Tomasz, S., Ewelina, W., Witold, L., Paul, B.: Comparison of two innovative precipitation systems for ZnO and Al-doped ZnO nanoparticle synthesis. Proc. Appl. Ceram. **4**, 107–114 (2010)

Ashok, C. H., Venkateswara Rao, K.: ZnO/TiO_2 nanocomposite rods synthesized by microwave-assisted method for humidity sensor application. Superlattices Microstruct. **76**, 46–54 (2014)

Askarinejad, A., Alavi, M.A., Morsali, A.: Sonochemically assisted synthesis of ZnO nanoparticles: a novel direct method. Iran. J. Chem. Chem. Eng. **30**, 75–81 (2011)

Kankanit, P., Sineenart, S., Wanichaya, M., Wisanu, P.: Synthesis of CuO nanoparticles by precipitation method using different precursors. Energy Procedia. **34**, 740–745 (2013)

Khorsand, Z.A., Razali, R., Abd Majid, W.H., Darroudi, M.: Synthesis and characterization of a narrow size distribution of zinc oxide nanoparticles. Int. J. Nanomed. **6**, 1399–1403 (2011)

Meilong, H., Chenguang, B., Mei, S., Xuewei, L., Shengfu, Z., Guibao, Q.: Preparation of spherical monodispersed titanium dioxide by microwave assistance. IJRSA, **2**, 31–33 (2012)
Qi, Q., Zhang, T., Qingjiang, Y., Wang, R., Zeng, Y., Liu, L., Haibin, Y.: Properties of humidity sensing ZnO nanorods-base sensor fabricated by screen-printing. Sens. Actuat. **133**, 638–643 (2008)
Suhendro, P.P., Rosari, S.: Hydrogen incorporation in undoped ZnO nanoparticles. World J. Condens. Matter Phys. **1**, 130–136 (2011)
Suresh, C.P., Declan, E.M., Pradeepan, P., Nigel, L., John, C., David, C.: Rapid microwave synthesis of mesoporous TiO_2 for electrochromic displays. J. Mater. Chem. **20**, 3650–3655 (2010)
Yadav, B.C., Richa, S., Dwivedi, C.D.: Synthesis and characterization of ZnO nanorods by the hydroxide route and their application as humidity sensors. Synth. React. Inorg., Metal-Org. Nano-Metal Chem. **37**, 417–423 (2007)

Part III
Pollution Control

Assessment of Ambient Air Quality Parameters in Various Industries of Uttarakhand, India

Abhishek Nandan, S.M. Tauseef and N.A. Siddiqui

1 Introduction

Uttarakhand industrial cluster comprises of more than 7000 Micro, Small and Medium Enterprises (MSME's) of which more than 800 MSMEs units are located in Dehradun. Since Uttarakhand was carved out of Uttar Pradesh, it has grown leaps and bounds in terms of both industrialization and urban development. Rapid industrialization has resulted in the deterioration of ambient air quality with consequent health effects, which are of great concern, and should be dealt with guardedly to contain and mitigate the consequences. Incentives given by the State Infrastructure and Industrial Development Corporation of Uttarakhand Ltd. (SIDCUL) since 2002 has led to influx of industries in various cities of Uttarakhand such as Dehradun, Haridwar, Sitarganj and Pantnagar due to increasing demand fueled by rapid urbanization. This has played a major role in deterioration of the ambient air quality.

The recent technological advancements have enabled manufacturing of chemicals, in larger quantities than was possible previously, which boosts the economy of developing nations. The Indian subcontinent is experiencing considerable troubles because of air pollution from anthropogenic activities, rapid urbanization and industrialization. Previous studies have suggested multiple health effects due to air pollution. Most common health effects are cardiopulmonary health impact, cardiovascular deaths from increasing blood coagulability, and lung diseases. In spite of air pollution the industrial activities cannot be halted as growing demands of increasing population is directly met by them (Varma et al. 1994).

Industrialization has resulted in release of large quantities of air pollutants—Respirable Suspended Particulate Matter (RSPM), Sulphur dioxide (SO_2), Suspended Particulate matter (SPM) and Nitrogen dioxide (NO_2) (Reddy and

A. Nandan (✉) · S.M. Tauseef · N.A. Siddiqui
University of Petroleum and Energy Studies, Dehradun, India
e-mail: anandan@ddn.upes.ac.in

R. Mohan B. et al. (eds.), *Materials, Energy and Environment Engineering*,
DOI 10.1007/978-981-10-2675-1_33

Suneela 2001). These air pollutants result in infection and several respiratory disorders (Chauhan and Johnston 2003). Suspended particulate matter (SPM), a complex mixture of different particle sizes generates in environment due to volcanic eruptions, geochemical sources, windblown dust, soil re-suspension, spray from marine sources, power plant emissions, agricultural and industrial emissions and vehicular emissions, plays critical role in air quality (Abam and Unachukwu 2009).

It has been observed in the study involving particle transport and transformation that ambient air SPM commonly occur in two forms, namely the fine (aerodynamic diameter two and half or less, $PM_{2.5}$) mode and the coarse (aerodynamic diameter 2.5–10 μm, PM_{10}) (Ngele and Onwu 2015). Respirable suspended particulate matter or PM_{10} are those particulate matters which have diameter ≤ 10 μm. While RSPMs are of three types—ultra-fine (size range less than 0.1 μm), fine (0.7–1 μm) and coarse (1–10 μm) (Mohanraj and Azeez 2004).

Detrimental effects of air quality have compelled the scientists to study the significant adverse health effects of various SPM and other pollutant present in the ambient air. A number of experiments have been conducted in the past to study the detrimental health effect of air pollutants. Jahn et al. (2013) monitored $PM_{2.5}$, with exposure over 24-hr periods, simultaneously at different locations in various districts of Guangzhou, China. Jahn et al. (2013) found that the quality of air exceeded the permissible limit during 24 h monitoring period and consequently adverse effects on the inhabitant's health were observed and the results indicated that the indoor activities/sources have significant influence on particulate matter generation. Saha et al. (2014) conducted an experiment to determine the effect of air pollution on meteorological parameters during Deepawali festival over an Indian urban metropolis and observed the effect of atmospheric pollutants (PM_{10}, CO, surface O_3, NO_2, SO_2, BC) emitted during fireworks display in air over Kolkata. He (Saha et al. 2014) concluded that both long-term and short-term pollutant concentration of PM_{10} were significant during Diwali. Chauhan (2010) in his work which deals with the effect of industrialization, urbanization and automobile emission on ambient air quality in Haridwar and Dehradun City for air pollutants (SPM, RSPM, SO_2 and NO_2), observed that the PM and SPM levels at all the sites (excluding Roshnabad) exceeds the prescribed limits suggested by Central Pollution Control Board (CPCB), New Delhi, India. Robert and Pirro (2014) during evaluation of air quality status performed over the period 2002–2011, in the major cities of Albania, calculated air quality index (AQI) and the air quality sub-indices (AQSI) for each of the pollutants, both aggregated AQI and AQSI methods showed that air quality during the whole study period varies between poor quality (heavy pollution) and very poor quality (severe pollution). The AQSI values of PM and SPM indicated that they are the main air pollutants in the cities of Albania, causing a high or severe pollution.

Although multiple studies have been done to assess the ambient air quality of Uttarakhand (Tripathi and Gautam 2007; Amit et al. 2001; Chauhan et al. 2010) but most of the studies emphasize on vehicular emission and its impact on health, this work is perhaps the first-ever attempt to assess the environmental impact of various industries in Dehradun, Haridwar, Kashipur and Rudrapur industrial belt of Uttarakhand state.

The objective of the present study is to assess the ambient air quality of various industries located in Uttarakhand and its impact on the human health. Ambient air quality monitoring equipment were positioned in 13 different industries and the results obtained was used to calculate the Air Quality Index in order to transmutes weighted values of individual air pollutant (such as SO_2, NO_2, SPM and RSPM) into a single number for better and easy understanding of the quality of ambient air and to minimize its long and short term impact on the human health.

2 Materials and Methods

2.1 *Identification of Industries for Air Quality Monitoring*

Different industries from various cities of Uttarakhand were identified for monitoring of air borne emissions harmful to humans and environment. Different industrial sector included thermal power plants, pulp and paper industries, sugar and fertilizer industries etc.

2.2 *Survey and Analysis of Various Industries*

The inspection were performed at various industries such as pharmaceutical, chemical, fertilizer, electronics, leather, food processing, automobile, paper and pulp, textile, cement, cosmetic, furniture, printing and sugar. Identifying their source of emission and sampling of air quality for pollutants such as, SPM, RSPM, SO_2, and NO_2. The results were compared with the standards given by Central Pollution Control Board (CPCB), New Delhi, India.

2.3 *Data Collection/Sampling*

Different techniques were used for sampling of pollutants. For this study, ambient air quality was monitored for 13 industrial sectors over the period of July–August, September–October and November–December 2014. Respirable Dust Sampler (Make Envirotech), India was used for 24 h sampling using specified flow rate (LPM) as per the guidelines given by Central Pollution Control Board (Table 1). Sampling and analysis of pollutants SPM, RSPM, SO_2, and NO_2 is carried out as per the standard methods given in Table 2.

Table 1 Permissible limit of ambient air quality parameters as per CPCB, New Delhi, India

S. No.	Pollutant	Time weighted average	Concentration in ambient air ($\mu g/m^3$)
1	Sulphur dioxide (SO_x)	24 h	80
2	Nitrogen dioxide (NO_2)		80
3	RSPM		100
4	SPM		500

Source National ambient air quality standards, 2009

Table 2 Standard test method for ambient air quality sampling and analysis

S. No.	Test parameter	Test method
1	RSPM or PM_{10}	IS: 5182 (part 23): Gravimetric
2	SPM	IS: 5182 (part 04): Gravimetric
3	Sulphur dioxide (SO_2)	Improved West and Gaeke
4	Oxides of nitrogen (NO_x)	Modified Jacobs and Hochheiser

3 Results and Discussion

The ambient air quality was analysed through monitoring of four potential pollutants i.e. SO_2, NO_2, SPM and RSPM over the period of July–August, September–October and November–December 2014 focusing on 13 different industrial sectors located in Uttarakhand. Sampling of dust particles and gaseous pollutants were carried out using Respirable Dust Sampler (RDS, APM 460BL) Envirotech, India for 24 h period using specified flow rate (LPM) as per the guidelines given by Central Pollution Control Board (Table 1). All the samples were analyzed using standard methods (Table 2) for various pollutants and the results are summarized in Table 3. Average concentration of each of the pollutants was determined and Air Quality Indices was calculated (Fenger 1999) (Table 5). Emission patterns for SO_2, NO_2, SPM and RSPM over the assessment periods is shown in Figs. 1–12.

Comparison of the calculated values of AQI for different parameters (Table 5) and the standard AQI values (Table 4), shows that the air quality indices is within permissible range (41–100) indicating that the quality of emission as measured by adopting standard methods covering 13 different industrial sectors is within the permissible limit prescribed by the Central Pollution Control Board (CPCB), New Delhi, India. The only plausible reason will be good environmental protection practices followed by the cluster of industries resulting in a clean and a healthy environment.

Table 3 Average concentration of ambient air quality parameters for the year 2014

Type of industry	SO_2 (μg/Nm^3)			NO_2 (μg/Nm^3)			RSPM (μg/Nm^3)			SPM (μg/Nm^3)		
	July-Aug	Sep-Oct	Nov-Dec	July-Aug	Sep-Oct	Nov-Dec	July-Aug	Sep-Oct	Nov-Dec	July-Aug	Sep-Oct	Nov-Dec
Pharmaceutical	23.2	24.2	25.6	39.05	42.30	45.1	64.7	65.3	67.7	178	180.2	184.04
Chemical	41.57	42.57	44.7	74.2	86.5	87.2	58	62.9	64.9	189	193.5	197.1
Fertilizer	10.16	15.16	17.16	12.74	14.34	16.74	72	75.3	78.35	174	179.8	210.81
Electronics	3.41	7.41	9.21	39.65	41.65	43.65	60	66.6	69.2	163	167.3	173.3
Food processing	17	19	21.4	21	23.9	25.8	63	62.5	63.9	112	117	127
Automobile	53	58	62.1	38	42.1	44.5	78	65.1	69	123	125.3	128.4
Pulp and paper	13	15	17.3	11	13	16.1	49	51.1	54.3	93	95.7	99.2
Textile	38	40.5	42.1	29	31.3	34.1	57	60.1	64	119	120.4	124
Cement	48	51	53.1	39	42.4	43	63	68	71.2	120	124.2	130.2
Cosmetic	32	34	37	28	32.7	35.3	64	69.1	72	152	155.1	160.4
Furniture	44	45.3	48.3	41	43.1	44.6	45	48	52	120	121.3	125.1
Printing	36	37.5	38.2	28	31	34	67	69	73.7	136	138.1	139.7
Sugar	28	30.1	32.1	31	33	35	72	74.4	77.3	141	152.1	176

Table 4 Standard air quality index scale

AQI value	Nature of air pollution
0–50	Good
51–100	Satisfactory
101–200	Moderately polluted
201–300	Poor
301–400	Very poor
>401	Severe air pollution

Source National air quality index, CPCB

Table 5 Ambient air quality indices

Name of industry	Calculated AQI values
Pharmaceutical	45.8
Chemical	63.7
Fertilizer	76.8
Electronics	38.8
Food processing	60.3
Automobile	38.3
Pulp and paper	26.8
Textile	32
Cement	48.5
Cosmetic	45.3
Furniture	45
Printing	45.1
Sugar	45.7

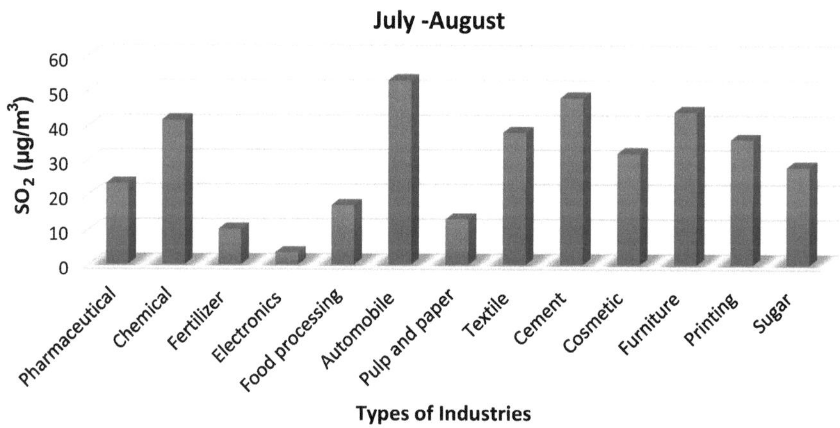

Fig. 1 Ambient SO_2 concentration in the months of July–August

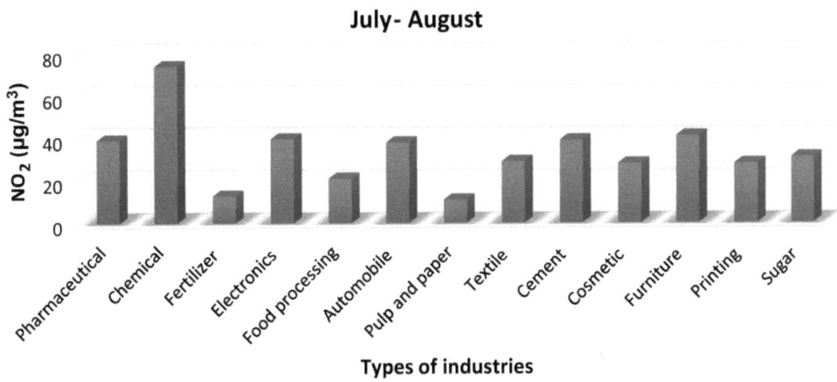

Fig. 2 Ambient NO_2 concentration during the months of July–August

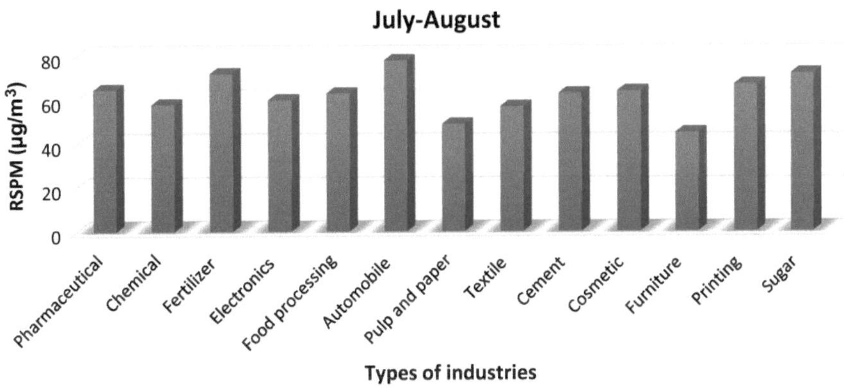

Fig. 3 Ambient RSPM concentration during the months of July–August

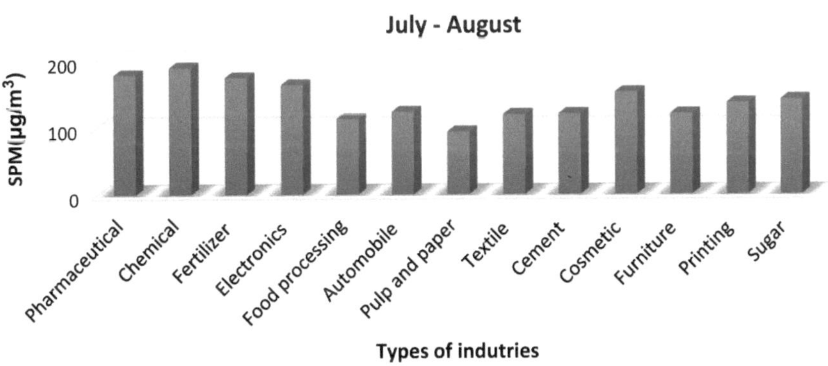

Fig. 4 Ambient SPM concentration during the months of July–August

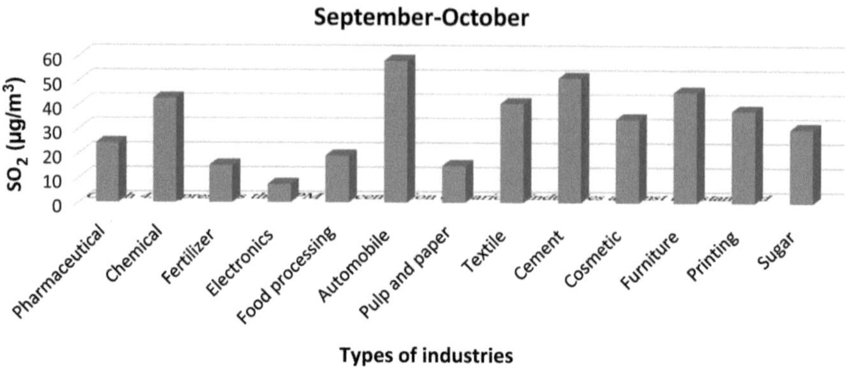

Fig. 5 Ambient SOx concentration during the months of September–October

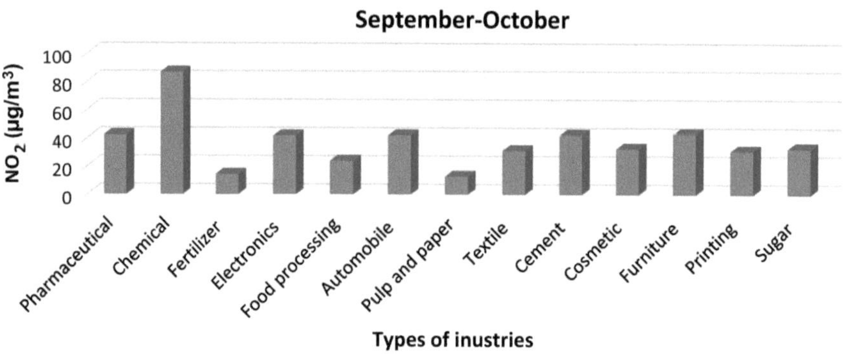

Fig. 6 Ambient NO_2 concentration during the months of September–October

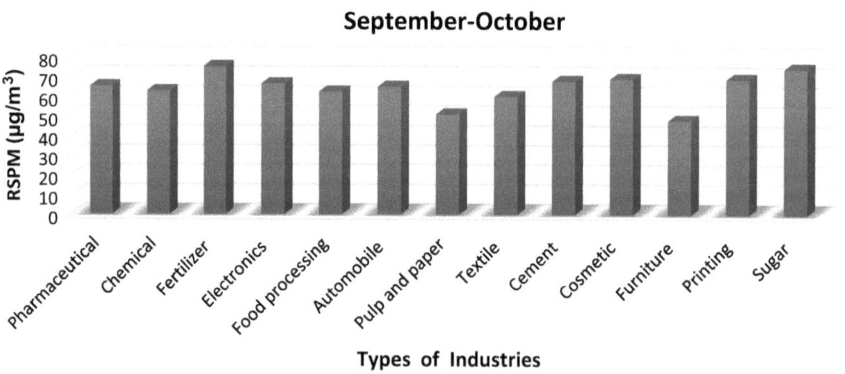

Fig. 7 Ambient RSPM concentration during the months of September–October

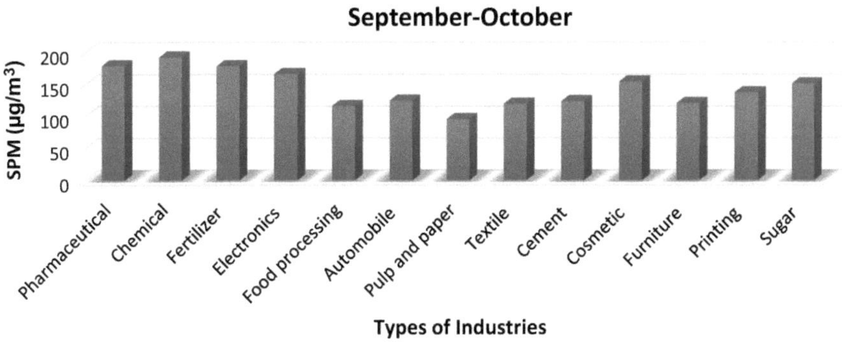

Fig. 8 Ambient SPM concentration during the months of September–October

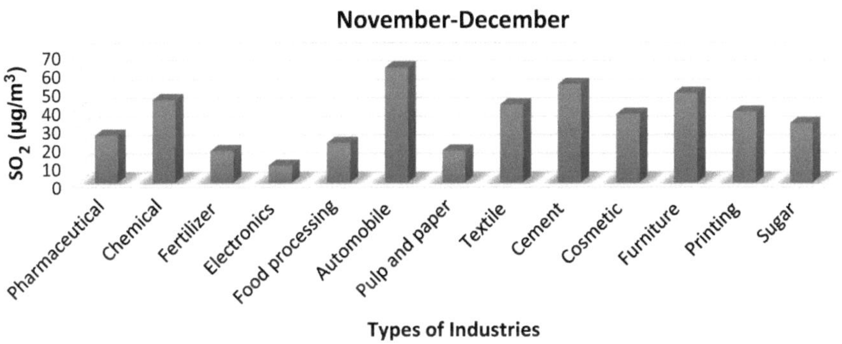

Fig. 9 Ambient SO_2 concentration during the months of November–December

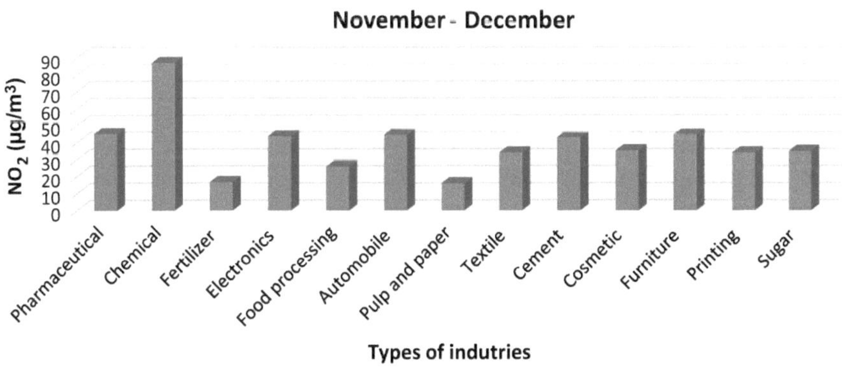

Fig. 10 Ambient NO_2 concentration during the months of November–December

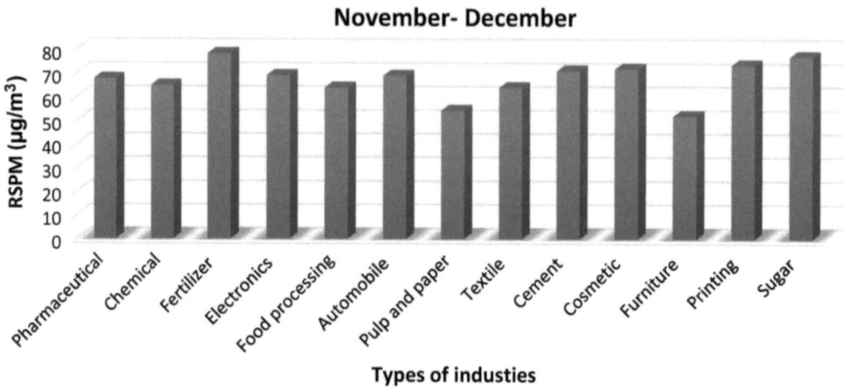

Fig. 11 Ambient RSPM concentration during the months of November–December

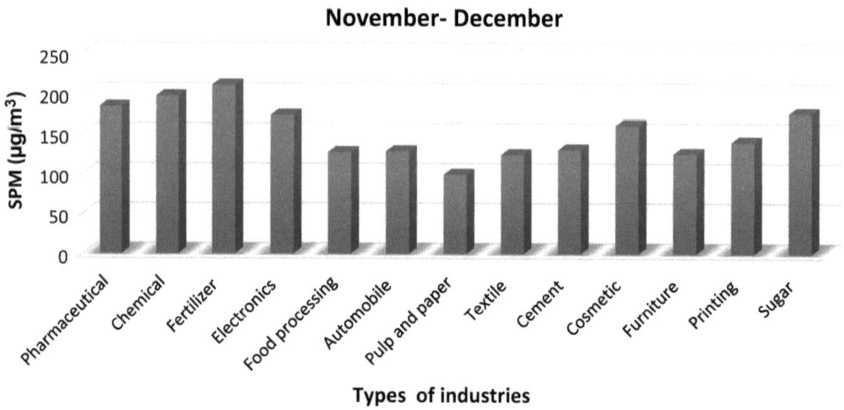

Fig. 12 Ambient SPM concentration during the months of November–December

4 Conclusion

A general assessment of ambient air quality of Dehradun using Air Quality Index (AQI) has been done in this study. An analysis of the results reveals the fact that the quality of ambient air in various industrial clusters with high pollution potential is within the standard limits prescribed by the Central Pollution Control Board, New Delhi, India. Evaluation of the ambient air quality using air quality index (AQI) indicates that the values of air quality indices falls within the acceptable range (41–100) on AQI rating scale indicating that the quality of emission as measured by adopting standard methods covering 13 different industrial sectors is adequate and within the permissible limit.

The only plausible reason will be good environmental protection practices followed by the cluster of industries resulting in a clean and healthy environment. However, for ensuring continuous improvement in the quality of air or to maintain the same air quality in future, the industry should develop, implement and maintain principles of sustainable development.

Acknowledgments The authors would like to thank professors and lab assistants of Health, Safety and Environment Engineering Department of University of Petroleum and Energy Studies, Dehradun, Uttarakhand for their extended motivational and infrastructure support in carrying out this research work and for completing it successfully.

References

Abam, F., Unachukwu, G.: Vehicular emissions and air quality standards in Nigeria. Eur. J. Sci. Res. **34**(4), 550–560 (2009)

Amit, G. et al.: Sub-region (district) and sector level SO_2 and NOx emissions for India: assessment of inventories and mitigation flexibility. Atmos. Environ. **35**(4), 703–713 (2001)

Brunekreef, B., Forsberg, B.: Epidemiological evidence of effects of coarse airborne particles on health. Eur. Respir. J. **26**, 309–318 (2005)

Burnett, R.T., Brook, J., Dann, T., Delocla, C., Philips, O., Cakmak, S., et al.: Association between particulate- and gas-phase components of urban air pollution and daily mortality in eight Canadian cities. Inhal. Toxicol. **12**, 15–39 (2000)

Chauhan, A.: Photosynthetic pigment changes in some selected trees induced by automobile exhaust in Dehradun, Uttarakhand. N.Y. Sci. J. **3**(2), 45–51 (2010)

Chauhan, A., et al.: Ambient air quality status in Uttarakhand (India): a case study of Haridwar and Dehradun using air quality index. J. Am. Sci. **6**(9), 565–574 (2010)

Chauhan, A.J., Johnston, S.L.: Air pollution and infection in respiratory illness. Br. Med. Bull. **68** (1), 95–112 (2003)

CPCB.: Guidelines for ambient air quality monitoring. Report Under Central Pollution Control Board, Ministry of Environment & Forests (2003)

Dominici, F., Peng, R.D., Barr, C.D., Bell, M.L.: Protecting human health from air pollution shifting from a single-pollutant to a multipollutant approach. Epidemiology **21**, 187–194 (2010)

Fenger, J.: Urban air quality. Atmos. Environ. **33**, 4877–4900 (1999)

Guidelines for the Measurement of Ambient Air Pollutants Volume-I, Central Pollution Control Board (Ministry of Environment, Forests & Climate Change) Government of India, May 2011

Jahn, H.J., Kraemer, A., Chen, X.-C., Chan, C.-Y., Engling, G., Ward, T.J.: Ambient and personal PM 2.5 exposure assessment in the Chinese megacity of Guangzhou. Atmos. Environ. **74**, 402–411 (2013)

Joshi, S.G.: Air quality indices proceedings of WHO training programme on air pollution (1992)

Mohanraj, R., Azeez, P.A.: Health effects of airborne particulate matter and the Indian scenario. Curr. Sci. **87**(6), 741–748 (2004)

Monitoring and Management at IIT, Bombay, India (1984)

National Air Quality Index, Central Pollution Control Board (Ministry of Environment, Forests & Climate Change) Government of India, Oct 2014

Ngele, S., Onwu, F.: Comparison of particulate matter levels in major urban centres in Eastern Nigeria. Int. J. Environ. Sci. **5**, 765 (2015)

Ott: Environmental Indices, Theory and Practice. Ann Arbor Science Publishers, Michigan (1976)

Reddy, M.K., Suneela, M.: Status of ambient air quality at Hazira with reference to modified air quality index. IJEP **21**(8), 707–712 (2001)

Robert, D., Pirro, I.: A comparative study of ambient air quality status in major cities of Albania. The Annals of "VALAHIA" University of Targoviste, **8**, (2014)

Saha, U., Talukdar, S., Jana, S., Maitra, A.: Effects of air pollution on meteorological parameters during Deepawali festival over an Indian urban metropolis. Atmos. Environ. **98**, 530–539 (2014)

Tripathi, A.K., Gautam, Mukesh: Biochemical parameters of plants as indicators of air pollution. J. Environ. Biol. **28**(1), 127 (2007)

Varma, P.C., Jha, C., Mian, S., Sinha, S.B., Gupta, V.S.: Air quality studies in an industrial and urban township of Jharia coal field, Dhanbad. IJEP **14**(4), 301–306 (1994)

Zereini, F., Wiseman, C.L.S.: Urban airborne particulate matter: origin, chemistry, fate, and health impacts. Springer, Heidelberg, New York (2010)

Urban Air Pollution Impact and Strategic Plans—A Case Study of a Tier-II City

N. Lokeshwari, Keshava Joshi, G. Srinikethan and V.S. Hegde

1 Introduction

The trend in the increase in urbanisation has been witnessed in the most of the developing countries in the 20th century. Presently most of the cities in the world are facing pollution and congestion as two major problems due to rapid urbanization (Ashmore 2005). In 1960, the urban development was less than 22 % in developing countries and it proportionally increased to 34 % by 1990. In the coming year's urbanisation will still increase and may reach to 50 % of the total world's population by 2020. According to World Bank's report the air quality is deteriorating in many cities due to the increased industrial activities and vehicular movement in Asia (World Bank 1997). The major sources of air contamination are vehicles, many industries including power plants and many among them vehicles are the major contributors to the regional air pollution. In India, the numbers of vehicles are increasing from 21, 54 and 70 million in 1991, 2000 and 2003, respectively. The World Health Organization has reported that around 800,000 deaths annually occur due to air pollution (WHO 2002). In India urban areas are highly polluted due to the major contribution from vehicles and power plants (Partheeban et al. 2012). World Health Organization reported that air contamination is one of the major causes of mortality in the world (WHO 2009) and deaths due to increased cardiovascular diseases (Araujo and Nel 2009) in the exposed population, particularly those of ischemic nature (Pope et al. 2004). The past evidences of air disasters like London smog, Meuse valley, Bhopal gas tragedy etc. show that

N. Lokeshwari (✉) · K. Joshi · V.S. Hegde
Department of Chemical Engineering, S.D.M. College of Engineering and Technology, Dharwad, India
e-mail: lokeshwarinavalgund@gmail.com

G. Srinikethan
Department of Chemical Engineering, National Institute of Technology Karnataka, Surathkal, India

R. Mohan B. et al. (eds.), *Materials, Energy and Environment Engineering*,
DOI 10.1007/978-981-10-2675-1_34

ambient air pollution can seriously be much harmful to public health. Recently, it is investigated and reported that premature deaths have increased significantly due to higher particulate matters in the ambient air (Singh 2008).

In India, air pollution is increasing significantly in most of the metropolitan cities. About 72 % ambient air pollution is due to vehicular and 20 % due to industrial emissions (Garg et al. 2001). The unplanned growth of the cities have serious impact on the natural resources and local ecology (Ramachandra et al. 2008). India's state air monitoring centres have reported that the parameters like Respirable Suspended Particulate Matter (RSPM) in Delhi (Tier-I city) are comparable to that of highly polluted cities of the world like Beijing. RSPM levels in most of the Tier-I cities like Bangalore, Kolkata, Chennai, Mumbai etc. are well above the limits of National Ambient Air Quality System (NAAQS) and the reasons are: increase in population, rapid urbanization, increase in the number of vehicles with poor quality of fuel, commercial developments, industrialization. In Tier-I cities, because of increased population and improper urban planning have reached almost the saturation level. This is evident from lack of basic facilities, water contamination, shortage of water and electricity, higher amount of pollutants in environment, over congestion due to inadequate infrastructure (Sudhir and Prasad 2003). These studies reported and suggested that there is necessity to plan for Tier II cities in India having population less than one million, which are urbanizing rapidly and mainly to avoid the environmental and congestion problems as that of Tier-I cities. There is a necessity to understand, plan and implement strategies for effective environmental managements in Tier-II cities. Hubli-Dharwad a Tier-II city in Karnataka state, India, with a typical tropical climate has seen serious rapid changes and the rate of urbanization is exceeding the global average.

In a common man's perspective, air quality is represented by Air Quality Index (AQI) which is a cumulative effect of all the pollutants (Lohani 1984). The level of air pollution and health concern depends on the value of the AQI. The AQI studies have been carried out for the Indian cities of Kanpur (Sharma et al. 2003), Delhi (Sengupta et al. 2000) and Mumbai (Sharma 1999). This AQI helps to understand the local level air contamination, and its health effects on people and with those existing with cardiovascular and respiratory diseases. The government of India, Ministry of environment and forest, has launched AQI, as huge initiative under 'Swachh Bharat' on 17th October 2014 as 'One Number- One Colour-One Description' for the common person to understand the air quality within his vicinity. Under the new measurement process, AQI includes a comprehensive set of parameters.

Air dispersion modelling is one of the frequently used tools throughout the world to understand the level of accumulation of pollutants (Bluett et al. 2004). Meteorology is one of primary factor for determining the diluting effect of the atmosphere in dispersion modelling (Jacobson 2002). Air flow and dispersion around buildings in urban area are dependent on the atmospheric stability. Giri et al. (2008), studied air pollution problem in Kathmandu, Nepal, which is serious since it is in the valley topography where in the role of meteorological parameters is significant. Pollutant dispersion is a complex process as it is the net effect of topography, land use/land cover, meteorological factors, trends and pattern of urbanization, industry and transport. Along with preventive measures, proper siting (Zoning atlas) of industrial estate in a country like India is required where the process of planned industrialization is still in the initial stage.

2 Methodology

The air quality with respect to RSPM, SO_2 and NO_x were monitored for 2012 and 2013 at the selected locations in the study region as per the standard methods for both ambient and stack monitoring. The AQI of the region was determined using the Oak Ridge Index method and the dispersion pattern for particular industries was studied using Gaussian dispersion model of Air Resource laboratory. To understand the urbanization pattern, change in land use land cover pattern viz area under settlements, agriculture land, barren land, water bodies etc. were studied from satellite images of land sat MSS (1975), land sat TM (1989) and land sat ETM 2000 and 2009 covering the study area. The images for the years 1975, 1989, and 2000, 2009 were downloaded from the website http://www.landsat.org. Geo-referencing of images is done selecting 25 permanent ground reference points selected carefully both in Landsat and images, then resampled in World Geodetic System (WGS 84) datum and Universal Transverse Mercator (UTM) projection system. During the Geo-referencing process Root Mean Square Error (RMSE) calculated for geo-corrected image was less than 0.4 pixels. Band 4 (Near Infrared Region of the spectrum) of the geo-corrected satellite images are processed in ERDAS 9.0 version software for clear delineation. These images, after geo rectification in ERDAS Imagine software 9.0, were used for supervised classification, and thematic maps have been generated, and area under each category of land is quantified and maps are generated. For the supervised classification ground truth data especially on soil, land use pattern etc. were recorded along with the corresponding ground coordinate systems using Garmin make GPS having spatial resolution of ± 9 m. Population data has been collected from Census, Government of India, vehicle data from Road transport office, Dharwad for the period from 2004 to 2014.

3 Results and Discussion

The AQI range calculated by Oak Ridge Index method for Hubli-Dharwad ranges from 19.1 in green vegetated area to 73.7 in high traffic zones and industrial areas in 2013. It is observed that 50 % of the area shows the good air quality and falls under the good health condition and other 50 % are in the range of red category as per Oak Index method and falls under unhealthy conditions. Based on the linear analysis of the urbanization, population, number of vehicles and the trend of air pollutants, the 2030 AQI predicted shows that almost 80 % of the region will be under the dangerous level.

Numerous specialists have reported the progressions seen in urban area use/land spread example (Stow and Chen 2002) and such reports especially are critical for considerate the different effects of human action on the general environmental state of the urban development (Yeh 1999). The level of emissions has been increased due to the conversion of grasslands, forest, and farmland to industrial complexes, large commercial centre and residential housing. The Hubli-Dharwad region has shown a drastic change in LULC pattern, the total urbanized area development was from 92.1 km^2 (1975) to 271.07 km^2 (2009) among the 644 km^2 total area. The agricultural land has decreased by 42 % from 1989 to 2009. The 2030 predictions of the urbanization with linear correlation indicate that around 500 km^2 will be in urban area. The effect of urbanization on air quality was correlated by superimposing the present level of pollutant concentration in air on the land use pattern in Hubli-Dharwad region to understand the effects and where these pollutants are getting accumulated.

The spatial exhibiting of air quality in the structure of GIS demonstrate the geospatial coordinate system, for delineation of surface things maintained by raster and TIN surface data, and vector subjects for representation of air contamination inputs (Matejicek 2005). Digital elevation model data helps for analysing pollutant distribution in the region, and to develop measuring and modelling strategies. Digital elevation model represented the spatial distribution of air pollutants throughout the Burdwan town (Chattopadhyay et al. 2010). Using AQI of Hubli-Dharwad city, GIS based air pollution elevation models were developed by means of Inverse Distance Interpolation (IDINT) technique. The dispersion and dilution of industrial emissions due to the meteorological parameters of the Hubli-Dharwad, indicated that maximum mixing height was around 1700 m in summer season with unstable conditions as shown in Fig. 1 The lower mixing height in rainy and winter leads to the accumulation of pollutants, because of the topographical effects of the region. The dispersion sensitivity map of Hubli-Dharwad shows the dispersion and dilution of air pollutants with respect to its vegetative cover and topography of the area.

The dispersion of pollutants in a region is of more concern with high inversion and extreme climatic conditions and the impact area in different seasons are studied. The effect of the impact on environment due to air contamination is studied using the change in land pattern. Normally, the impact on the receiving environment due to small air polluting industries extends up to a maximum of 500 m from the source. The impact from medium air polluting industries is normally less than 2 km and that of high air polluting industries is 5 km. The impact of large scale industries units like cement, thermal power plants fertilizer, petrochemical, steel industries are more than 10 km (CPCB 1995). Impact zone created based on the Gaussian models output and existing meteorological data for Hubli-Dhawad region with the change

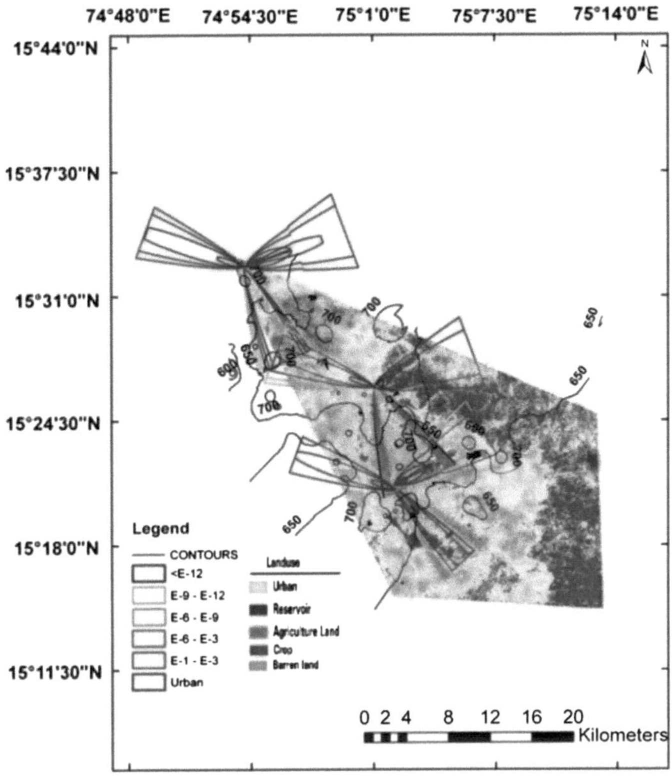

Fig. 1 Dispersion pattern derived from Gaussian model superimposed on land use and land cover

in land use/land cover, suggest that risk zone area is becoming larger which leads to the effect on the urban population and other environmental problems. The 2030 predicted impact zone for Hubli-Dharwad shows that 80 % of region is going to be in the risk zone as shown in Fig. 2.

The management strategy and policies for a particular region or city like Hubli-Dharwad (Tier-II city) is proposed for sustainable development. Strategy like creating the industrial buffer zones, proper monitoring and control measures to be adopted by industries, compliance by industries on emissions, EIA based clearance, adoption of cleaner technologies by industries, introduction of mass transport system, clean fuel policies may enhance the air quality in a region and control the health effects. Sustainable urban transport systems with GPS adoption in emergency management and for rerouting, encouraging public transportation formation, pedestrian zones and incentives for use of bicycles and eco-friendly fuels, taxation on new vehicles, regular monitoring of vehicle emissions may lead to sustainability.

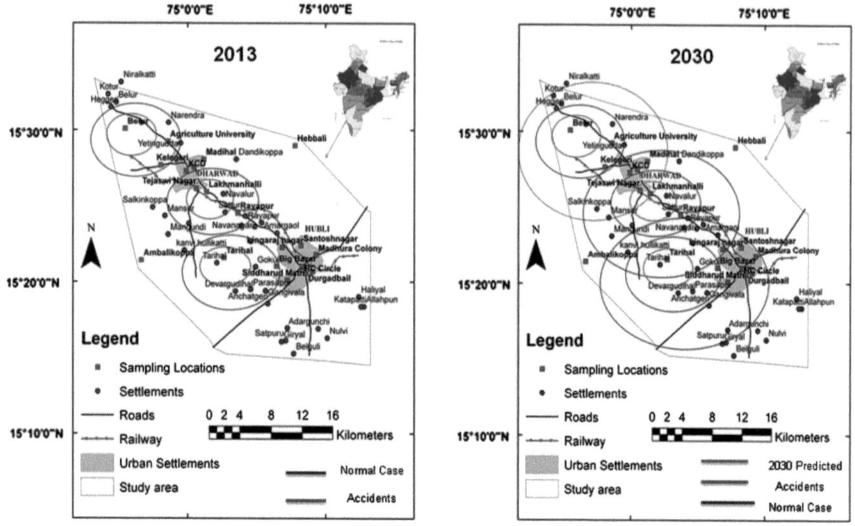

Fig. 2 Impact zone for the air quality of Hubli-Dharwad region for 2013 and 2030

4 Conclusion

The output of the remote sensing and GIS based air quality atlas specifies the level of concentration of pollutants in Hubli-Dharwad city for the present level of pollutants and for 2030 predictions. The impact zone of the study area in the present condition, during any failure of equipment's and the predicted impact zone is indicated by the dispersion sensitivity map. Hence the buffer zone can be created between industrial and residential area. This method can be used by environmental managers and local authorities to plan for short term and long term programs and formulate strategies for better air quality management.

Acknowledgments The authors wish to thank, Department of Chemical Engineering, National Institute of Technology Surathkal, Karnataka, Department of Chemical Engineering, SDMCET, Dharwad and NRE coke Industry, Dharwad for providing research facilities. One of the authors, Lokeshwari. N is thankful to the Management and Principal, SDM College of Engineering and Technology Dharwad for their support and encouragement.

References

Araujo, J.A., Nel, A.E.: Particulate matter and atherosclerosis: role of particle size, composition and oxidative stress. Part Fibre Toxicol. **6**, 24 (2009)

Ashmore, M.R.: Assessing the future global impacts of ozone on vegetation. Plant, Cell Environ. **28**, 949–964 (2005)

Bluett, N., Gimson, G., Fisher, C., Heydenrych, T., Freeman J., Godfrey, J.: Good practice guide for atmospheric dispersion modeling (2004)
Chattopadhyay, S., Gupta, S., Saha, R.N.: Spatial and temporal variation of urban air quality: a GIS approach. J. Environ. Protect. **1**(03), 264 (2010)
CPCB: Manual for Preparation of Zoning Atlas for Siting of Industries. Project report submitted to Ministry of Environment and Forest (1995)
Garg, A., Shukla, P.R., Bhattacharya, S., Dadhwal, V.K.: Sub-region (district) and sector level SO_2 and NOx emissions for India: assessment of inventories and mitigation flexibility. Atmos. Environ. **35**, 703–713 (2001)
Giri, N., Aryal, R.K., Lee, B.K., Karki, R., Gurung, A., Kandasamy, J., Pathak, B.K.: Seasonal PM dynamics in Kathmandu Valley. Atmos. Environ. **42**(37), 8623–8633 (2008)
Jacobson, M.Z.: Effect of meteorology on air pollution. In: Atmospheric Pollution: History, Science, and Regulation, pp. 145–176. Cambridge University Press, New York, USA (2002)
Lohani, B.N.: Environmental Quality Management. South Asian Publishers, New Delhi (1984)
Matejicek, L.: Spatial modelling of air pollution in urban areas with GIS: a casestudy on integrated database development. Adv. Geosci. **4**, 63–68 (2005)
Partheeban, P., Rani, H.R., Prasad Raju, H.: Vehicular emission monitoring using internet GIS, GPS and sensors. In: 2012 International Conference on Environment Energy and Biotechnology IPCBEE.33 (2012)
Pope, C.A., Burnett, R.T.: Cardiovascular mortality and long-term exposure to particulate air pollution: epidemiological evidence of general pathophysiological pathways of disease. Circulation **109**(1), 71–77 (2004)
Ramachandra, T.V., Kumar, U.: Wetlands of greater Bangalore, India: automatic delineation through pattern classifiers. Electron. Green J. **1**(26) (2008)
Sengupta, B., Sharma, M., Shukla, P., Maheshwari, M.: Air quality index for data interpretation and public information. In: Proceedings of International Conference on Air Quality Index and Emission inventory for Delhi. Centre for Science and Environment, New Delhi, India (2000)
Sharma, V.K.: Development of air quality indices for Mumbai, India. Int. J. Environ. Pollut. **11**, 141–146 (1999)
Sharma, M., Maheshwari, M., Sengupta, B., Shukla, B.P.: Design of a website for dissemination of an air quality index in India. Environ. Modell. Softw. **18**, 405–411 (2003)
Singh, S.K.: Transport policy for better air quality in urban India. In: Bhatt, K.N. (ed.) Population Environment and Health Emerging Issues, Govind Ballabh Pant Social Science Institute, Allabahad. Rawat Publication, Jaipur (2008)
Stow, D.A., Chen, D.M.: Sensitivity of multi-temporal NOAA AVHRR data of an urbanizing region to land use/cover changes and misregistration. Remote Sens. Environ. **80**, 297–307 (2002)
Sudhir, P., Prasad, C.E.: Prevalence of exercise-induced bronchospasm in school children: an urban–rural comparison. J. Trop. Pediatr. **49**(2), 104–108 (2003)
World Bank: Urban air quality management strategy in Asia guidebook. URBAIR (1997)
World Health Organization: Regional Office for Europe, and World Health Organization. 2009. Air quality guidelines: global update 2005: particulate matter, ozone, nitrogen dioxide, and sulfur dioxide. (2009)
World Health Organization: World health report 2002: reducing risk, Promoting healthy life. World Health Organization. Geneva, Switzerland. Available from www.who.int/whr/2002/en/ (2002)
Yeh, A.: Urban planning and GIS. Geogr. Inform. Syst. (Principles technical issues). **2**, 877–887 (1999)

Optimization of Engineering and Process Parameters for Electro-Chemical Treatment of Textile Wastewater

Sachin Koshti, Abhinav Rai, S. Arisutha, Prerna Sen and S. Suresh

1 Introduction

The textile industry is one of the prevailing industry practicing enormous amount of dyes, fibers and other chemicals in different production practices including dyeing, bleaching and finishing. Approximately 100,000 dyes are commercially available with an estimated annual production exceeding 700,000 tons, and nearly 30 % of these dyes are gone astray along with wastewater. So it becomes a necessity to treat the wastewater before discharge. More than 2500 textile weaving industries and 4200 textile finishing factories are established in India. The major pollutants from textile wastewater mainly including surfactants, finishing agents, metal complexes, inhibitor compounds, ionic compounds, dyeing substances, phosphates, dissolved and suspended solids. Textile wastewater has been observed to have high COD values, strong color, immense TDS and varying pH. Therefore, due to large quantities of pollutants present in wastewater, textile industries suffer a lot of problems in discharging the same. Various methods have been proposed for effluent treatment, including Fentons method, ozonation, photochemical treatment, coagulation, adsorption, biological method and electrochemical treatment (Khandegar and Saroha 2013). Due to use of numerous chemicals in the physical, chemical and biological methods these are inefficient for effective treatment. Hence the industrial operators are searching for better methods to solve their wastewater treatment problems.

S. Koshti · A. Rai · P. Sen · S. Suresh (✉)
Advanced Industrial Pollution Abatement Laboratory and Analytical and Simulation Laboratory, Department of Chemical Engineering, Maulana Azad National Institute of Technology, Bhopal 462 003, India
e-mail: sureshshpecchem@gmail.com

S. Arisutha
Energy Centre, Maulana Azad National Institute of Technology, Bhopal 462 003, India

R. Mohan B. et al. (eds.), *Materials, Energy and Environment Engineering*,
DOI 10.1007/978-981-10-2675-1_35

In recent years, electrochemical treatment is gaining attention due to its potential towards degrading and removing both carbonaceous and non-carbonaceous pollutants from wastewater without using any expensive chemicals to waste water. So the process can be considered to be environment friendly using 'electron' as the main reagent. Electrochemical treatment yields less sludge, which comprises metal oxides capable of passing the leachability test and is more easily filterable and further finds application to be used as a soil additive. The apparatus obligatory for electrochemical treatment is simple, compact and easy to operate.

RSM is based upon assimilation of investigational approaches, numerical methodology along with statistical supposition for determining the best suited result. It helps in reducing the investigational runs necessary for examining numerous factors and their relations, hence is less arduous and is time saving as compared to other methodologies. RSM related approach has been broadly used for optimization in various fields including production, biomedical, metallurgical and industrial engineering process (Elboughdiri et al. 2015). In this paper, RSM is employed to examine the variations due to pH, current density and also time of electrochemical treatment for COD reduction (mg/l). MINITAB14 software used for RSM.

2 Materials and Method

2.1 Materials Used

20 l of textile wastewater (diluted to 25 %), Aluminium electrodes, Rheostat, DC power supply, 1000 ml glass beaker, pH meter, spectrophotometer, COD digester, Chemicals-potassium dichromate, sulfuric acid, silver sulfate, mercuric sulphate were used for the experimental study.

2.2 Apparatus Design

The electrochemical treatment cell used was 1000 ml glass reactor with 2 electrode arrangement and was equipped with a magnetic stirrer with temperature control. Aluminium electrodes with dimensions 8.7 cm × 6.5 cm were used for electrochemical treatment. For the entire process the current was regulated using a rheostat. A set of current densities was tested at ambient temperature of 25 °C. The whole setup is shown in following Fig. 1.

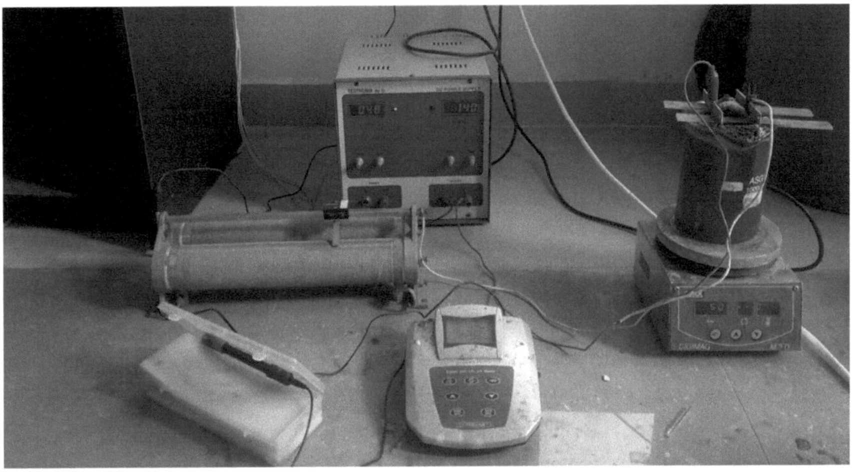

Fig. 1 Electrochemical setup

2.3 Experimental Procedure

In each run, 1 l of textile wastewater sample was treated by an Electro-chemical process. For each run parameters (pH, current density and electrochemical treatment time) were adjusted according to the experimental trial table made by using RSM. The gap between the anode and the cathode was fixed at 3 cm. The controlled direct current was supplied using a DC power supply. For a particular experimental trial the current was kept steady using a rheostat (Wheatstone Type) and was measured with an ammeter. The sample water in the electrochemical treatment cell was continuously stirred using a magnetic plate stirrer with a needle at 150 RPM. Electrochemical treatment was performed at different current densities for different time periods. Treated wastewater samples were collected after settling for 60 min to estimate its COD reduction. In each experimental setup, 3 ml sample was collected for analysis.

2.4 Analytical Method

COD was estimated using COD digester and visible spectrophotometer according to the standard approaches for analysis of water (Ghanim 2014). Digesting reagent was prepared by mixing potassium di-chromate solution with concentrated sulfuric acid reagent (in 1:3 ratio). Sulfuric acid reagent was prepared by adding silver sulfate to strong sulfuric acid. Then 3 ml of reagent was used for digestion of an equal volume of sample. After digestion absorbance was measured using a visible spectrophotometer (Li et al. 2009).

2.5 *Data Analysis*

The textile wastewater treatment in a batch wise approach by electrochemical treatment was developed and optimized by RSM. The RSM is applied to augment the three most important operating parameters; current density, pH and operation time having a significant influence on the electrochemical treatment process and also to maximize process efficiency. MINITAB14 was used for RSM. As COD is one of the most important measures of adverse contaminants in textile wastewater. The COD percentage reduction was chosen to be an indicator of effectiveness of electrochemical treatment. The optimized results are shown by response, main effect, interaction, surface and contour plots.

3 Results and Discussion

Interaction of three factor variables (pH, current density, time of electrochemical treatment) with the response (% COD reduction) for electrochemical treatment was examined using RSM. The Central Composite Design (CCD) (Khuri and Cornell 1996) endorsed in formulating various equations, where the % COD reduction was expressed in terms of factor variables. This study involved 48 runs of experiments for getting a response.

Response Surface Regression: percentage COD reduction versus pH, current density, time of electrochemical treatment.

The analysis was done using uncoded units.

$$S = 1834 \quad \text{R-Sq} = 99.2\,\% \quad \text{R-Sq(adj)} = 99.0\,\%$$

"Polynomial equation" developed through design of experiments by RSM (MINITAB14 software) is:

$$\begin{aligned} \text{COD} = {} & 56.9362 + 4.1539 * X + 0.9924 * Y + 0.7793 * Z - 1.4797 * X^2 - 0.0231 * Y^2 \\ & - 0.0163 * Z^2 + 0.1047 * X * Y + 0.1194 * X * Z - 0.0194 * Y * Z \end{aligned}$$

Here, X = pH, Y = Current Density, Z = Time of electro-coagulation.

"*p*" values are used as a measure of significance of each factor. Lower the value of *p*, the higher is the significance of the factor (Ravikumar et al. 2006). From Table 1 it can be observed that pH, quadratic factor, interaction factor of pH and time of electrochemical treatment and interaction factor of pH and current density are more significant as compared to others. Typically the factors having p values less than or equal to 0.05 are regarded as significant factors. The response surface model established in this study has an R-Sq value higher than 99 %. Coefficients of above polynomial equation are taken from Table 2. The observations which do not fit in above polynomial equation and their corrected values are listed in Table 3.

Table 1 Factor setting table

pH	Current density (mA/cm^2)	Time of electro-coagulation (min)	% COD reduction	pH	Current density (mA/cm^2)	Time of electro-coagulation (min)	% COD reduction	pH	Current density (mA/cm^2)	Time of electro-coagulation (min)	% COD reduction
4	10	5	61.3600	6	10	5	51.0920	8	10	5	20.5300
4	10	10	70.0200	6	10	10	54.0980	8	10	10	20.5300
4	10	15	77.1390	6	10	15	56.3530	8	10	15	32.8100
4	10	20	78.1420	6	10	20	59.4140	8	10	20	36.5650
4	15	5	71.5050	6	15	5	57.3540	8	15	5	22.5400
4	15	10	72.6320	6	15	10	58.3560	8	15	10	28.5500
4	15	15	77.8910	6	15	15	61.6120	8	15	15	37.0672
4	15	20	78.8935	6	15	20	63.1150	8	15	20	40.0720
4	20	5	73.4330	6	20	5	58.6100	8	20	5	27.8000
4	20	10	75.6375	6	20	10	61.1100	8	20	10	33.3100
4	20	15	78.3900	6	20	15	64.8700	8	20	15	40.5700
4	20	20	79.3998	6	20	20	65.8700	8	20	20	43.0780
4	25	5	73.6800	6	25	5	60.6100	8	25	5	31.3063
4	25	10	76.0100	6	25	10	61.6120	8	25	10	37.8100
4	25	15	78.4900	6	25	15	66.6210	8	25	15	43.5700
4	25	20	79.8950	6	25	20	67.6223	8	25	20	46.0830

Table 2 Estimated regression coefficients for percentage COD reduction

Term	Coefficient	SE coefficient	T	*p*
Constant	56.9362	7.18450	7.925	0.000
pH	4.1539	1.80367	2.303	0.027
Current density	0.9924	0.42553	2.332	0.025
Time of electro-coagulation	0.7793	0.35294	2.208	0.033
pH * pH	−1.4797	0.14093	−10.540	0.000
Current density * Current density	−0.0231	0.01059	−2.186	0.035
Time of electro-coagulation * Time of electro-coagulation	−0.0163	0.01059	−1.544	0.131
pH * Current density	0.1047	0.02900	3.611	0.001
pH * Time of electro-coagulation	0.1194	0.02900	4.116	0.000
Current density * Time of electro-coagulation	−0.0194	0.00847	−2.292	0.028

Table 3 Residuals values

Observation no.	% COD reduction	Fit	Residual
1	61.360	66.578	−5.218
21	57.354	53.304	4.050

The main effect plots are shown in Fig. 2 which indicated that the maximum COD reduction was obtained at pH 4 and % COD reduction is nearly constant over current density range of 20–25 mA/cm^2 and between electrochemical treatment times of 15–20 min. The 2D contour plots are shown in Fig. 3. Interaction of two factors is shown on each plot and shaded area represents the response values, whereas darkness represents a higher COD reduction.

The plots shown in Figs. 4, 5 and 6 represent 3D response surface plots for COD reduction. 3D response surface plots for COD reduction on z-axis gives the 3

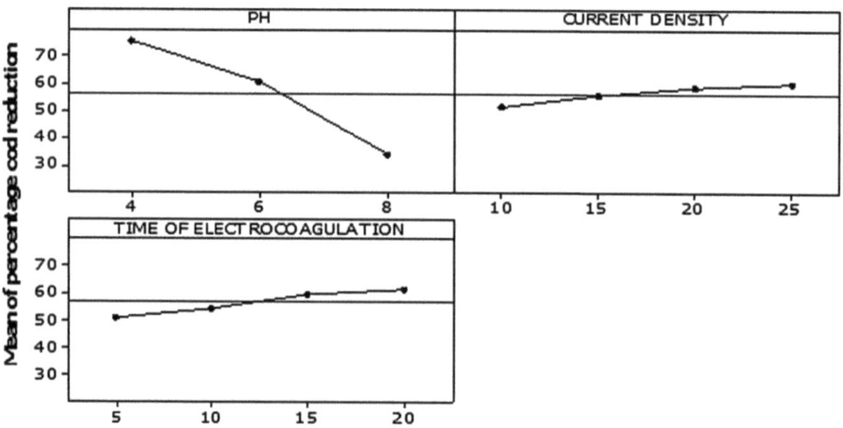

Fig. 2 Main effects plot for percentage reduction in COD

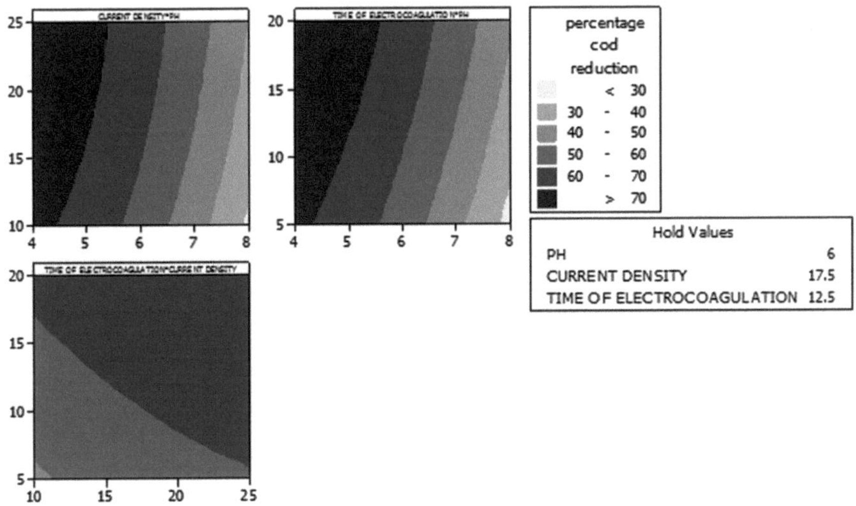

Fig. 3 Contour plots of percentage COD reduction

dimensional relations with the factor variables pH, current density and electrochemical treatment time.

Figure 7 shows the optimum COD removal percent, which occurs at pH 5.0729, current density 22.5501 and time of electrochemical treatment 17.717 min. The predicted percentage COD reduction was 74.16 % at composite desirability of 0.9846.

Fig. 4 Surface plot of percentage COD reduction versus time of electrochemical treatment, pH

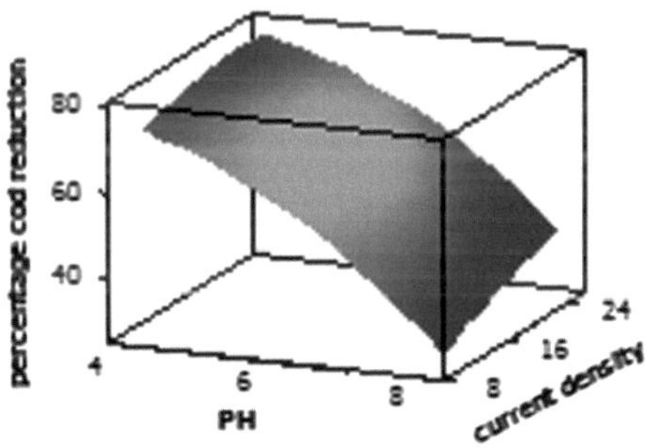

Fig. 5 Surface plot of percentage COD reduction versus current density, pH

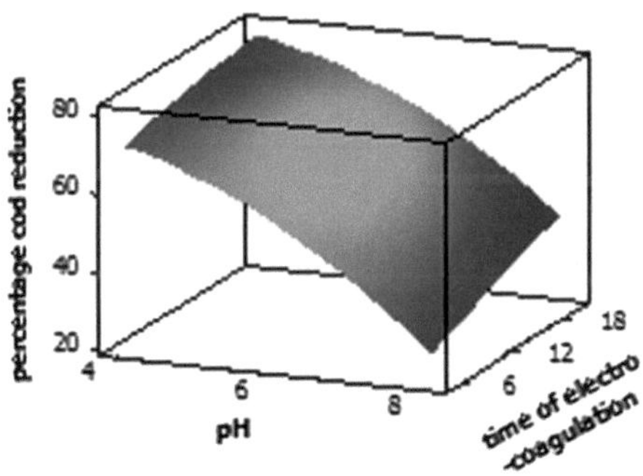

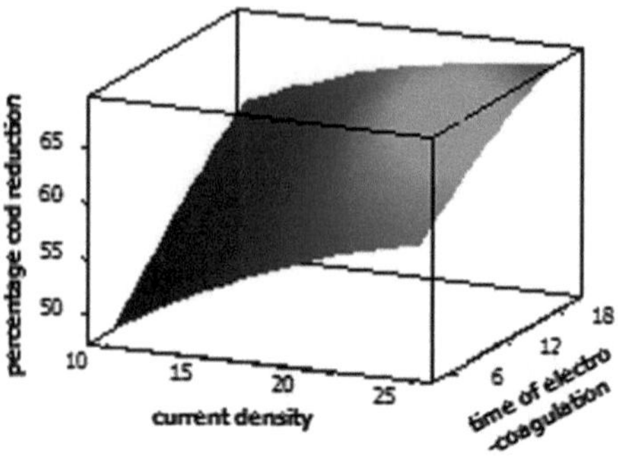

Fig. 6 Surface plot of percentage COD reduction versus current density, time of electrochemical treatment

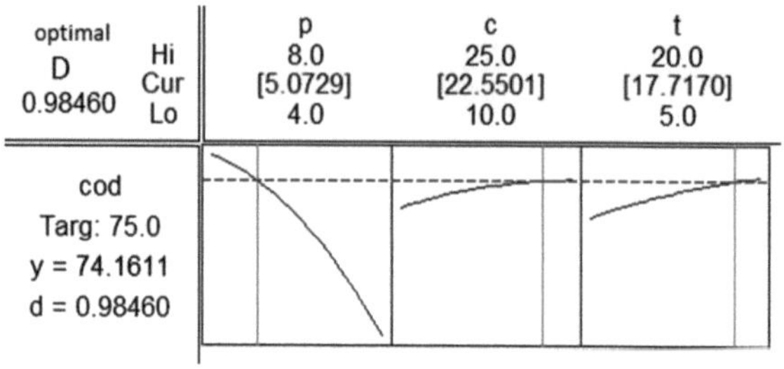

Fig. 7 Response optimization plot of maximum % COD reduction

Interaction plots are shown in Fig. 8 which indicates the effects of interacting factors on percentage cod reduction.

Ghamin (2014) have done a similar experiment and obtained predicted maximum % cod reduction of 85.7538 % with R^2 value of 0.9946 and with pH 4.52, current density 20 mA/cm^2 and operation time 56.5403 min. The results obtained by Ghamin are very comparable to our results.

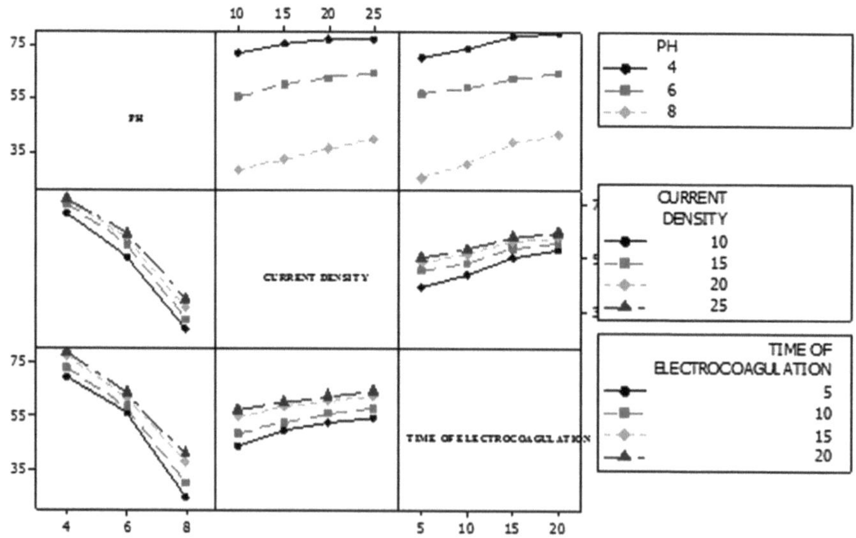

Fig. 8 Interaction plot for percentage COD reduction

4 Conclusion

In this study RSM has been applied to optimize the experimental data of textile wastewater treatment by electrochemical treatment process. The response surface model was developed with R^2 value 0.992 for the % COD reduction and the model was represented by 3D surface plots, 2D contour plots, main effects plots and interaction plots. The predicted optimal % cod reduction was found to be 74.1611 % with pH 5.0729, current density 22.55 mA/cm^2 and time of electro-coagulation 17.710 min.

Acknowledgments The authors would like to thank the Department of Chemical Engineering, Maulana Azad National Institute of Technology (MANIT) Bhopal, India, for providing necessary facilities and established the Advanced Industrial Pollution Abatement Laboratory and Analytical and Simulation Laboratory through Research grant.

References

Elboughdiri, N., Mahjoubi, A., Shawabkeh, A., Khasawneh, H.E., Jamoussi, B.: Optimization of the degradation of hydroquinone, resorcinol and catechol using response surface methodology. Adv. Chem. Eng. Sci. **5**, 111–120 (2015)

Ghanim, A.N.: Optimization of pollutants removal from textile wastewater by electrochemical treatment through RSM. J. Eng. Sci. **22**, 1–13 (2014)

Khandegar, V., Saroha, A.K.: Electrochemical treatment for the treatment of textile industry effluent—A review. J. Environ. Manag. **128**, 949–963 (2013)

Khuri, A.I., Cornell, J.A.: Response surfaces, 2nd edn. Marcel Dekker, New York (1996)

Li, J., Tao, T., Li, X.-B., Zuo, J.-L., Li, T., Lu, J., Li, S.-H., Chen, L.-Z., Xia, C.-Y., Liu, Y., Wang, Y.-L.: A spectrophotometric method for determination of chemical oxygen demand using home-made reagents. Desalination **239**, 139–145 (2009)

Ravikumar, K., Ramalingam, S., Krishnan, S., Balu, K.: Application of response surface methodology to optimize the process variables for reactive red and acid brown dye removal using a novel adsorbent. Dyes Pigm. **70**, 18–26 (2006)

Secondary Treatment of Dairy Effluents with Trickle Bed

M. Ramananda Bhat, Shivaprasad Nayak, Akshay Pariti and Sahil Dhawan

1 Introduction

Trickling filters are used to provide secondary treatment of primary settled wastewater (Metcalf and Eddy 2003). They are usually 3–8 ft long and are packed with either plastic or rock over which a biofilm grows. Due to the continuous flow of wastewater over packed medium, microbes present in the wastewater attach to the surface of the medium and a film begins to develop. The biofilm contains different species of bacteria, fungi, algae and protozoa. These microorganisms oxidize the organic materials in the wastewater as a food source in the presence of oxygen. Thus, it is a non-submerged, fixed film process (Lin 2007). The effluent from primary treatment is allowed to trickle through the porous bed (usually rock pieces). Air is passed from the bottom of the bed, in the counter current direction to the flow of liquid. The biomass on the rock surfaces will consume nutrients (from wastewater) and oxygen from the air stream. The secondary treatment of wastewater takes place between three phases—biofilm, wastewater and air (Logan et al. 1987).

Trickling filters are currently being studied for treatment of dairy wastewater (Shahriari and Shokouhi 2015). The dairy waste water has considerable amounts of

M. Ramananda Bhat (✉) · S. Nayak · A. Pariti · S. Dhawan
Department of Biotechnology, Manipal Institute of Technology,
Manipal University, Manipal 576104, India
e-mail: bhatchem@gmail.com

S. Nayak
e-mail: nayakshivaprasad@yahoo.co.in

A. Pariti
e-mail: go4akshaypariti@gmail.com

S. Dhawan
e-mail: dhawansahil@gmail.com

R. Mohan B. et al. (eds.), *Materials, Energy and Environment Engineering*,
DOI 10.1007/978-981-10-2675-1_36

chemical oxygen demand (COD), biochemical oxygen demand (BOD), total Kjeldahl nitrogen (TKN), total phosphorus (TP) and nitrate and the levesl of these pollutants need to be brought down before recycle or discharge (Abdulgader et al. 2009) and bio-trickle filters are capable of reducing the pollutant levels to desirable levels. Shahriari and Shokouhi (2015) have utilized active sludge as seeding for the trickle bed filters for the treatment of dairy wastewater. Amal Raj and Murthy (1999) have carried out dairy effluent treatment using trickle bed filters with the hydraulic loadings employed in the study in the range of 5–17 m^3/m^2 d for the influent COD concentrations of 427 to 1384 mg/L and it was observed that the relationship between organic removal rate and organic loading rate showed two regions of linear behavior with the point of inflection at an organic loading rate of 26 g COD/m^2 d. The microorganisms grown in the trickling filters were able to effectively reduce COD levels up to 2750 mg/L, under aerobic conditions at pH values between 6.8 and 7.2 at temperatures as low as 10 to 13 °C (Mehrdadi et al. 2012). As reported by the website edie.net (2003), Crest super dairy at Stroud Severnside in Gloucestershire, UK, is treating its wastewater by trickle bed filters since 2003, consisting of high technology Marpak cross-flow media capable of reducing BOD from 904 mg/L and Ammonia from 20 mg/L down to 25 mg/L BOD and 5 mg/L Ammonia respectively.

2 Materials and Methods

In the present study, the laboratory scale unit of trickling filter (vertical flow system) was fabricated from PVC pipe of internal diameter of 10 cm and 3 ft in height. It is attached to a firm wooden frame as shown in Fig. 1. Internal segmentation into equal parts of 1 ft each was made by fitting water redistributors. Water redistributors were fixed to ensure even distribution of fluid flow and to bear the load of the packing material. The 1st and 3rd compartments were filled with gravels and the 2nd compartment with aquarium pebbles. The pre-treated effluent of dairy wastewater was used as influent for the trickle bed filter. The pretreatment was given by the SBR (Sequencing Batch Reactor) technique. The size of the gravels and that for pebbles chosen in the present study ranged from 10 to 30 mm and 5 to 15 mm respectively. The porosity of gravels and pebbles are 46.96 and 41.7 % respectively. The influent wastewater from the dosing tank was allowed to trickle through the sprinkler over the filter media. The filtered water was then made to flow through the pipe system and collected through the control valve fitted at the bottom. The details of design and operation of trickle bed filter used for the present study is given in Table 1.

The filter media was soaked in pretreated dairy wastewater for 3 weeks. This filter media was used to fill the trickle bed column. Aeration for the biofilm growth was provided by an air compressor. A liquid seal is used at the bottom part of the setup (shown with shaded portion in the outlet end) so that air does not escape from the bottom but only enters the porous bed.

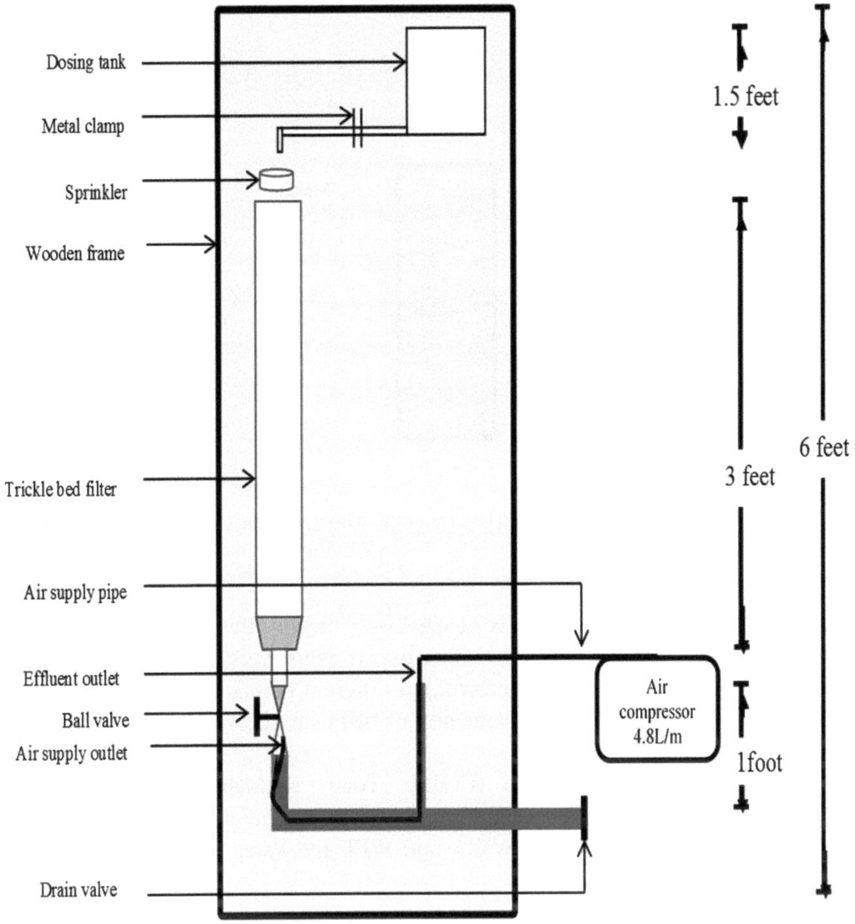

Fig. 1 Schematic diagram of trickle bed filter

Table 1 Characteristics of trickle bed filter

Trickle bed characteristic	Value
Volume of dosing tank	9 L
Volume of trickle bed filter	7.18 L
Average porosity of trickle bed filter	45.20 %
Number of passes per trial	4
Flow rate of wastewater	2.25 L/h
Air flow rate	4.8 L/min

The flow pattern through the porous bed (Kister 1990) of the trickle bed filter was studied using a Powdered Carbon (PC) solution (1 % by weight). Initially, 1 ft pipe column (with internal diameter 10 cm), was used to study the flow pattern

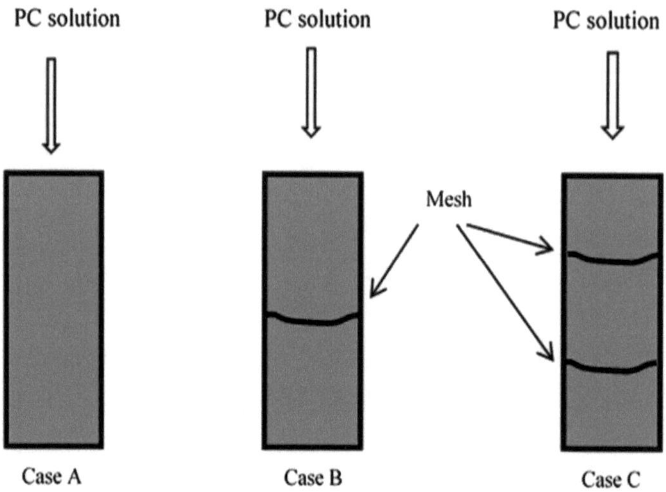

Fig. 2 Flow pattern of powdered carbon (*PC*) solution within the pipe packed with gravels of size 10–30 mm and placing meshes. Pipe diameter is 10 cm and pipe length is 3 ft

within the pipe packed with gravels of size 10 to 30 mm. Based on the observation in the experiment for 1 ft column, similar experiments were conducted by placing meshes (of sieve size 2 mm, GI material) at different points on a 3 ft column pipe for testing the flow pattern through the porous bed using Powdered Carbon solution as shown in Fig. 2.

Chemical Oxygen Demand (COD), Total Kjeldahl Nitrogen (TKN), Ammoniacal Nitrogen and Organic Nitrogen were determined as per the standard procedure outlined by APHA, AWWA and WEF for water and wastewater.

3 Results and Discussions

The gravel and pebbles that were soaked in effluent of SBR for three weeks had developed slime layer on them. Figure 3 shows the flow pattern in a trickle bed with experiments conducted using PC solution. Placing meshes will help to avoid channelling of water passage. The dark arrows show the pattern of flow of water. In case A, there is lot of channelling where as in case C, there very less channelling. Table 2 shows the comparison of the three cases that are shown in Fig. 3.

It is the general phenomenon of a liquid to follow the path of least resistance while trickling. Hence channelling is more in case A and least in case C. The meshes tend to interfere with the least resistance path of the fluid passage and hence there is better redistribution of flow through the bed.

The pretreated dairy wastewater was subjected to secondary treatment in trickle bed filter. Reduction in COD, TKN, Ammonia nitrogen and Organic nitrogen was

Table 2 Comparison of the flow pattern of three cases that are shown in Fig. 3

Observation	Case A	Case B	Case C
Flow path of PC solution	Reaches the wall of the pipe before 1 ft of its downward passage and continues through the wall rather than filter bed	Reaches the wall of the pipe by 1 ft of its downward passage but is redistributed by the mesh	Only very small fraction of PC solution reaches the walls of the pipe. Rest of the PC solution is distributed throughout the bed
Distribution of PC powder (after the flow experiments)	Major fraction of powder on the walls of the pipes	Small fraction of PC powder was on the filter media at the liquid entry and near the mesh. Rest of the powder was on the walls of the pipe	Major fraction of PC powder was found on the filter media of the bed
Channeling effects	High channeling, resulting in poor distribution of liquid in the filter bed	Moderate channeling	The liquid is well distributed throughout the filter media in the bed

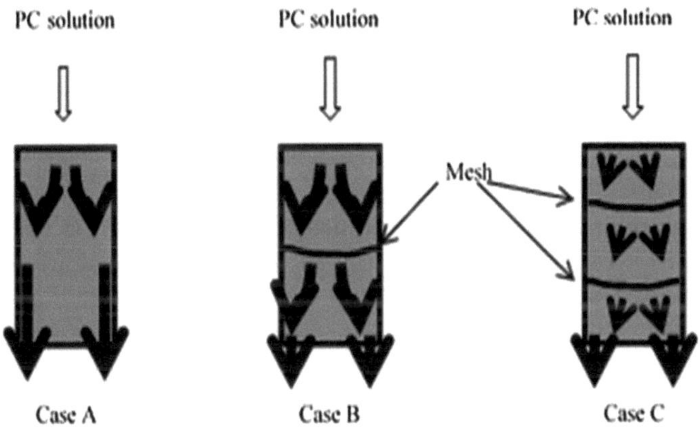

Fig. 3 Flow pattern of PC solution within the pipe packed with gravels of size 10–30 mm and placing meshes. Pipe diameter is 10 cm and pipe length is 3 ft

determined to study the performance of trickle bed filter. The percentage reduction of the COD, TKN, Ammonia Nitrogen and Organic Nitrogen is depicted in Fig. 4.

From Fig. 4, it can be observed that there is steady increase in the percentage reduction in COD, TKN, Ammonia nitrogen and Organic nitrogen content of the wastewater treated in trickle bed. This is due to the improved performance of the biomass in the trickle bed filter media. As observed by Shahriari and Shokouhi (2015), the liquid recirculation enhanced mass transfer between the dairy wastewater and the surface of solid medium, thereby accelerating the startup

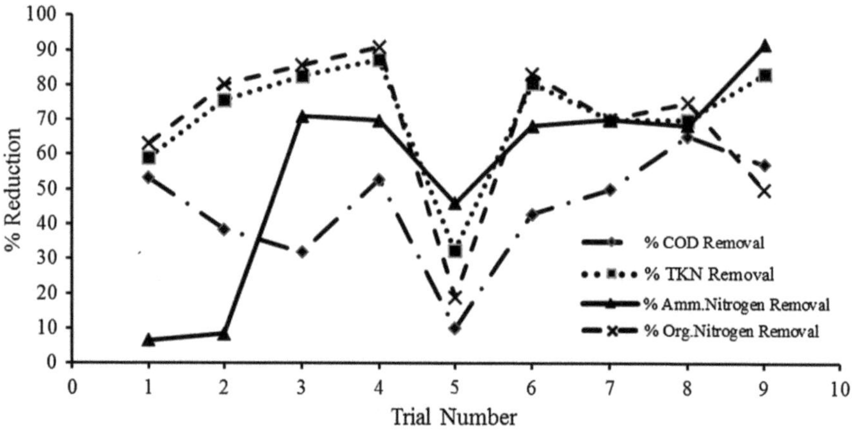

Fig. 4 Reduction (%) in COD, TKN, Ammonia nitrogen and Organic nitrogen in trickle bed

process. However there is significant drop in the performance of the trickle bed filter in 5th trial. This may be due to sloughing (Metcalf and Eddy 2003) or loss of activity of the biofilm on the media of trickle bed filter.

The dark bold lines in each of Figs. 5 and 6 are the true variation in the percentage removal of the content of wastewater. The dashed lines in the graphs are the linear fit for the percentage removal. From Fig. 5, we can infer that there is always an overall increasing trend in the percentage reduction in the removal of contents from the wastewater except organic nitrogen. The slope of linear trend is positive for all the contents except for organic nitrogen removal. The overall negative trend in organic nitrogen could be due to the microbial flora variation in the process of trickling bed filter over a long period of time. Amal Raj and Murthy (1999) concluded that the trickle bed reactor must operate in a batch mode for 20 days to initiate the growth of nitrifies and for an additional period of 40 days (Amal Raj and Murthy 1998) to enhance the attachment of nitrifiers to the media so as to assist the removal on organic nitrogen. It can also be inferred from Figs. 4 and 5, that the present trickle bed filter removes TKN and its components more efficiently than that of COD. Also the efficiency for removal of ammonia nitrogen is the highest. This indicates that the microbial flora that remove nitrogenous compounds are more dominating and active than the microbial flora for COD/BOD removal. The ratio of percentage TKN removal to percentage COD removal is always above 1 for the present trickle bed filter. The variation of ratio of percentage TKN removal to percentage COD removal is shown in Fig. 6.

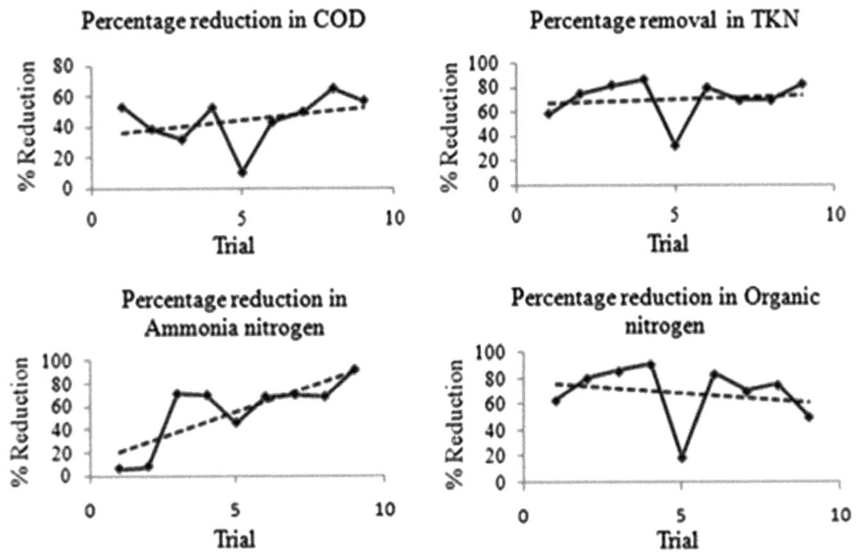

Fig. 5 Fitting linear trend line to study the trend in the percentage removal of COD, TKN, Ammonia nitrogen and Organic nitrogen

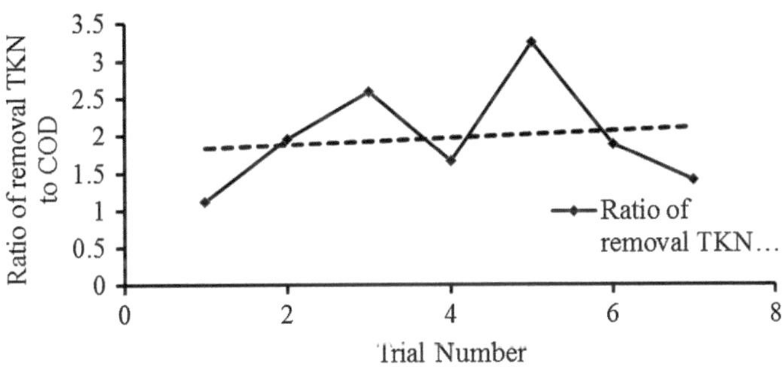

Fig. 6 Variation of ratio of percentage TKN removal to percentage COD removal

4 Conclusion

Pretreated dairy wastewater was subjected to secondary treatment using a trickle filter with three packings and the results showed that COD, TKN (Total Kjeldahl Nitrogen) and organic nitrogen undergoes reduction, with a steady rise in percentage reduction of COD and TKN. However, percentage reduction in organic nitrogen showed a negative trend probably due to variation in microbial flora. Also, significant drop in the performance of the trickle bed filter in 5th trial was observed, probably due to sloughing or due to loss of activity of the biofilm on the media of trickle bed filter

References

Abdulgader, M., Yu, Q.J., Zinatizadeh, A., Williams, P.: Biological treatment of milk processing wastewater in a sequencing bath flexible fibre biofilm reactor. Asia-Pac. J. Chem. Eng. **4**(5), 698–703 (2009)

Amal Raj, S., Murthy, D.V.S.: Synthetic dairy wastewater treatment using cross flow medium trickling filter. J. Environ. Sci. Health Part A **34**(2):357–369 (1999)

Amal Raj, S., Murthy, D.V.S.: Nitrification of synthetic wastewater in a cross flow medium trickling filter. Bioprocess Eng. **19**(2), 149–154 (1998)

Kister, H.Z.: Distillation Operation, pp. 211–228. McGraw Hill Inc. New York (1990)

Lin, S.D.: Water and Wastewater Calculation Manual, pp. 692–695. Tata McGraw-Hill Publishing Company Limited, New York (2007)

Logan, B.E., Hermanowicz, S.W., Parker, D.S.: A fundamental model for trickling filter process design. J. Water Pollut. Control Federation **59**, 1029–1046 (1987)

Mehrdadi, N., Bidhendi, G.R., Shokouhi, M.: Determination of dairy wastewater treatability by bio-trickling filter packed with lava rocks—case study PEGAH dairy factory. Water Sci. Technol. **65**(8), 1441–1447 (2012)

Metcalf, E., Tchobanoglous, G., Burton Franklin, L., Stesel, D.H.: Wastewater Engineering, Treatment and Reuse. Tata McGraw-Hill Publishing Company Limited, New Delhi (2003)

Shahriari, T., Shokouhi, M.: Assessment of bio-trickling filter startup for treatment of industrial wastewater. Int. J. Environ. Res. **9**(2), 769–776 (2015)

Website: http://www.edie.net/news/0/Trickle-filter-system-extends-water-treatment-capability-at-midlands-super-dairy/7683/ (2003)

MIX
Papier aus verantwortungsvollen Quellen
Paper from responsible sources
FSC® C105338

If you have any concerns about our products,
you can contact us on
ProductSafety@springernature.com

In case Publisher is established outside the EU,
the EU authorized representative is:
Springer Nature Customer Service Center GmbH
Europaplatz 3, 69115 Heidelberg, Germany

Printed by Libri Plureos GmbH
in Hamburg, Germany